T0275695

# Evolution, Explanation, Ethics, and Aesthetics

Academic Press is an imprint of Elsevier
125 London Wall, London EC2Y 5AS, United Kingdom
525 B Street, Suite 1800, San Diego, CA 92101-4495, United States
50 Hampshire Street, 5th Floor, Cambridge, MA 02139, United States
The Boulevard, Langford Lane, Kidlington, Oxford OX5 1GB, United Kingdom

**Notices**
Knowledge and best practice in this field are constantly changing. As new research and
experience broaden our understanding, changes in research methods, professional practices,
or medical treatment may become necessary.

Practitioners and researchers must always rely on their own experience and knowledge in
evaluating and using any information, methods, compounds, or experiments described
herein. In using such information or methods they should be mindful of their own safety and
the safety of others, including parties for whom they have a professional responsibility.

To the fullest extent of the law, neither the Publisher nor the authors, contributors, or editors,
assume any liability for any injury and/or damage to persons or property as a matter of
products liability, negligence or otherwise, or from any use or operation of any methods,
products, instructions, or ideas contained in the material herein.

**British Library Cataloguing-in-Publication Data**
A catalogue record for this book is available from the British Library

**Library of Congress Cataloging-in-Publication Data**
A catalog record for this book is available from the Library of Congress

ISBN: 978-0-12-803693-8

For information on all Academic Press publications
visit our website at https://www.elsevier.com/

Working together
to grow libraries in
developing countries

www.elsevier.com • www.bookaid.org

*Publisher:* Sara Tenney
*Acquisition Editor:* Kristi A.S. Gomez
*Editorial Project Manager:* Pat Gonzalez
*Production Project Manager:* Kirsty Halterman and Karen East
*Designer:* Matthew Limbert

Typeset by TNQ Books and Journals

# Contents

## 10.    Reduction and Emergence

## 11.    Microevolution and Macroevolution: A New Evolutionary Synthesis?

# Part III
# Ethics, Aesthetics, and Religion

# Preface

Theodosius Dobzhansky, credited as one of the founders of the modern theory of evolution, wrote in 1972 that "Nothing in Biology Makes Sense Except in the Light of Evolution," the title of his address to a convention of the National Association of Biology Teachers.[1] The theory of evolution is, indeed, the central organizing concept of modern biology. Evolution scientifically explains why there are so many different kinds of organisms, and it accounts for their similarities and differences. It accounts for the appearance of humans on Earth, and reveals humans' biological connections with other living beings. It provides an understanding of constantly evolving viruses, bacteria, and other pathogens and enables the development of effective ways to protect ourselves against the diseases they cause. Knowledge of evolution has made possible advances in agriculture, medicine, and biotechnology.

Michael Ruse, the distinguished and prolific philosopher of biology, extended Dobzhansky's statement with a simple modification: "Nothing makes sense except in the light of evolution," a statement he always attaches to his signature. What does Ruse mean? Does he mean that evolution encompasses other scientific disciplines, such as astrophysics, physics, and chemistry? Does he mean that the concept of evolution goes beyond science and extends to art, literature, religion, and social and political institutions? And to human history and family life, including love and care? I suspect that Ruse means all of these: indeed, that as his statement says, "*nothing* makes sense except in the light of evolution."

Dobzhansky would have agreed. At the very end of his *Mankind Evolving*, Dobzhansky quotes the Jesuit anthropologist Pierre Teilhard de Chardin: "Is evolution a theory, a system, or a hypothesis? It is much more—it is a general postulate to which all theories, all hypotheses, all systems must henceforward bow and which they must satisfy in order to be thinkable and true. Evolution is a light which illuminates all facts, a trajectory which all lines of thought must follow—this is what evolution is."[2]

In the introduction to Part I of his 1988 magnificent collection of previously published papers, *Toward a New Philosophy of Biology*, Ernst Mayr

---

1. Dobzhansky, T., 1972. Nothing in biology makes sense except in the light of evolution. The American Biology Teacher 35, 125–129.
2. Dobzhansky T., 1962. Mankind Evolving. The Evolution of the Human Species. Yale University Press, New Haven, CT and London.

points out his "special concerns" about the neglect of biology in the works of philosophers of science. "From the 1920s to the 1960s, the logical positivists and physicalists who dominated the philosophy of science had little interest in and even less understanding of biology, because it simply did not fit their methodology."[3] In 1961–1964, I was a PhD student at Columbia University in New York, under the mentorship of Dobzhansky. Earlier, I had three years of formal philosophical training in Spain and remained interested in the philosophical implications of my scientific research, focused on genetics and evolution. I read at the time two recently published books, *The Biological Way of Thought* by Morton Beckner (1959), a Ph.D. student of the eminent philosopher of science Ernest Nagel at Columbia University, and *The Ascent of Life. A Philosophical Study of the Theory of Evolution* (1961) by T. A. Goudge, a professor of philosophy at the University of Toronto. At Columbia, I befriended and occasionally attended the lectures of Ernest Nagel and read the two chapters dedicated to biological issues in his *The Structure of Science. Problems in the Logic of Scientific Explanation* (1961).

Most important were my frequent conversations with Dobzhansky, concerned as he was about the little attention philosophers of science were paying to the multitude of fundamental and ever-mounting philosophical issues emerging from evolutionary biology. Dobzhansky often expressed his frustration at the limited influence of biology on the thinking of contemporary philosophers. He saw that evolutionary biology raises important new philosophical issues and illuminates old ones. In 1971, we decided to convene a conference where in eight days of intense discussion such issues would be jointly explored by concerned philosophers and philosophically savvy scientists. The conference took place in Villa Serbeloni, a magnificent palace with beautiful gardens owned by the Rockefeller Foundation, in northern Italy, overlooking Lake Como, on September 9–15, 1972, including Karl Popper among the philosophers, and four Nobel Laureates, John C. Eccles, Gerald M. Edelman, Peter Medawar, and Jacques Monod. The papers presented at the conference were published as *Studies in the Philosophy of Biology* (1974). Two relatively short books were published at about the same time by two authors, Michael Ruse (*The Philosophy of Biology*, 1973) and David Hull (*Philosophy of Biological Science*, 1974), who would become over the ensuing decades two leaders among the increasing number of philosophers dedicated to the philosophy of biology since the last quarter of the 20th century.

The 17 chapters included in the present volume should be of interest to philosophy students as a textbook for a university course on the philosophy of biology, but the chapters are not intended as an encompassing treatise. The book is intended for scientists and philosophers concerned with the issues at hand, as

---

3. Mayr, E., 1988. Toward a New Philosophy of Biology. Harvard University Press, Cambridge, MA.

well as for interested general readers. Technical language has been avoided to the extent possible. The chapters can be read in any sequential order and are largely self-contained, so that an issue mostly considered in a particular chapter may be brought up elsewhere, if only briefly, whenever necessary for understanding a particular topic under consideration.

There are many scientists, philosophers, and other scholars to whom I am indebted, both intellectually and personally. Theodosius Dobzhansky, my Ph.D. mentor at Columbia University in New York, deserves particular mention. I want to express my gratitude also to Denise Chilcote, my executive assistant, who, for nearly three decades, has never failed to do whatever needed to be done as well as it could be done. I have tremendously benefited from her dedication and perfectionism.

**Francisco J. Ayala**
*University of California, Irvine, CA, United States*

# Part I

# Evolution

# Chapter 1

# The Darwinian Revolution

## INTRODUCTION

The publication in 1859 of Darwin's *Origin of Species* opened a new era in the intellectual history of mankind. The discoveries by Copernicus, Kepler, Galileo, and Newton had gradually led to a conception of the universe as a system of matter in motion governed by natural laws. The Earth was found to be not the center of the universe but a small planet rotating around an average star; the universe appeared as immense in space and in time; and the motions of the planets around the Sun could be explained by simple laws—the same laws that accounted for the motion of objects in our planet. These and many other discoveries greatly expanded human knowledge. But the conceptual revolution that went on through the 17th and 18th centuries was the realization that the universe obeys immanent laws that can account for natural phenomena. The workings of the universe were brought into the realm of science—explanation through natural laws. Physical phenomena could be reliably predicted whenever the causes were adequately known. Darwin completed the Copernican Revolution by drawing out for biology the ultimate conclusions of the notion of nature as a lawful system of matter in motion. The adaptations and diversity of organisms, the origin of novel and highly organized forms, even the origin of man himself could now be explained by an orderly process of change governed by natural laws.

All human cultures, including primitive ones, have developed their own explanations for the origin of the world and of human beings and other creatures. Traditional Judaism and Christianity explain the origin of living beings and their adaptations to their environments—wings, gills, hands, flowers—as the handiwork of an omniscient God. Among the Western cultures, the earliest systematic attempts to account for the origin of the world, humans, and other creatures were formulated by Greek philosophers (Mayr, 1982; Moore, 1993).

## GREEK ANTIQUITY

Thales of Miletus (c.620–555 BCE) was an astronomer, geographer, and meteorologist who considered water the first principle of *creation*, but did not

Evolution, Explanation, Ethics, and Aesthetics. http://dx.doi.org/10.1016/B978-0-12-803693-8.00001-3

advance any specific ideas about the origins of organisms. His disciple Anaximander (c.611–547), also an astronomer and geographer, proposed that the world is composed of four elements: earth, water, air, and fire. He speculated that the first animals were generated in an aquatic environment and would later migrate to the dry land. Humans also were formed first as fish-like creatures which would eventually develop and live independently on the land. Anaximander's account of human origins was not a theory about *evolution* in the modern sense, but an *ontogenetic* account of the early stages of each individual who would eventually develop into a full human. The doctrine of the four elements—earth, water, air, and fire—was adopted a century later by Empedocles (c.490–430 BCE) and, another century later, by Aristotle, through whom it became central to Western thought for the next 2,000 years. Anaximenes (c.555–500 BCE) and Parmenides (c.515–445 BCE) accepted similar ideas, propounding the spontaneous generation of organisms from mud or slime.

Anaxagoras (c.500–428 BCE) and Democritus (c.460–370 BCE) were concerned with what we now would call "*adaptation*" of organisms to the environment. Their explanations were thoroughly materialistic or, as we could say, physicalist: the composition and functional structures of organisms are a necessary consequence of the properties of atoms.

More interesting in some ways were the origination ideas of Empedocles, because one could see in them a (gross) anticipation of *natural selection*. Body parts, such as heads, limbs, eyes, and mouths, originate spontaneously and separately. These parts would attract each other in various combinations, only some of which would survive, namely those that would become functional organisms, while all others would be eliminated.

The three most important philosophers of classical Greece in terms of their posterior influence, some extending to the present, are Socrates, Plato, and Aristotle. Socrates (469–399 BCE), born in Athens, is known through the writings ("dialogues") of Plato, particularly the *Apology*, the *Crito*, and the *Phaedo*, where Plato immortalized the story of Socrates' trial and last days. Beyond his trial and death, associated with his criticisms of the moral distortions of the Athenians, Socrates is particularly recognized for the "Socratic method": asking for definitions, most importantly about moral values, such as piety and justice, and subjecting them to criticism and analysis, showing the contradictions often elicited by the responses.

Plato (c.428–347 BCE) was born, probably in Athens, into a wealthy and aristocratic *family*. After the self-executed death sentence of his mentor Socrates, Plato traveled extensively through Greece, Italy, and Egypt. He returned to Athens around 387 BCE and founded the Academy, which became a distinguished center for philosophical, mathematical, and scientific research. Plato's ideas are well known through his 30 dialogues, mostly philosophical. He replaced the spontaneous generation of the previous Greek philosophers by a creative power, or demiurge, author of all the ideal forms, or categories that account for all realities in the world, organisms and otherwise, which are limited and imperfect instantiations of the ideas in the mind of the Creator demiurge. Plato was a pagan polytheist. His demiurge is very different form of the Creator–God of the great monotheistic religions. According to Plato, for example, the idea of a horse exists only in the mind of the demiurges; the horses that exist on Earth are imperfect instantiations of the ideal horse.

Aristotle (384–322 BCE) was a disciple of Plato and later a teacher at the Academy for 20 years, from 367 to 347 BCE. Aristotle, born in Stagira in Macedon, was the son of Nichomachus, court physician to Amyntas III of Macedonia, grandfather of Alexander the Great, who was tutored from age 13 to 16 (343–340 BCE) by Aristotle. Aristotle spend three years on the Aegean island of Lesbos, close to the western coast of today's Turkey, where he developed his interest in and study of biology. In 335 BCE Aristotle returned to Athens, where he founded the Lyceum, his own school, and taught there for the next 12 years.

Aristotle is fittingly recognized as one of the great philosophers of Greek antiquity, known for his works on logic, *epistemology*, metaphysics, *ethics*, and politics. He also qualifies as a great scientist, the first biologist in the history of the world, as he dissected a variety of animals and described in detail the *development* of the chicken embryo from egg to hatching. During his three years in Lesbos he investigated all sorts of marine and lacustrine organisms, from coral reefs and molluscs to fish, as well as land animals. In his numerous zoological works, such as *The Generation of Animals* and others, Aristotle mentions 500 animal *species*, a large number relative to the knowledge of the time (Leroi, 2014). Aristotle rejected the classification of animals based only on their external structures (for example, winged or wingless), basing his classification on the principle of organization, consisting of structure and function (including mode of reproduction), and introduced the binomial method of nomenclature. To define each animal, Aristotle used the *genus* (which encompassed interrelated animals or organisms in general) and their *difference* (the distinctive characteristics of the *species* or the specific organism). This binomial two-term method of identifying each particular organism would be modified a few centuries later by Porphyry (234–305), a prolific author, mainly of philosophical treatises, and disciple of Plotinus (considered the father of Neoplatonism in Rome). Porphyry introduced the *category* of *species* combined with *genus* to identify organisms, thus refining the combination of *genus* and *difference* used by Aristotle and thereby originating the binomial system of nomenclature that persists to the present day and is often erroneously credited to Linnaeus (1707–78) (Ayala, 2014).

Aristotle developed a system of classification of animals, recognizing nine groups within two large categories: animals with blood and without blood. He classified the animals with blood into five groups—mammals, birds, reptiles, amphibians, and fishes—and divided the animals without blood into cephalopods, crustaceans, insects, and a fourth group that would include the rest of the animals. Aristotle's classification compares favorably with the classification by Linnaeus, 21 centuries later, who recognized only six kinds of animals: mammals, birds, reptiles, fishes, insects, and worms. Aristotle considered some animals more complex or advanced than others, but he believed in the immutability of *species*, not in their *evolution*, and thus he did not see the difference between lower and higher animals as the outcome of a process of change or *progress* over time.

## CHRISTIANITY AND EUROPE UNTIL DARWIN

Among the early Fathers of the Church, Gregory of Nyssa (335–394) and Augustine (354–430) maintained that not all of *creation*, all *species* of plants and animals, were initially created by God; rather, some had evolved in historical times from God's creations.

According to Gregory of Nyssa, the world came about in two successive stages. The first stage, the creative step, was instantaneous; the second stage, the formative step, is gradual and develops through time. According to Augustine, many plant and animal *species* were not directly created by God, but only indirectly, in their potentiality (their *rationes seminales*), so they would come about by natural processes later in the *development* of the world. Gregory's and Augustine's motivation was not scientific but theological. For example, Augustine was concerned that it would have been impossible to hold representatives of all animal *species* in a single vessel, such as Noah's Ark; some *species* must have come into existence only after the Flood.

The notion that organisms may change by natural processes was not investigated as a biological subject by Christian theologians of the Middle Ages, but it was, usually incidentally, considered as a possibility by many, including Albertus Magnus (1200–80) and his student Thomas Aquinas (1224–74). Aquinas concluded, after consideration of the arguments, that the *development* of living creatures, such as maggots and flies, from nonliving matter, such as decaying meat, was not incompatible with Christian faith or philosophy, but he left it to others (scientists, in current parlance) to determine whether this actually happened.

The issue of whether living organisms could spontaneously arise from dead matter was not settled until four centuries later by the Italian Francesco Redi (1626–97), one of the first scientists to conduct biological experiments with proper controls. Redi set up flasks with various kinds of fresh meat: some were sealed, others covered with gauze so that air but not flies could enter, and others left uncovered. The meat putrefied in all flasks, but maggots appeared only in the uncovered flasks which flies had entered freely. Redi was a poet as well as a physician, chiefly known for his *Bacco in Toscana* (1685, *Bacchus in Tuscany*).

The cause of putrefaction was discovered two centuries later by Darwin's younger contemporary, the French chemist Louis Pasteur (1822–95), one of the greatest scientists of all time. Pasteur demonstrated that fermentation and putrefaction were caused by minute organisms that could be destroyed by heat. Food decomposes when placed in contact with germs present in the air. The germs do not arise spontaneously within the food. We owe to Pasteur the process of pasteurization, the destruction by heat of microorganisms in milk, wine, and beer, which can thus be preserved if kept out of contact with the microorganisms in the air. Pasteur also demonstrated that cholera and rabies are caused by microorganisms, and he invented vaccination, treatment with attenuated (or killed) infective agents that would stimulate the *immune system* of animals and humans, thus protecting them against infection.

In the 18th century Pierre-Louis Moreau de Maupertuis (1698–1759) proposed the spontaneous generation and extinction of organisms as part of his theory of origins, but he advanced no theory about the possible transformation of one *species* into another through knowable natural causes. One of the greatest naturalists of the time, Georges-Louis Leclerc, Comte de Buffon (1707–78), explicitly considered—and rejected—the possible descent of several distinct kinds of organisms from a common ancestor. However, he made the claim that organisms arise from organic molecules by spontaneous generation, so there could be as many kinds of animals and plants as there are viable combinations of molecules.

The Swedish botanist Carolus Linnaeus (1707–88) devised the hierarchical system of plant and animal classification that is still in use in a modernized form. Although he insisted on the fixity of *species*, his classification system eventually contributed much to the acceptance of the concept of *evolution* by common descent.

The first broad theory of *evolution* was proposed by the French naturalist Jean-Baptiste de Monet, Chevalier de Lamarck (1744–1829). In his *Philosophie zoologique* (1809, *Zoological Philosophy*), Lamarck held the enlightened view, shared by the intellectuals of his age, that living organisms represent a progression, with humans as the highest form. Lamarck's theory of *evolution* asserts that organisms evolve through eons of time from lower to higher forms, a process still going on, always culminating in human beings. The remote ancestors of humans were worms and other inferior creatures, which gradually evolved into more and more advanced organisms, ultimately humans.

The "inheritance of acquired characters" is the theory most often associated with Lamarck's name. Yet this theory was actually a subsidiary construct of his theory of *evolution*: that *evolution* is a continuous process, and that today's worms will yield humans as their remote descendants. It stated that as animals become adapted to their environments through their habits, modifications in their body plans occur by "use and disuse." Use of an organ or structure reinforces it; disuse leads to obliteration. Lamarck's theory further asserted that the characteristics acquired by use and disuse would be inherited.

This assumption would later be called the inheritance of acquired characteristics (or Lamarckism). It was disproved in the 20th century.

Lamarck's evolutionary theory was metaphysical rather than scientific. He postulated that life possesses the innate property to improve over time, so progression from lower to higher organisms would continually occur, and always follow the same path of transformation from lower organisms to increasingly higher and more complex organisms. A somewhat similar evolutionary theory was formulated a century later by another Frenchman, the philosopher Henri Bergson (1859—1940) in his *L'Evolution créatrice* (1907, *Creative Evolution*).

Erasmus Darwin (1731—1802), a physician and poet and the grandfather of Charles Darwin, proposed, in poetic rather than scientific language, a theory of the transmutation of life forms through eons of time (*Zoonomia, or the Laws of Organic Life*, 1794—96). More significant for Charles Darwin was the influence of his older contemporary and friend, the eminent geologist Sir Charles Lyell (1797—1875). In his *Principles of Geology* (1830—33), Lyell proposed that Earth's physical features were the outcome of major geological processes acting over immense periods of time, incomparably greater than the few thousand years since *Creation* that was commonly believed at the time.

## CHARLES DARWIN

Charles Robert Darwin (1809—82) was born in Shrewsbury, Shropshire, UK. His father, Robert Darwin, was a physician, like his own father Erasmus, but he was also a very successful financier and left a substantial inheritance to his son Charles, who thus never had to work for a living. Darwin's mother, Susannah Wedgwood, was the daughter of the famous Wedgwood potter. Charles Darwin went to school in Shrewsbury. In 1825 he was enrolled as a medical student at the University of Edinburgh. He left after two years and moved to Christ's College at Cambridge, where he graduated in 1831 with a BA degree. Darwin was not an exceptional student, but he was deeply interested in natural history.

On December 27, 1831, a few months after his graduation from Cambridge, Darwin sailed as a naturalist aboard the HMS *Beagle* on a round-the-world trip that lasted until October 1836 (Moorehead, 1969). On that voyage, Darwin often disembarked for extended trips ashore to collect natural specimens. The discovery of *fossil* bones from large extinct mammals in Argentina and the observation of numerous *species* of finches in the Galápagos Islands were among the events credited with stimulating Darwin's interest in how *species* originate.[1]

Darwin's observations in the Galápagos Islands are often described as the most influential on his thinking. The islands, which lie on the equator, 500 miles off the west coast of South America, had been named Galápagos (the Spanish word for tortoises) by the Spanish explorers who discovered them because of the abundance of giant tortoises, different ones on different islands and all of them different from those known anywhere else in the world. The Galápagos tortoises sluggishly clanked their way around, feeding on the vegetation and seeking the few pools of fresh water. They would have been vulnerable to predators had there been any on the islands.

Darwin also found large lizards, marine iguanas that feed, unlike any others of their kind, on seaweed, and mockingbirds quite different from those found on the South American mainland. His observations of several kinds of finches that varied from island to island in their features, notably their distinctive beaks that were adapted to disparate feeding habits—crushing nuts, probing for insects, grasping worms—are now part of the canon of science history.

In addition to *The Origin of Species* (1859), Darwin published many other books, notably *The Descent of Man, and Selection in Relation to Sex* (1871), which extends the theory of *natural selection* to human *evolution*.

## COPERNICUS AND DARWIN

There is a version of the history of the ideas that sees a parallel between the Copernican and the Darwinian Revolutions. In this view, the Copernican Revolution consisted of displacing the Earth from its previously accepted *locus* as the center of the universe, moving it to a subordinate place as just one more planet revolving around the Sun. In congruous manner, the Darwinian Revolution is viewed as consisting of the displacement of humans from their exalted position as the center of life on Earth, with all other *species* created for the service of humankind. According to this version of intellectual history, Copernicus had accomplished his revolution with the heliocentric theory of the solar system. Darwin's achievement emerged from his theory of organic *evolution*.[2]

This version of the two revolutions is inadequate. What it says is true, but it misses what is most important about these two intellectual revolutions, namely that they ushered in the beginning of science in the modern sense of the word. The two revolutions may jointly be seen as one Scientific Revolution with two stages, the Copernican and the Darwinian (Ayala, 2007).

The Copernican Revolution was launched with the publication in 1543, the year of Nicolaus Copernicus's death, of his *De revolutionibus orbium celestium* (*On the Revolutions of the Celestial Spheres*), and bloomed with the publication in 1687 of Isaac Newton's *Philosophiae naturalis principia mathematica* (*The Mathematical Principles of Natural Philosophy*). The discoveries by Copernicus, Kepler, Galileo, Newton, and others in the 16th and 17th centuries had gradually ushered in a conception of the universe as matter in motion governed by natural laws, and that the motions of the planets around the Sun can be explained by the same simple laws that account for the motion of physical objects on our planet—laws such as $f = m \times a$, *force* $=$ *mass* $\times$ *acceleration*, or the inverse-square law of attraction, $f = g(m_1 m_2)/r_2$, the force of attraction between two bodies is directly proportional to their masses, but inversely related to the square of the distance between them.

These discoveries greatly expanded human knowledge. The conceptual revolution they brought about was more fundamental yet: a commitment to the postulate that the universe obeys immanent laws that account for natural phenomena. The workings of the universe were brought into the realm of science: explanation through natural laws. All physical phenomena could be accounted for as long as the causes were adequately known.

The advances of physical science brought about by the Copernican Revolution drove mankind's conception of the universe to a split-personality state of affairs which persisted well into the mid-19th century. Scientific explanations derived from natural laws dominated the world of nonliving matter, on the Earth as well as in the heavens. Supernatural explanations, such as John Ray's (1691) and William Paley's (1802) explanation of *design*, which depended on the unfathomable deeds of the Creator, accounted for the origin and configuration of living creatures—the most diversified, complex, and interesting realities of the world.

It was Darwin's genius to resolve this conceptual schizophrenia. Darwin completed the Copernican Revolution by drawing out for biology the notion of nature as a lawful system of matter in motion that human reason can explain without recourse to supernatural agencies. The conundrum faced by Darwin can hardly be overestimated. The strength of the *argument from design* to demonstrate the role of the Creator had been forcefully set forth by Paley. Wherever there is function or *design*, we look for its author. It was Darwin's greatest accomplishment to show that the complex organization and functionality of living beings can be explained as the result of a natural process—*natural selection*—without any need to resort to a Creator or other external agent. The origin and *adaptation* of organisms in their profusion and wondrous variations were thus brought into the realm of science.

Darwin accepted that organisms are "designed" for certain purposes; that is, they are functionally organized. Organisms are adapted to certain ways of life, and their parts are adapted to perform certain functions. Fish are adapted to live in water, kidneys are designed to regulate the composition of blood, and the human hand is made for grasping. But Darwin went on to provide a natural explanation of the *design*. The seemingly purposeful aspects of living beings could now be explained, like the phenomena of the inanimate world, by the methods of science, as the result of natural laws manifested in natural processes (Reznick, 2010).

## DARWIN'S THEORY

The conclusion that Darwin considered *natural selection* (rather than his demonstration of *evolution*) to be his most important discovery emerges from consideration of his life and works. Darwin himself treasured *natural selection* and designated it as "my theory," a term he never used when referring to the *evolution* of organisms. The discovery of *natural selection*, Darwin's aware-ness that it was a greatly significant discovery because it was science's answer to Paley's *argument from design*, and Darwin's designation of it as "my the-ory" can be traced in his "Red Notebook" and "Transmutation Notebooks B to E," which he started in March 1837, not long after returning (on October 2, 1836) from his voyage on the *Beagle*, and completed in late 1839.[3]

The *evolution* of organisms was commonly accepted by naturalists in the middle decades of the 19th century, and the distribution of exotic *species* in South America, in the Galápagos Islands and elsewhere, and the discovery of *fossil* remains of long-extinguished animals confirmed the reality of *evolution* in Darwin's mind. The intellectual challenge was to explain the origin of distinct *species* of organisms, how new ones adapted to their environments, that "mystery of mysteries," as it had been labeled by Darwin's older contemporary, the prominent scientist and philosopher Sir John Herschel (1792–1871).

Early in the Notebooks of 1837–39, Darwin registers his discovery of *natural selection* and repeatedly refers to it as "my theory." From then until his death in 1882, Darwin's life was dedicated to substantiating *natural selection* and its companion postulates, mainly the pervasiveness of hereditary variation and the enormous fertility of organisms, which much surpassed the capacity of available resources. *Natural selection* became for Darwin "a theory by which to work." He relentlessly pursued observations and performed experiments to test the theory and resolve presumptive objections (Mayr, 1982; Ghiselin, 1969).

## DARWIN AND WALLACE

Alfred Russel Wallace (1823–1913) is famously given credit for discovering, independently of Darwin, *natural selection* as the process accounting for the *evolution* of *species*. On June 18, 1858, Darwin wrote to Charles Lyell that he had received by mail a short essay from Wallace such that "if Wallace had my [manuscript] sketch written in [1844] he could not have made a better ab-stract." Darwin was thunderstruck.

Darwin and Wallace started occasional correspondence in late 1855. At the time, Wallace was in the Malay archipelago collecting biological spec-imens. In his letters, Darwin offered sympathy and encouragement to the occasionally dispirited Wallace for his "laborious undertaking." In 1858 Wallace came upon the idea of *natural selection* as the explanation for evolutionary change and he wanted to know Darwin's opinion about this hypothesis, since Wallace, as well as many others, knew that Darwin had been working on the subject for years, had shared his ideas with other scientists, and was considered by them as the eminent expert on issues concerning biological *evolution*.

Darwin was uncertain how to proceed about Wallace's letter. He wanted to credit Wallace's discovery of *natural selection*, but he did not want altogether to give up his own earlier independent discovery. Eventually, Sir Charles Lyell and Joseph Hooker proposed, with Darwin's consent, that Wallace's letter and two of Darwin's earlier writings would be presented at a meeting of the Linnean Society of London. On July 1, 1858, three papers were read by the society's undersecretary, George Busk, in the *order* of their date of composition: Darwin's abbreviated abstract of his 230-page essay from 1844; an "abstract of abstract" that Darwin had written to the American botanist Asa Gray on September 5, 1857; and Wallace's essay, "On the Tendency of Varieties to Depart Indefinitely from Original Type; Instability of Varieties Supposed to Prove the Permanent Distinctness of *Species*."[4] The meeting was attended by some 30 people, who did not include Darwin or Wallace. The papers generated little response and virtually no discussion, their significance apparently lost to those in attendance. Nor was it noticed by the president of the Linnean Society, Thomas Bell, who, in his annual address the following May, blandly stated that the past year had not been enlivened by "any of those striking discoveries which at once revolutionize" a branch of science.

Wallace's independent discovery of *natural selection* is remarkable. Wallace, however, was not interested in explaining *design*, but rather in accounting for the *evolution* of *species*, as indicated in his paper's title: "On the Tendency of Varieties to Depart Indefinitely from Original Type." Wallace thought that *evolution* proceeds indefinitely and is progressive.[5] Darwin, in contrast, did not accept that *evolution* would necessarily represent *progress* or advancement, nor did he believe that *evolution* would always result in morphological change over time; rather, he knew of the existence of "living fossils," organisms that had remained unchanged for millions of years. For example, "some of the most ancient Silurian animals, as the Nautilus, Lingula, etc., do not differ much from living *species*."[6]

In 1858 Darwin was at work on a multivolume treatise, intended to be titled "On Natural Selection." Wallace's paper stimulated Darwin to write *The Origin*, which was published the following year. Darwin intended this as an abbreviated version of the much longer book he had meant to write. As noted earlier, Darwin's focus, in *The Origin* as elsewhere, was the explanation of *design*, with *evolution* playing the subsidiary role of supporting evidence.

## THE DARWINIAN AFTERMATH

The publication of *The Origin* produced considerable public excitement. Scientists, politicians, clergymen, and notables of all kinds read and discussed the book, defending or deriding Darwin's ideas. The most visible actor in the controversies immediately following publication was the English biologist T. H. Huxley, who later became known as "Darwin's bulldog," and who defended the theory of *evolution* with articulate and sometimes mordant words on public occasions as well as in numerous writings.

A younger English contemporary of Darwin, with considerable influence over the public during the latter part of the 19th and in the early 20th century, was Herbert Spencer. A philosopher rather than a biologist, he became an energetic proponent of evolutionary ideas, popularized a number of slogans, such as "survival of the fittest" (which was taken up by Darwin in later editions of *The Origin*), and engaged in social and metaphysical speculations. Unfortunately, his mistaken ideas considerably damaged proper understanding and acceptance of the theory of *evolution* by *natural selection*. Darwin wrote of Spencer's speculations that "his deductive manner of treating any subject is wholly opposed to my frame of mind ... His fundamental generalizations are of such a nature that they do not seem to me to be of any strictly scientific use." Most pernicious was the crude extension by Spencer and others of the notion of the "struggle for existence" to human economic and social life that became known as Social Darwinism (Moore, 1993).

## DARWIN AND MENDEL

The most serious difficulty facing Darwin's evolutionary theory was the lack of an adequate theory of inheritance that would account for the preservation through the generations of the variations on which *natural selection* was supposed to act. Contemporary theories of "blending inheritance" proposed that offspring merely struck an average between the characteristics of their parents. But as Darwin became aware, blending inheritance (including his own theory of "pangenesis," in which each organ and tissue of an organism throws off tiny contributions of itself that are collected in the sex organs and determine the configuration of the offspring) could not account for the conservation of variations, because differences between variant offspring would be halved each generation, rapidly reducing the difference between any new variation (which by itself might enhance *adaptation*) and the preexisting characteristics.

The missing link in Darwin's argument was provided by Mendelian genetics. About the time *The Origin of Species* was published, the Augustinian monk Gregor Mendel was starting a long series of experiments with peas in the garden of his monastery in Brünn, Austria–Hungary (now Brno, Czech Republic). These experiments and the analysis of their results are by any standard an example of masterly scientific method. Mendel's paper, "Experiments on Plant *Hybridization*," originally published in German in 1866 in the proceedings of the Natural Science Society of Brünn, formulated the fundamental principles of the theory of heredity that is still current. Mendel's theory accounts for biological inheritance through particulate factors (now known as genes) inherited one from each parent, which do not mix or blend in the progeny, but segregate in the formation of their sex cells or gametes.

Mendel's discoveries remained unknown to Darwin, however, and, indeed, did not become generally known until 1900, when they were almost simultaneously rediscovered by several scientists in Europe. In the meantime, Darwinism in the latter part of the 19th century faced the alternative evolutionary theory known as neo-Lamarckism. This hypothesis shared with Lamarck's the importance of use and disuse in the *development* and obliteration of organs, and added the notion that the environment acts directly on organic structures, which explained their *adaptation* to the way of life and environment of the organism. Adherents of neo-Lamarckism theory discarded *natural selection* as an explanation for *adaptation* to the environment.

Prominent among the defenders of *natural selection* was the German biologist August Weismann, who in the 1880s published his germ plasm theory. He distinguished two substances that make up an organism: the soma, which comprises most body parts and organs, and the germ plasm, which contains the cells that give rise to the gametes and hence to progeny. Early in the *development* of an egg, the germ plasm becomes segregated from the somatic cells that give rise to the rest of the body. This notion of a radical separation between germ plasm and soma—that is, between the reproductive tissues and all other body tissues—prompted Weismann to assert that inheritance of acquired characteristics was impossible, and it opened the way for his championship of *natural selection* as the only major process that would account for biological *evolution*. Weismann's ideas became known after 1896 as neo-Darwinism.

## THE SYNTHETIC THEORY

The rediscovery in 1900 of Mendel's theory of heredity by the Dutch botanist and geneticist Hugo de Vries and others led to an emphasis on the role of heredity in *evolution*. De Vries proposed a new theory of *evolution* known as mutationism, which essentially did away with *natural selection* as a major evolutionary process. According to de Vries (who was joined by other geneticists such as William Bateson in England), two kinds of variation take place in organisms. One is the "ordinary" variability observed among individuals of a *species*, which is of no lasting consequence in *evolution* because, according to de Vries, it could not "lead to a transgression of the *species* border [ie, to establishment of new *species*] even under conditions of the most stringent and continued *selection*." The other consists of the changes brought about by mutations, spontaneous alterations of genes that result in large modifications of the organism and give rise to new *species*: "The new *species* thus originates suddenly; it is produced by the existing one without any visible preparation and without transition."

Mutationism was opposed by many naturalists and in particular by the so-called biometricians, led by the English statistician Karl Pearson, who defended Darwinian *natural selection* as the major cause of *evolution* through the cumulative effects of small, continuous, individual variations (which the biometricians assumed passed from one generation to the next without being limited by Mendel's laws of inheritance).

The controversy between mutationists (also referred to at the time as Mendelians) and biometricians approached a resolution in the 1920s and 1930s through the theoretical work of geneticists. These scientists used mathematical arguments to show, first, that continuous variation (in such characteristics as body size, number of eggs laid, and the like) could be explained by Mendel's laws, and, second, that *natural selection* acting cumulatively on small variations could yield major evolutionary changes in form and function. Distinguished members of this group of theoretical geneticists were R. A. Fisher and J. B. S. Haldane in Britain and Sewall Wright in the United States. Their work contributed to the downfall of mutationism and, most important, provided a theoretical framework for the integration of genetics into Darwin's theory of *natural selection*. Yet their work had a limited impact on contemporary biologists for several reasons—it was formulated in a mathematical language that most biologists could not understand; it was almost exclusively theoretical, with little empirical *corroboration*; and it was limited in scope, largely omitting many issues, such as *speciation* (the process by which new *species* are formed), that were of great importance to evolutionists.

A major breakthrough came in 1937 with the publication of *Genetics and the Origin of Species* by Theodosius Dobzhansky, a Russian-born American naturalist and experimental geneticist. Dobzhansky's book advanced a reasonably comprehensive account of the evolutionary process in genetic terms, laced with experimental evidence supporting the theoretical arguments. *Genetics and the Origin of Species* (1937; third edition 1951) may be considered the most important landmark in the formulation of what came to be known as the *synthetic theory* of *evolution*, effectively combining Darwinian *natural selection* and Mendelian genetics. It had an enormous impact on naturalists and experimental biologists, who rapidly embraced the new understanding of the evolutionary process as one of genetic change in populations. Interest in evolutionary studies was greatly stimulated, and contributions to the theory soon began to follow, extending the synthesis of genetics and *natural selection* to a variety of biological fields.

The main writers who, with Dobzhansky, may be considered the architects of the *synthetic theory* were the German-born American zoologist Ernst Mayr (1942), the English zoologist Julian Huxley (1942), the American paleontologist George Gaylord Simpson (1944), and the American botanist George Ledyard Stebbins (1950). These scientists contributed to a burst of evolutionary studies in the traditional biological disciplines and some emerging ones—notably *population* genetics and, later, evolutionary ecology. By 1950 acceptance of Darwin's theory of *evolution* by *natural selection* was universal among biologists, and the *synthetic theory* had become widely adopted.

## MOLECULAR BIOLOGY

The most important line of investigation after 1950 was the application of molecular biology to evolutionary studies. In 1953 the American geneticist James Watson and the British biophysicist Francis Crick deduced the molecular structure of *DNA* (deoxyribonucleic acid), the hereditary material contained in the chromosomes of every cell's *nucleus*. The genetic information is encoded within the sequence of nucleotides that make up the chainlike *DNA* molecules. This information determines the sequence of amino acid building blocks of *protein* molecules, which include, among others, structural proteins such as collagen, respiratory proteins such as hemoglobin, and numerous enzymes responsible for the organism's fundamental life processes. Genetic information contained in the *DNA* can thus be investigated by examining the sequences of amino acids in the proteins.

In the mid-1960s laboratory techniques such as electrophoresis and selective assay of enzymes became available for the rapid and inexpensive study of differences among enzymes and other proteins. The application of these techniques to evolutionary problems made possible the pursuit of issues that earlier could not be investigated—for example, exploring the extent of genetic variation in natural populations (which sets bounds on their evolutionary potential) and determining the amount of genetic change that occurs during the formation of new *species*.

Comparisons of the amino acid sequences of corresponding proteins in different *species* provided quantitatively precise measures of the divergence among *species* evolved from common ancestors, a considerable improvement over the typically qualitative evaluations obtained by comparative anatomy and other evolutionary subdisciplines. In 1968 the Japanese geneticist Motoo Kimura proposed the neutrality theory of *molecular evolution*, which assumes that at the level of the sequences of nucleotides in *DNA* and amino acids in proteins, many changes are adaptively neutral; they have little or no effect on the molecule's function and thus on an organism's *fitness* within its environment. If the neutrality theory is correct, there should be a *"molecular clock"* of *evolution*; that is, the degree to which amino acid or *nucleotide* sequences diverge between *species* should provide a reliable estimate of the time since the *species* diverged. This would make it possible to reconstruct an evolutionary history that would reveal the *order* of branching of different lineages, such as those leading to humans, chimpanzees, and orangutans, as well as the time in the past when the lineages split from one another. During the 1970s and 1980s it gradually became clear that the *molecular clock* is not exact; nevertheless, into the early 21st century it continued to provide the most reliable evidence for reconstructing evolutionary history.

The laboratory techniques of *DNA cloning* and sequencing have provided a new and powerful means of investigating *evolution* at the molecular level. The fruits of this technology began to accumulate during the 1980s following the *development* of automated DNA-sequencing machines and the invention of the polymerase chain reaction, a simple and inexpensive technique that obtains, in a few hours, billions or trillions of copies of a specific *DNA* sequence or *gene*. Major research efforts such as the Human *Genome* Project further improved the technology for obtaining long *DNA* sequences rapidly and inexpensively. By the first few years of the 21st century the full *DNA* sequence—that is, the full genetic complement, or *genome*—had been obtained for more than 20 higher organisms, including human beings, the house mouse (*Mus musculus*), the rat *Rattus norvegicus*, the vinegar fly *Drosophila melanogaster*, the mosquito *Anopheles gambiae*, the nematode worm *Caenorhabditis elegans*, the malaria parasite *Plasmodium falciparum*, the mustard weed *Arabidopsis thaliana*, and the yeast *Saccharomyces cerevisiae*, as well as for numerous microorganisms. By the middle of the second decade of the 21st century the *DNA* sequences of more than 1,000 organisms were known and made available to investigators in GenBank and other publicly accessible electronic databases. The Human *Genome* Project was initiated in 1989 with the goal of completing the sequence of one human *genome* in 15 years at a cost of $3 billion. A draft of the *genome* sequence was completed ahead of schedule in 2001. In 2003 the Human *Genome* Project was completed, ahead of schedule and ahead of budget. One decade later, a human *genome* can be sequenced in less than one month and at a cost of $10,000.

## EARTH SCIENCES, BIOGEOGRAPHY, AND ECOLOGY

The Earth sciences also experienced, in the second half of the 20th century, a conceptual revolution with considerable consequence to the study of *evolution*. The theory of plate tectonics, which was formulated in the late 1960s, revealed that the configuration and position of the continents and oceans are dynamic, rather than static, features of Earth. Oceans grow and shrink, while continents break into fragments or coalesce into larger masses. The continents move across the Earth's surface at rates of a few centimeters a year. Over millions of years of geologic history this movement profoundly alters the face of the planet, causing major climatic changes along the way. These previously unsuspected massive modifications of the Earth's past environments are, of necessity, reflected in the evolutionary history of life. *Biogeography*, the evolutionary study of the geographic distribution of plant and animal *species*, has been revolutionized by the knowledge, for example, that Africa and South America were part of a single landmass some 200 million years ago and the Indian subcontinent was not connected with Asia until geologically recent times.

Ecology, the study of the interactions of organisms with their environments, has evolved from descriptive studies—"natural history"—into a vigorous biological discipline with a strong mathematical component, both in the *development* of theoretical models and in the collection and analysis of quantitative data. Evolutionary ecology is an active field of evolutionary biology; another is evolutionary ethology, the study of the *evolution* of animal behavior. *Sociobiology*, the evolutionary study of social behavior, is perhaps the most active subfield of ethology; it is also the most controversial, because of its extension to human societies.

## THE ORIGIN OF SPECIES

It deserves emphasis that Darwin accomplished something much more important for intellectual history than demonstrating *evolution* (Ghiselin, 1969; Mayr, 1982; Reznick, 2010; Ayala, 2007). Indeed, accumulating evidence for common descent with diversification may very well have been a subsidiary objective of Darwin's masterpiece. The main claim advanced here is that Darwin's *The Origin of Species* is, first and foremost, a sustained effort to solve the problem of how to account scientifically for the *design* of organisms. Darwin seeks to explain the *design* of organisms, their complexity, diversity, and marvelous contrivances as the result of natural processes. The evidence for *evolution* is brought in because *evolution* is a necessary consequence of his theory of *design*.

The introduction, this chapter, and 2–8 of *The Origin* explain how *natural selection* accounts for the adaptations and behaviors of organisms, their "*design*." The extended argument starts in this chapter, where Darwin describes the successful *selection* of domestic plants and animals with desired characteristics by farmers and, in considerable detail, the success of pigeon fanciers seeking exotic "sports." The success of plant and animal breeders manifests how much *selection* can accomplish by taking advantage of spontaneous variations that occur in organisms but happen to fit the breeders' objectives. A sport (*mutation*) that first appears in an individual can be multiplied by selective breeding, so after a few generations that sport becomes fixed in a breed, or "race." The familiar breeds of dogs, cattle, chickens, and food plants have been obtained by this process of *selection* practiced by people with particular objectives.

The ensuing 2–8 of *The Origin* extend the argument to variations propagated by *natural selection* for the benefit of the organisms themselves, rather than by artificial *selection* of traits desired by humans. As a consequence of *natural selection*, organisms exhibit *design*—that is, adaptive organs and functions. The *design* of organisms as they exist in nature, however, is not "*intelligent design*," imposed by God as a Supreme Engineer or by humans; rather, it is the result of a natural process of *selection*, promoting the *adaptation* of organisms to their environments. This is how *natural selection* works. Individuals that have beneficial variations—that is, variations that improve their probability of survival and reproduction—leave more descendants than individuals of the same *species* that have less beneficial variations. The beneficial variations will consequently increase in frequency over the generations, while less beneficial or harmful variations will be eliminated from the *species*. Eventually, all individuals of the *species* will have the beneficial features; new features will continue arising over eons of time.

Organisms exhibit complex *design*, but it is not *"irreducible complexity,"* emerging all of a sudden in its current elaboration. Rather, according to Darwin's theory of *natural selection*, the *design* has arisen gradually and cumulatively, step by step, promoted by the reproductive success of individuals with incrementally more beneficial elaborations. If Darwin's explanation of the adaptive organization of living beings is correct, *evolution* necessarily follows as a consequence of organisms becoming adapted to different environments in different localities and to the ever-changing conditions of the environment over time, and as hereditary variations become available at a particular time that improve the organisms' chances of survival and reproduction. *The Origin*'s evidence for biological *evolution* is central to Darwin's explanation of *"design,"* because this explanation implies that biological *evolution* occurs, which Darwin seeks to demonstrate in most of the remainder of the book (Chapters 9–13).

In the concluding Chapter 14 of *The Origin*, Darwin returns to the *dominant* theme of *adaptation* and *design*. In an eloquent final paragraph, Darwin asserts the "grandeur" of his vision: "It is interesting to contemplate an entangled bank, clothed with many plants of many kinds, with birds singing on the bushes, with various insects flitting about, and with worms crawling through the damp earth, and to reflect that these *elaborately constructed* forms, *so different* from each other, and dependent on each other *in so complex a manner*, have all been produced by laws acting around us ... Thus, from the war of nature, from famine and death, the most exalted object which we are capable of conceiving, namely, the production of the higher animals, directly follows. There is grandeur in this view of life, with its several powers, having been originally breathed into a few forms or into one; and that, whilst this planet has gone cycling on according to the fixed law of gravity, from so simple a beginning *endless forms most beautiful and most wonderful* have been, and are being, evolved."

Darwin's *The Origin of Species* addresses the same issue as William Paley and earlier authors: how to account for the adaptive configuration of organisms and their parts, which are so obviously "designed" to fulfill certain functions. Darwin argues that hereditary adaptive variations ("variations useful in some way to each being") occasionally appear, and that these are likely to increase the reproductive chances of their carriers. The success of pigeon fanciers and animal breeders clearly shows the occasional occurrence of useful hereditary variations. In nature, over the generations, Darwin's argument continues, favorable variations will be preserved, multiplied, and conjoined; injurious ones will be eliminated. In one place, Darwin avers: "I can see no limit to this power [*natural selection*] in slowly and beautifully *adapting* each form to the most complex relations of life."

In his autobiography written in 1876, Darwin wrote, "The old argument of *design* in nature, as given by Paley, which formerly seemed to me so conclusive, falls, now that the law of *natural selection* has been discovered. We can no longer argue that, for instance, the beautiful hinge of a bivalve shell must have been made by an intelligent being, like the hinge of a door by a man" (Darwin, 1887).

*Natural selection* was proposed by Darwin primarily to account for the adaptive organization, or "*design,*" of living beings; it is a process that preserves and promotes *adaptation*. Evolutionary change through time and evolutionary diversification (multiplication of *species*) often ensue as byproducts of *natural selection* fostering the *adaptation* of organisms to their milieux. Evolutionary change is not directly promoted by *natural selection*, however, and therefore it is not its necessary consequence. Indeed, some *species* may remain unchanged for long periods of time. Nautilus, lingula, and other so-called "living fossils," for example, have remained unchanged in their appearance for millions of years.

## Endnotes

1. The Galápagos Islands were discovered on March 10, 1535, by the Spanish Dominican priest Fray Bartolomé de Berlanga, Bishop of Panama, who narrates the discovery in a letter to the Spanish Emperor Carlos V dated April 21, 1535. Curiously, this discovery occurred almost exactly 300 years before Darwin's visit, which lasted from September 15 to October 20, 1835.
2. Sigmund Freud (1856–1939) refers to these two revolutions as "outrages" inflicted upon humankind's self-image: "Humanity in the course of time had to endure from the hands of science two great outrages upon its naïve self-love. The first was when it realized that our earth was not the center of the universe, but only a tiny speck in a world-system of a magnitude hardly conceivable; this is associated in our minds with the name of Copernicus, although Alexandrian doctrines taught something very similar. The second was when biological research robbed man of his peculiar privilege of having been specially created, and relegated him to a descent from the animal world, implying an ineradicable animal nature in him: this transvaluation has been accomplished in our own time upon the instigation of Charles Darwin, Wallace, and their predecessors, and not without the most violent opposition from their contemporaries." S. A. Freud, *A General Introduction to Psycho-Analysis* (first published in English in 1920), in *Great Books of the Western World*, vol. 54, *Freud*, ed. M. J. Adler (Chicago: Encyclopædia Britannica, Inc., 1993), 562. Freud proceeds to assert that "the third and most bitter blow" upon "man's craving for grandiosity" was meted out in the 20th century by psychoanalysis, revealing that man's *ego* "is not even master in his own house." Ibid.
3. For a review of the Red and Transmutation B to E Notebooks see Chapter 3 in N. Eldredge, *Darwin* (New York: Norton, 2005), 71–138.
4. Wallace's essay was published in the *Journal of the Proceedings of the Linnean Society of London (Zoology)* 3 (1858): 53–62. Darwin's two papers were also published in the same issue. Wallace throughout his life expressed admiration for Darwin and dedicated to him his book *Malay Archipelago* (1869). Wallace titled his book of 1989 *Darwinism: An Exposition of the Theory of Natural Selection with Some of Its Applications. Macmillan, London.*

5. In his paper, Wallace writes: "We believe that there is a tendency in nature to the continued progression of certain classes of varieties further and further from the original type—a progression to which there appears no reason to assign any definite limits. This progression, by minute steps, in various directions" (Berra, 1990, p. 53).
6. Darwin, *The Origin of Species*, 306. The Silurian period spans from 416 to 444 million years ago.

# Chapter 2

# Evolution Is a "Fact"

## INTRODUCTION

The *evolution* of organisms is at the core of biological disciplines, such as genetics, molecular biology, biochemistry, neurobiology, physiology, and ecology, and makes sense of the *emergence* of new diseases, the *development* of antibiotic resistance in bacteria, and other matters of public health. *Evolution* explains the agricultural relationships among wild and domestic plants and animals. It is used in informatics and computer science, in the *design* of chemical compounds, and in various other industries.

The term *"evolution"* means change over time. It usually refers to the *evolution* of living things, but is also used in other scientific contexts, particularly in astronomy to refer to the processes by which the whole physical universe—galaxies, stars, and planets—forms and changes.

Contrary to popular opinion, neither the term nor the idea of biological *evolution* began with Charles Darwin and his foremost work, *On the Origin of Species* (1859). The *Oxford English Dictionary* (1933) tells us that the word *"evolution,"* to unfold or open out, derives from the Latin *evolvere*, which applied to the "unrolling of a book." It first appeared in the English language in 1647 in a nonbiological connection, and became widely used in English to refer to all sorts of progressions from simpler beginnings. *Evolution* was first used as a biological term in 1670, to describe the changes observed in the maturation of insects. It was not until the 1873 edition of *Origin of Species* that Darwin first employed the term to refer to what we now call biological *evolution*. He had earlier used the expression "descent with modification," which is still a good brief definition of the process.

## THE UNIVERSE

It is widely believed that the universe started about 15 billion years ago in the Big Bang, a monumental explosion that sent matter and energy expanding in all directions. As the universe expanded, matter collected into clouds that gradually condensed into galaxies, such as our own Milky Way. In these galaxies gravitational attraction compressed the material, which in many cases condensed into stars where nuclear reactions took place. In the case of our

Evolution, Explanation, Ethics, and Aesthetics. http://dx.doi.org/10.1016/B978-0-12-803693-8.00002-5

Sun, gas and dust collided and aggregated, forming very small planets, which in successive stages coalesced into the eight (or nine, if Pluto is included) planets of our solar system and their numerous satellites.

The age of the Earth is estimated at 4.54 billion years. The oldest known rocks, dated at 3.96 billion years, are found in northwestern Canada, although rocks found in other places, such as Western Australia, encase zircon crystals dated at 4.3 billion years, older than the rocks themselves. The origins of life on Earth may have been as early as 4 billion years ago. There is evidence that organisms similar to today's bacteria lived 3.5 billion years ago. All *species* living today, estimated to number at least 10 million, derive by *evolution* from those early simple organisms.

## BIOLOGICAL EVOLUTION

*Evolution* is the process of "descent with modification," as Charles Darwin named it, through which all *species* now living, and the even more numerous ones that became extinct in the past, came about.

The virtually infinite variations on life are the fruit of *evolution*. All living creatures are related by descent from common ancestors. Humans and other mammals descend from shrew-like creatures that lived more than 150 million years ago; mammals, birds, reptiles, amphibians, and fish share as ancestors aquatic worms that lived 600 million years ago; and all plants and animals derive from bacteria-like microorganisms that originated more than 3 billion years ago. Lineages of organisms change through generations; diversity arises because the lineages that descend from common ancestors diverge over time as they adapt to different environments.

Evolutionary research consists mainly of three different, though related, issues: the fact of *evolution*—that is, that organisms are related by common descent with modification; evolutionary history—that is, the details of when lineages split from one another and the changes that occurred in each lineage; and the mechanisms or processes by which evolutionary change occurs.

The fact of *evolution* is the most fundamental issue and the one established with utmost certainty. Darwin gathered much evidence in its support, but the evidence has accumulated continuously ever since, derived from all biological disciplines. The evolutionary origin of organisms is today a scientific conclusion established with the kind of certainty attributable to such concepts as the roundness of the Earth, the motions of the planets, and the molecular composition of matter. This degree of certainty beyond reasonable doubt is what is implied when biologists say that *evolution* is a "fact": the evolutionary origin of organisms is accepted by virtually every biologist.

The second and third issues go far beyond the general affirmation that organisms evolve. The theory of *evolution* seeks to ascertain the relationships between particular organisms and the events of evolutionary history, and explain how and why *evolution* takes place. These are matters of active scientific investigation. Many conclusions are well established; for example, that chimpanzees and gorillas are more closely related to humans than are any of those three *species* to baboons or other monkeys; or that *natural selection* explains the adaptive configuration of such features as the human eye and the wings of birds. Some other matters are less certain, others are conjectural, and still others—such as precisely when life originated on Earth and the characteristics of the first living things (see Chapter 3)—remain largely unresolved.

However, uncertainty about these issues does not cast doubt on the fact of *evolution*. Similarly, we do not know all the details about the configuration of the universe and the origin of the galaxies, but this is not a reason to doubt that the galaxies exist, to throw out all we have learned about their characteristics, or to reject the notion that the Earth revolves around the Sun. Biological *evolution* is one of the most active fields of scientific research at present, and significant discoveries continually accumulate, supported in great part by advances in other disciplines.

The study of biological *evolution* has transformed our understanding of life on this planet. *Evolution* provides a scientific explanation for why there are so many different kinds of organisms on Earth and how they are all related because they are part of a single evolutionary lineage. It demonstrates why some organisms that look quite different are in fact closely related, while others that may look similar may be more distantly related. It accounts for the appearance of humans on Earth, and reveals our *species'* biological connections with other living things. It details how different groups of humans are related to each other and how we acquired many of our features. It enables the *development* of effective new ways to protect ourselves against constantly evolving bacteria and viruses.

## THE "FACT" OF BIOLOGICAL EVOLUTION

Darwin and other 19th-century biologists found compelling evidence for biological *evolution* in the comparative study of living organisms, in their geographic distribution, and in the *fossil* remains of extinct organisms. Since Darwin's time, the evidence from these sources has become considerably stronger and more comprehensive, while biological disciplines that have emerged more recently—genetics, biochemistry, physiology, ecology, animal behavior (ethology), and especially molecular biology—have supplied powerful additional evidence and detailed confirmation. The amount of information about evolutionary history stored in the deoxyribonucleic acid (*DNA*) and proteins of living things is virtually unlimited; scientists can reconstruct any detail of the evolutionary history of today's organisms by investing sufficient time and laboratory resources.

Evolutionists no longer are concerned with obtaining evidence to support the fact of *evolution*, but rather are concerned with what sorts of knowledge can be obtained from different sources of evidence. We can explore the most productive of these sources and illustrate the types of information they have provided.

## THE FOSSIL RECORD

Paleontology was a rudimentary science up to the 18th century, and gradually matured early in the 19th century in Darwin's time. Large parts of the geological succession of stratified rocks were unknown or inadequately studied up to the middle of the 19th century. Darwin therefore worried about the rarity of intermediate forms between major groups of organisms. Anti-evolutionists then and now have seized on this as a weakness in evolutionary theory. Although gaps in the paleontological record remain even today, many have been filled since Darwin's time. Paleontologists have recovered and studied the *fossil* remains of many thousands of organisms that lived in the past. This *fossil* record shows that many kinds of extinct organisms were very different in form from any now living. It also shows successions of organisms through time, manifesting their transition from one form to another.

When an organism dies, it is usually decomposed by bacteria and other forms of life, and by weathering processes. On rare occasions some body parts—particularly hard ones, such as shells, teeth, or bones—are preserved by being buried in mud or protected in some other way from predators and weather. Eventually, they may become petrified and preserved indefinitely with the rocks in which they are embedded. Methods such as radiometric dating—measuring the amounts of natural radioactive atoms that remain in certain minerals to determine the elapsed time since they were constituted—make it possible to estimate the time period when the rocks, and the fossils associated with them, were formed.

Radiometric dating indicates that Earth was formed about 4.5 billion years ago. The earliest known fossils resemble microorganisms, such as bacteria and archaea (see Chapter 3); the oldest of these fossils appear in rocks 3.5 billion years old. Numerous fossils belonging to many living phyla and exhibiting mineralized skeletons appear in rocks about 540 million years old. These organisms are different from organisms living now and from those living at intervening times. Some are so radically different that paleontologists have created new phyla (phyla are the largest taxonomic *category* of organisms) to classify them.

Microbial life of the simplest type (ie, prokaryotes, cells whose nuclear matter is not bounded by a nuclear membrane) was already in existence more than 3 billion years ago. The oldest evidence suggesting the existence of more complex organisms (ie, eukaryotic cells with a true *nucleus*) has been discovered in fossils sealed in flinty rocks approximately 1.4 billion years old, but such complex cells are thought to have originated as early as 2000 million years ago. More advanced forms, such as true algae, fungi, higher plants, and animals have been found only in younger geological strata. The first multicellular animals appeared around 670 million years ago, during the Ediacaran period. The first vertebrates (animals with backbones) appeared somewhat before 500 million years ago; the first mammals about 250 million years ago (Table 2.1).

The sequence of observed forms and the fact that all except the first on the list (prokaryotes) are constructed from the same basic cellular type strongly implies that all these major categories of life (including animals, plants, and fungi) have a common ancestry in the first eukaryotic cell. Moreover, there have been so many discoveries of intermediate forms between fish and amphibians, between amphibians and reptiles, between reptiles and mammals, and even along the *primate* line of descent from apes to humans that it is often difficult to identify categorically where the transition occurs from one to another particular *genus* or from one to another particular *species*. Nearly all fossils can be regarded as intermediates in some sense; they are life forms that come between ancestral forms that preceded them and those that followed.

The *fossil* record thus provides compelling evidence of systematic change through time—descent with modification. From this consistent body of evidence it can be predicted that no reversals will be found in future paleontological studies: that is, amphibians will not appear before fish or mammals before reptiles, and no complex life will occur in the geological record before the oldest eukaryotic cells.

Although some creationists (proponents of the direct *creation* by God of each and every one of all living *species*) have claimed that the geological record, with its orderly succession of fossils, is the product of a single universal flood a few thousand years ago that lasted a little longer than a year and covered the highest mountains to a depth of some about 7 m, there is clear evidence in the form of intertidal and terrestrial deposits that at no recorded time in the past has the entire planet been under water. Moreover, a universal flood of sufficient magnitude to deposit the existing strata, which together are many scores of kilometers thick, would require a volume of water far greater than has ever existed on and in the Earth, at least since the formation of the first known solid crust about 4 billion years ago. The belief that all this sediment with its fossils was deposited in an orderly sequence in only a year defies all geological observations and physical principles. There were periods of unusually high rainfall, and extensive flooding of inhabited areas has occurred, but there is no scientific support for the hypothesis of a universal mountain-topping flood.

It was Darwin, above all others, who first marshaled convincing evidence for biological *evolution*, but earlier scholars had recognized that organisms on Earth had changed systematically over long periods of time. For example, in 1799 an engineer named William Smith reported that in undisrupted layers of rock, fossils occurred in a definite sequential order, with more modern ones appearing closer to the top. Because bottom layers of rock logically were laid down earlier and thus are older than top layers, the

**TABLE 2.1** The Geologic Time Scale From 700 Million Years Ago to the Present, Showing Major Evolutionary Events

| million years ago | Era | Period | | Events |
|---|---|---|---|---|
| 0 / 1.8 | Cenozoic | Quaternary | | Evolution of humans |
| 50 | Cenozoic | Tertiary | | Mammals diversify |
| 100 | Mesozoic | Cretaceous | | Extinction of dinosaurs<br>first primates<br>first flowering plants |
| 150 | Mesozoic | Jurassic | | Dinosaurs diversify<br>first birds |
| 200 | Mesozoic | Triassic | | First mammals<br>first dinosaurs |
| 250 | Paleozoic | Permian | | Major extinctions<br>reptiles diversify |
| 300 | Paleozoic | Carboniferous | Pennsylvanian | First reptiles<br>scale trees<br>seed ferns |
| 350 | Paleozoic | Carboniferous | Mississippian | |
| 400 | Paleozoic | Devonian | | First amphibians<br>jawed fishes diversify |
| 450 | Paleozoic | Silurian | | First vascular land plants |
| 500 | Paleozoic | Ordovician | | Sudden diversification<br>of metazoan families |
| 550 | Paleozoic | Cambrian | | First fishes<br>first chordates |
| 600 / 650 / 700 | Precambrian | Ediacaran/Cryogenian | | First skeletal elements<br>first soft-bodied metazoans<br>first animal traces |

sequence of fossils also could be given a chronology from oldest to youngest. Smith's findings were confirmed and extended in the 1830s by the paleontologist William Lonsdale, who recognized that *fossil* remains of organisms from lower strata were more primitive than the ones above. Georges Cuvier is often considered the founding father of paleontology. As a member of the faculty at the National Museum of Natural Sciences in Paris in the early 19th century, he had access to the most extensive collection of fossils available at the time. However, he was adamant that fossils of animals different from those now living represented extinct *species* rather than ancestors of current *species*. The older contemporary of Darwin, Sir Charles Lyell, in his *Principles of Geology* (1830–1833) proposed that Earth's physical features were the outcome of major geological processes acting over immense periods of time, incomparably greater than the few thousand years since *Creation* that was commonly believed at the time. Today, many thousands of ancient rock deposits have been identified that show corresponding successions of *fossil* organisms.

Thus the general sequence of fossils had already been recognized before Darwin conceived of his theory of descent with modification, or *evolution* by *natural selection* as we would more usually say. The paleontologists and geologists before Darwin (including Charles Lyell) used the sequence of fossils in rocks not as proof of biological *evolution*, but mostly for working out the original sequence of rock strata that had been structurally disturbed by earthquakes and other forces. Nevertheless, in Darwin's time paleontology was still a rudimentary science. Large parts of the geological succession of stratified rocks were unknown or inadequately studied.

The *fossil* record is still incomplete. Of the small proportion of organisms preserved as fossils, only a tiny fraction have been recovered and studied by paleontologists, although in numerous cases the succession of forms over time has been reconstructed in considerable detail. One example is the *evolution* of the horse, which can be traced to an animal the size of a dog with several toes on each foot and teeth appropriate for browsing (eating tender shoots, twigs, and leaves of trees and shrubs); this animal, called the dawn horse (scientific name Hyracotherium), lived more than 50 million years ago. The most recent form, the modern horse (Equus), is much larger, is one-toed, and has teeth appropriate for grazing (eating growing herbage). Transitional forms are well preserved as fossils, as are other kinds of extinct horses that evolved in different directions and left no living descendants (Fig. 2.1).

Using recovered fossils, paleontologists have reconstructed examples of radical evolutionary transitions in form and function. For example, the lower jaw of reptiles consists of several bones, but that of mammals has only one. The other bones in the reptilian jaw evolved into bones now found in the mammalian ear. At first, such a transition would seem unlikely—it is hard to imagine what function such bones could have had during their intermediate stages. Yet paleontologists have discovered two transitional forms of

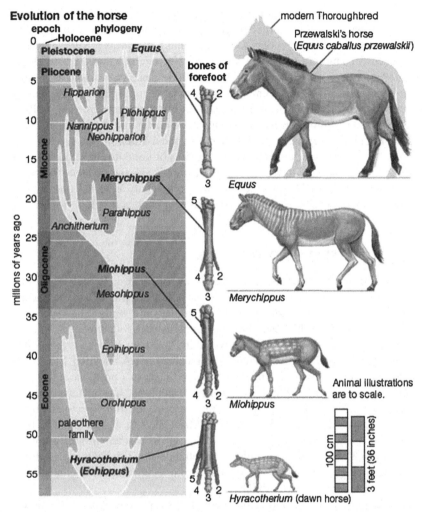

**FIGURE 2.1** Evolution of the horse. The earliest ancestor shown is Hyracotherium, which lived 50 million years ago and was small, about the size of a dog. Successive species became larger and had different dentition and fewer toes, as they adapted to different diets and ways of life. Branches on the left side of the figure represent species that lived at different times, most of which became extinct. The width of the branches corresponds to the abundance of the species. *Courtesy of Encyclopedia Britannica, Inc.*

mammal-like reptiles, called therapsids, that had a double jaw joint (ie, two hinge points side by side)—one joint consisting of the bones that persist in the mammalian jaw and the other composed of the quadrate and articular bones, which eventually became the hammer and anvil of the mammalian ear (Simpson, 1949, 1961).

For skeptical contemporaries of Darwin, the "missing link"—the absence of any known transitional form between apes and humans, or between fish and amphibians, or between reptiles (dinosaurs) and birds—was a battle cry, as it remained for uninformed people afterward. Not one but many creatures intermediate between living apes and humans have since been found as fossils. The oldest known *fossil* hominins—ie, primates belonging to the human lineage after it separated from lineages going to the apes—are 6–7 million years old, come from Africa, and are known as Sahelanthropus and *Orrorin* (or Praeanthropus). They were predominantly bipedal when on the ground, but had very small brains. Numerous *fossil* remains of hominins have been found over the last century and to the present which are intermediate between *Orrorin* and modern humans, notably with respect to brain size, which increases from about 300–400 cc in the earliest hominins to 1300–1400 cc in modern humans (Chapter 5).

## ARCHAEOPTERYX AND TIKTAALIK

In addition to the hominins, many fossils intermediate between diverse organisms have been discovered over the years. Two examples are Archaeopteryx, an animal intermediate between reptiles and birds, and Tiktaalik, intermediate between fish and tetrapods (animals with four limbs).

The first Archaeopteryx was discovered in Bavaria in 1861, two years after the publication of Darwin's *Origin*, and received much attention because it shed light on the origin of birds and bolstered Darwin's postulate of the existence of *missing links* (Fig. 2.2). Other Archaeopteryx specimens have been discovered in the past 100 years. The most recent, the 10th specimen so far recovered, was described in December 2005 (Mayr et al., 2005). The best preserved Archaeopteryx yet, it is now housed in a small privately owned museum in Thermopolis, Wyoming. The tetrapod-like fish Tiktaalik is also a very recent discovery, published only on April 6, 2006 (Daeschler et al., 2006; Shubin et al., 2006; Shubin, 2008).

Archaeopteryx lived during the Late Jurassic period, about 150 million years ago, and exhibited a mixture of both avian and reptilian traits. All known specimens are small, about the size of a crow, and share many anatomical characteristics with some of the smaller bipedal dinosaurs. The skeleton is reptile-like, but Archaeopteryx had feathers, clearly shown in the fossils, with a *skull* and a beak like those of a bird. Other features are intermediate between reptiles and birds. Unlike modern birds, Archaeopteryx had teeth and a long tail, similar in structure to that of the smaller dinosaurs. The hind legs are bird-like, but the well-preserved foot of the most recent specimen shows a hyperextensible second toe, similar to the killer claw of the dinosaur Velociraptor. The forelimbs retained primitive reptilian characteristics and had not yet completed their transformation into wings. Archaeopteryx may have been capable of flying, but it was not capable of sustained flight. The most recent

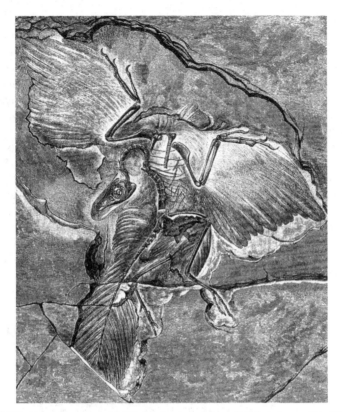

**FIGURE 2.2**    An Archaeopteryx fossil showing traits intermediate between reptiles (dinosaurs) and birds. *Courtesy of Museum für Naturkunde, Berlin.*

specimen indicates that Archaeopteryx lived mostly on the ground, rather than in trees (Mayr et al., 2005).

Paleontologists have known for more than a century that tetrapods (amphibians, reptiles, birds, and mammals) evolved from a particular group of fishes called lobe-finned. Until recently, Panderichthys was the closest known *fossil* fish to the tetrapods. It was somewhat crocodile-shaped and had a pectoral-fin skeleton and a shoulder girdle intermediate in shape between those of typical lobe-finned fish and those of tetrapods, which allowed it to "walk" in shallow waters, but probably not on land. In most features, however, it was more like a fish than a *tetrapod*. These transitional fish still had gill cover plates which pumped water over the gills, and the *skull* was fused to its shoulder girdle. Panderichthys is known from Latvia, where it lived some 385 million years ago (the Mid-Devonian period) (Shubin, 2008).

Until very recently the earliest *tetrapod* fossils that are more nearly fish-like were also from the Devonian, about 376 million years old. They have been found in Scotland and Latvia. Ichthyostega and Acanthostega from

Greenland, which lived more recently, about 365 million years ago, are un-ambiguous tetrapods, with limbs that bear digits, although they retain from their fish ancestors such characteristics as true fish tails with fin rays. Thus the time gap between the most tetrapod-like fish and the most fish-like tetrapods was nearly 10 million years, between 385 and 376 million years ago. The morphological gap was also substantial because none of these animals (all between 75 cm and 1.5 m in length) was truly an intermediate between fish and tetrapods (Daeschler et al., 2006; Shubin et al., 2006).

The recently discovered Tiktaalik goes a long way toward filling this gap; it is the most nearly intermediate between fish and tetrapods yet known. Several specimens have been found in Late Devonian river sediments, dated about 380 million years ago, on Ellesmere Island in Nunavut, Arctic Canada. *Tiktaalik* is Inuit for "big freshwater fish." It displays an array of features that are just about as precisely intermediate between fish and tetrapods as one could imagine, and exactly fits the time gap as well (Fig. 2.3; Shubin, 2008). The excavated fossils include three skulls, 10 jawbones, and two specimens with

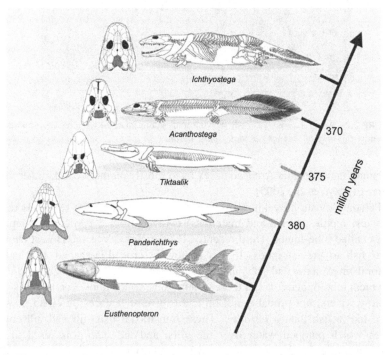

**FIGURE 2.3**  Tiktaalik and other fossil intermediates between fish and tetrapods. Other inter-mediate fossils, from closer to the fish to closer to the amphibians, are Eusthenopteron, Pander-ichthys, Acanthastega, and Ichthyostega, which lived between 385 and 359 million years ago. *Adapted from Ahlberg, E., Clark, J.A. Nature 440 2006, 747. Reprinted with permission from Macmillan Publishers Ltd.*

head and trunk in one piece. Tiktaalik was a flattened, superficially crocodile-like animal, with a *skull* about 20 cm in length. The pectoral fins are incipient forelimbs, with robust internal skeletons, but fringed with fin rays rather than digits. Fish-like features include small pelvic fins, the already-mentioned fin rays in their paired appendages, and well-developed gill arches, which suggest that they remained mostly aquatic. But the bony gill cover has disappeared, indicating reduced water flow through the gill chamber. The elongated snout suggests a shift from sucking up toward snapping up prey, mostly on land. The relatively large ribs indicate that Tiktaalik could support its body out of water (Daeschler et al., 2006; Shubin et al., 2006).

## EXTINCTION

The number of *species* living on Earth at present or at any time represents the balance between the *species* that originated in the past and those that became extinct. Many strange or bizarre animals of the past are revealed by the *fossil* record: dinosaurs, ammonites, trilobites, and many others. Paleontologists estimate that more than 99% and perhaps more than 99.9% of all *species* that lived in the past became extinct. That gives rise to a sobering thought. The number of *species* now living on Earth (excluding bacteria and archaea) is estimated at more than 10 million. If this number represents 1% of the total that ever existed, this would amount to 1 billion *species*; if it only represents 0.1%, the number of *species* that have lived on Earth would amount to 10 billion. Why did so many *species* become extinct (Simpson, 1949, 1953; Valentine, 1985, 2004; Raup, 1991; Sepkoski, 1997; Erwin, 2006; Erwin and Valentine, 2013)?

*Natural selection* acts to improve or maintain *adaptation* to prevailing conditions. If environmental change is very slow, *selection* may be able to maintain a high level of *adaptation*. However, adaptations are not selected for future conditions, but for present ones. Environmental change is environmental deterioration so far as organisms are concerned. Since environmental change is incessant, we expect extinctions to occur continuously, rising in intensity during periods of greater and more rapid change (Mayr, 1982; Valentine, 1973, 1985; Jablonski, 1999).

Paleontologists identify two patterns of extinction. One is the pattern of "background extinctions," which reflects the failure of *species* to adapt to the ongoing process of change in not only the physical environment but also the biotic environment, including competitors, parasites, and predators. But there are also "mass extinctions," when a very large number of taxa, even a majority in some cases, become extinct over a relatively short (in the geological scale) period of time.

Five mass animal extinctions are generally recognized; they occurred at the end of the Ordovician period, late in the Devonian, at the boundary between the Permian and the Triassic, at the end of the Triassic, and at the boundary between the Cretaceous and the *Tertiary* (K/T extinction). The corresponding times for the five mass extinctions, in millions of years ago, are approximately 440, 370, 260, 200, and 65. Not all were equally severe, nor were all animal taxa equally impacted. The most severe episode of mass extinction was the one at the boundary between the Permian and the Triassic, some 260 million years ago, when trilobites became extinct and corals, brachiopods, and crinoids very nearly so. Estimates of the taxa extinguished around the boundary between the Permian and the Triassic are 54% of marine families, 84% of genera, and 80−90% of all *species* (Valentine, 1973, 1985; Erwin, 2006; Raup, 1991; Sepkoski, 1987; Raup, 1991).

The causes of mass extinctions are for the most part conjectural. Massive volcanic eruptions throughout the planet are considered one likely cause of the event at the Permian/Triassic boundary, with associated global warming. Particularly interesting is the K/T extinction. It was suggested in 1980 that the cause was the impact of an extraterrestrial body, which would have extended a thick layer of dust throughout the atmosphere, darkening the sky and lowering temperatures. Large terrestrial animals, notably the dinosaurs, would have become extinct. The small mammals then in existence would have fared better and some survived. As the atmosphere recovered, the mammals diversified and larger taxa evolved, notably the primates and humans. Geologists have identified the Chicxulub crater off the coast of the Yucatan peninsula of Mexico as the place where the impact occurred (Alvarez et al., 1980).

The dinosaurs dominated the Earth, seemingly inhibiting the *evolution* of other large animals. Their extinction made possible the diversification in size and otherwise of the preexisting small mammals. If so, we humans owe our existence to a meteorite. Its impact made it possible for primates and eventually humans to evolve. A sobering thought, indeed, that a meteorite impact on the Yucatan peninsula may have made possible the *evolution* of our *species*, *Homo sapiens*.

## ANATOMICAL SIMILARITIES

The skeletons of turtles, horses, humans, birds, whales, and bats are strikingly similar (Fig. 2.4), in spite of the different ways of life of these animals and the diversity of their environments. The correspondence, bone by bone, can easily be seen in the limbs as well as in other parts of the body. From a purely pragmatic point of view, it seems incomprehensible that with forelimb structures built of the same bones a turtle and a whale should swim, a horse run, a person write, and a bird or bat fly. An engineer could *design* better limbs for each purpose. However, if we accept that these animals inherited their skeletal structures from a common ancestor and became modified only as they adapted to different ways of life, then the similarity of their structures makes sense.

Scientists call such structures *homologous* or inherited similarities, and have concluded that they are best explained by common descent; that is, the

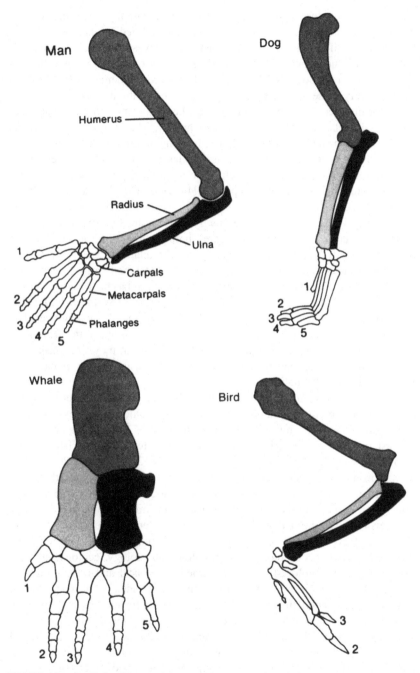

**FIGURE 2.4** Forelimb skeleton of four vertebrates, showing similar and similarly arranged bones, although used for different functions, in human, dog, whale, and bird.

*homologous* structure evolved in an ancestor common to all the *species* exhibiting the *homology* and was subsequently modified by *natural selection* to fit different lifestyles. Comparative anatomists investigate such homologies, not only in bone structure but also in other parts of the body, working out relationships from degrees of similarity. The correspondence of structures is typically very close among some organisms—the different varieties of song-birds, for instance—but becomes less so as the organisms being compared are less closely related in their evolutionary history. The similarities are fewer between mammals and birds than they are among mammalian *species*, and are fewer still between mammals and fish. Similarities in structure therefore not only manifest *evolution* but also help to reconstruct the *phylogeny*, or evolutionary history, of organisms (Simpson, 1944, 1953; Dobzhansky et al., 1977; Ayala, 2007).

Comparative anatomy also reveals why most organismic structures are not perfect. Similar to the forelimbs of turtles, horses, humans, birds, and bats, an organism's body parts are less than perfectly adapted because they are modified from an inherited structure rather than designed from scratch for a specific purpose. The anatomy of animals shows that it has been designed to fit their lifestyles, but it is "imperfect" *design*, accomplished by *natural selection*, rather than "intelligent" *design*, as it would be if designed by an engineer. The imperfection of structures is evidence for *evolution*, contrary to the arguments asserting that the *design* of organisms demonstrates that they were fashioned by a Creator, just as the *design* of a watch demonstrates that it has been fashioned by a watchmaker.

Separate lineages sometimes independently evolve similar features, known as *analogous* structures. Although these look like homologies, they result not from common ancestry but rather from "*convergence*," because the organisms, at least in some respects, have similar ways of life (Simpson, 1944, 1961; Dobzhansky et al., 1977; Valentine, 1985; Shubin, 2008; Gould, 2002; Conway Morris, 2003). For example, dolphins are aquatic mammals that have evolved from terrestrial mammals over the past 50 million years. In evolutionary terms they are as distant from fish as are mice or humans. But they have evolved streamlined bodies that closely resemble the bodies of fish, sharks, and even extinct dinosaurs known as ichthyosaurs. Evidence from many different fields of biology (morphology, physiology, ecology, etc.) allows evolutionary biologists to discern whether physical and behavioral similarities are the product of common descent or are independent responses to similar environmental challenges. Typically, *analogous* features are similar in overall configuration but not in the details, unlike *homologous* structures. Thus, for example, the skeletons of forelimbs of humans, dogs, whales, and birds consist of the same components arranged in similar ways: humerus, radius, and ulna. The number and arrangement of carpals, metacarpals, and phalanges are also the same in humans, dogs, and whales, although they are somewhat modified in birds. This precise correspondence in several organisms between the component parts of a limb or an organ is evidence of common descent from an ancestor that had a similar structure with the same components arranged in the same way; that is, the structures are *homologous*.

On the other hand, the anatomies of a dolphin and a tuna are clearly designed for swimming and living in the oceans, but the details of the *design* are quite different (for example, tuna, unlike dolphins, have gills for breathing), showing that they are *analogous*. Similarly, the wings of both a bird and a bat are designed for flying, but they are quite different in their anatomical detail; from this we can tell that they evolved independently rather than being inherited from the same winged ancestor.

## EMBRYONIC SIMILARITIES AND VESTIGES

Evolutionists, following Darwin, find evidence for *evolution* in the similarities in the *development* of embryos; that is, in the *development* of organisms, say reptiles, birds, and mammals, from fertilized egg to moment of birth or hatching. Vertebrates, from fish through lizards to humans, develop in ways that are remarkably similar during the early stages, but they become more and more differentiated as the embryo approaches maturity. The similarities persist longer between organisms that are more closely related (eg, humans and monkeys) than between those less closely related (such as humans and lizards) (Haeckel, 1866; Simpson, 1944, 1961; Dobzhansky et al., 1977; Valentine, 1985, 2004; Conway Morris, 1998; Ruse and Richards, 2009; Gould, 2002; Mayr, 1982; Vermeij, 1987).

Common developmental patterns reflect evolutionary kinship. Lizards and humans share a developmental pattern inherited from their remote common ancestor; the inherited pattern of each was modified only as the separate descendant lineages evolved in different directions. The common embryonic stages of the two creatures reflect the constraints imposed by this shared inheritance, which prevent changes that have not been necessitated by their diverging environments and ways of life.

The embryos of humans and other nonaquatic vertebrates exhibit gill slits even though they never breathe through gills. These slits are found in the embryos of all vertebrates because they share as common ancestors the fish in which these structures first evolved. Human embryos also exhibit, by the fourth week of *development*, a well-defined tail, which reaches its maximum length at six weeks. Similar embryonic tails are found in other mammals, such as dogs, horses, and monkeys; in humans, however, the tail eventually shortens, persisting only as a rudiment in the adult coccyx. Embryonic rudiments are inconsistent with claims of intentional *design*: why would a structure be designed to form during early *development* if it will disappear before birth? *Evolution* makes sense of embryonic rudiments (Dobzhansky et al., 1977; Shubin, 2008; Valentine, 1985).

A close evolutionary relationship between organisms that appear drastically different as adults can sometimes be recognized through their embryonic homologies. Barnacles, for example, are sedentary crustaceans with little apparent likeness to free-swimming crustaceans, such as lobsters, shrimps, or copepods. Yet they pass through a free-swimming larval stage, the nauplius, which is unmistakably similar to that of other crustacean larvae.

Embryonic rudiments that never fully develop, such as the gill slits in humans, are common in all sorts of animals. Some, however, persist as adult vestiges, reflecting evolutionary ancestry. A familiar rudimentary organ in humans is the vermiform appendix. This worm-like structure attaches to a short section of intestine called the caecum, which is located at the point where the large and small intestines join. The human vermiform appendix is a functionless vestige of a fully developed organ present in mammals, such as the rabbit and other herbivores, where a large caecum and appendix store vegetable cellulose to enable its digestion with the help of bacteria. Vestiges are instances of imperfections—like those seen in anatomical structures—that argue against *creation* by *design* but are fully understandable as a result of *evolution* by *natural selection* (Dobzhansky et al., 1977).

The discovery of genes as the carriers of biological heredity, and later of *DNA* as the chemical that encodes genetic information, raised the challenge of "ontogenetic decoding," or the "egg-to-adult transformation." I mean by these phrases the problem of how the linear information contained in the sequence of letters in the *DNA* becomes transformed into a three-dimensional organism that exists and changes through time. Scientific knowledge about this problem has increased enormously, particularly in the past four decades, made possible by the rapidly advancing discipline of molecular biology, as mentioned in Chapter 1 and to which we return later in this chapter (Ayala, 2007; see also Chapter 5).

## BIOGEOGRAPHY

The disparate geographic distribution of plants and animals throughout the world and the distinctive flora and fauna of island archipelagos were, for Darwin, evidence of *evolution*, eventually reinforced by later knowledge. The diversity of life is stupendous. Approximately 250,000 *species* of living plants, 100,000 *species* of fungi, and more than 1 million *species* of animals have been described and named, each occupying its own peculiar ecological setting or *niche*, and the census is far from complete (Simpson, 1944; Vermeij, 1987; Valentine, 1973). Some *species*, such as human beings and our companion the dog, can live under a wide range of environmental conditions. Others are amazingly specialized. One *species* of the fungus *Laboulbenia* grows exclusively on the rear portion of the covering wings of a single *species* of beetle (*Aphaenops cronei*) found only in some caves of southern France.

The larvae of the fly *Drosophila carcinophila* can develop only in specialized grooves beneath the flaps of the third pair of oral appendages of a land crab (*Gecarcinus ruricola*) found on certain Caribbean islands.

How can we make intelligible the colossal diversity of living beings and the existence of such extraordinary, seemingly whimsical, creatures as the fungus beetle and the fly described above? And why are island groups like the Galápagos so often inhabited by forms similar to those on the nearest mainland but belonging to different species? *Evolution* explains that biological diversity results from the descendants of local or migrant predecessors becoming adapted to their diverse environments. This explanation can be tested by examining present *species* and local fossils to see whether they have similar structures, which would indicate how one is derived from the other. There should also be evidence that forms without an established local ancestry have migrated into the locality.

Wherever such tests have been carried out, these conditions have been confirmed. A good example is provided by the mammalian populations of North and South America, where strikingly different *endemic* forms evolved in isolation until the *emergence* of the isthmus of Panama approximately 3 million years ago. Thereafter, the armadillo, porcupine, and opossum— mammals of South American origin—migrated north, along with many other *species* of plants and animals, while the mountain lion and other North American *species* made their way across the isthmus to the south.

Each of the world's continents has its own distinctive collection of animals and plants. In Africa, there are rhinoceroses, hippopotamuses, lions, hyenas, giraffes, zebras, lemurs, monkeys with narrow noses and nonprehensile tails, chimpanzees, and gorillas. South America, which extends over much the same latitudes as Africa, has none of these animals; instead it has pumas, jaguars, tapirs, llamas, raccoons, opossums, armadillos, and monkeys with broad noses and large prehensile tails. Australia boasts a great diversity of marsupial mammals, which lack placentas so much of the early *development* takes place in a mother's external pouch rather than inside the womb. Marsupials include kangaroos, Australian moles and anteaters, and Tasmanian wolves (Fig. 2.5).

The vagaries of *biogeography* are not due solely to the suitability of the different environments. There is no reason to believe that South American animals are not well suited to living in Africa or those of Africa to living in South America. When rabbits were intentionally introduced in Australia, so they could be hunted for sport, they prospered beyond the expectations of the introducers and became an agricultural pest. Hawaii lacks native land mammals, but when feral pigs and goats were brought to the islands in the 19th century for hunting, they multiplied to large numbers and are now endangering the native vegetation.

An interesting story is the case of Santa Catalina Island, some 25 miles southwest of Los Angeles. Bison were introduced to the island in the 1940s by some filmmakers. After their filming was completed, they did not bother

marsupials | placentals

Tasmanian wolf
(*Thylacinus*)

wolf
(*Canis*)

glider, or flying phalanger
(*Petaurus*)

flying squirrel
(*Glaucomys*)

marsupial mouse
(*Dasycercus*)

mouse
(*Mus*)

marsupial mole
(*Notoryctes*)

mole
(*Talpa*)

native cat
(*Dasyurus*)

ocelot
(*Felis*)

numbat, or banded anteater
(*Myrmecobius*)

anteater
(*Myrmecophaga*)

wombat
(*Phascolomys*)

groundhog
(*Marmota*)

**FIGURE 2.5**   Parallel evolution of marsupial mammals in Australia (left) and placental mammals on other continents (right).

removing the animals. Santa Catalina does not have any native mammal larger than a fox. Yet the bison prospered and successfully reproduced throughout the uninhabited parts of the island, threatening the vegetation. The bison are now regularly culled so that only two herds with a few dozen animals are kept, and these have become a distinctive tourist attraction.

The remarkable diversification of life in different parts of the world is evidence of *evolution* promoted by *natural selection*. Even though climate and other features of the environment may be comparable at similar latitudes, the flora and fauna are diverse in different continents and on different islands. This diversity occurs because *natural selection* depends on the opportunism of genetic mutations, which are random events. Moreover, *evolution* relies on previous changes, so diversification from one continent or island to another, or between continents and islands, is cumulative over time. Evolutionary change occurs in response to the environment, but it is conditioned by history: mammals do not evolve into fish, or insects into molluscs.

Darwin's observations of the flora and fauna of South America, so different from those of the Old World, convinced him of the reality of *evolution*. The evidence from *biogeography* is also apparent on a scale much smaller than continental: Darwin observed that different Galápagos islands had different kinds of tortoises and different *species* of finches, which in turn were different from those found in continental South America. He was startled by the Galápagos tortoises, giant lizards, mockingbirds, and finches, different as they were from mainland *species* and diverse among the islands as well.

## HAWAII'S BIOLOGICAL CAULDRON

The Galápagos Islands are on the equator, about 500 miles west of the South American country of Ecuador. More remote yet than the Galápagos are the Hawaiian Islands, about 2500 miles from the North American mainland (Fig. 2.6). Many sorts of plants and animals are lacking in Hawaii, whereas others are *endemic* (ie, native nowhere else on Earth) and extraordinarily diverse. Table 2.2 lists groups of organisms with numerous and very diverse *species* native to Hawaii.

Kohala, the oldest volcano on the large island of Hawaii, is somewhat less than a million years old; of the other volcanoes, Mauna Kea and Mauna Loa are much younger, and Kilauea is still active. The island of Kauai was formed less than 10 million years ago; other islands are of intermediate age, increasingly older from southeast to northwest. The gradual formation over millions of years of these volcanic islands has resulted in successive colonizations by plants and animals, and therefore *species* diversification. *Drosophila* fruitflies are favored by experimental geneticists because they can easily and inexpensively be cultured in the laboratory. The ecology, behavior, and genetics of Hawaiian fruitflies have been studied intensively. There are about 2000 known *species* of *Drosophila* flies in the world; nearly one-third of them live in Hawaii and nowhere else, although the total area of the archipelago is less than one-twentieth the area of California. Moreover, the morphological and behavioral diversity of Hawaiian *Drosophila* exceeds that of *Drosophila* in the rest of the world. There are more than 1000 *species* of

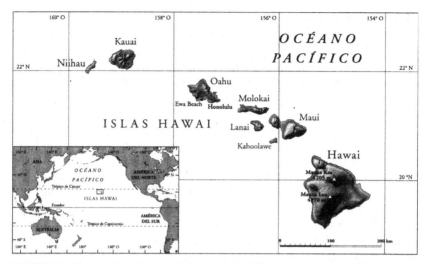

**FIGURE 2.6**   The Hawaiian Islands (with inset of the Pacific Ocean), more than 2000 miles away from the nearest continent. These volcanic islands formed between approximately 5 million (Kauai) and 500,000 (Hawaii) years ago. *Pacific Ocean inset courtesy of NOAA Coastal Services Center.*

**TABLE 2.2** Endemic Biota of Hawaii

|  | Number of Species | % Endemic |
|---|---|---|
| Ferns | 168 | 65 |
| Flowering plants | 1729 | 94 |
| Snails | 1064 | 99+ |
| *Drosophila* | 600 | 100 |
| Other insects | 3750 | 99+ |
| Land mammals | 0 | 0 |

land snails in Hawaii, all of which have evolved in the archipelago; and about 80 bird *species*, all but one of which exist nowhere else (Ayala, 2007).

Why has such "explosive" *evolution* occurred in Hawaii? The overabundance of fruitflies there contrasts with the absence of many other native insects, such as mosquitoes and cockroaches. Because of their remote isolation, the Hawaiian Islands have rarely been reached by colonizing plants and animals. Some that did reach the islands found suitable habitats without competitors or predators. The ancestors of Hawaiian fruitflies were passively transported to the archipelago by air currents or flotsam before other groups of insects reached it, and there they found a multitude of opportunities for living. They rapidly evolved and diversified by exploiting the available resources. It is known from genetic studies that several hundred *species* have derived from a single colonizing *species*; they adapted to the variety of opportunities available in diverse niches by evolving suitable adaptations, which range broadly from one *species* to another. In Hawaii some *Drosophila species* feed on decaying leaves on the forest floor, others feed on flowers, still others on fungi, and so on.

The geographic remoteness of the Hawaiian Islands is a more reasonable explanation for the explosive diversity of a few kinds of organisms—such as fruitflies, snails, and birds—than an inordinate preference on the part of the Creator for providing the archipelago with numerous flies, or a peculiar distaste for creating mosquitoes, cockroaches, and other insects there. There are no native land mammals in Hawaii; no mammals existed there until pigs and goats were introduced by humans. Hawaii was never colonized by mammals because none happened to reach the archipelago from the distant continents where mammals lived.

The Hawaiian Islands are no less hospitable than other parts of the world. The absence of many kinds of organisms, and the great multiplication of a few kinds, is due to the fact that many sorts of organisms never reached the islands because of their geographic isolation. Those that did diversified over time because of the absence of related organisms that would compete for resources.

## MOLECULAR BIOLOGY

Molecular biology provides the most detailed and convincing evidence available for biological *evolution*. In its unveiling of the nature of *DNA* and the workings of organisms at the level of enzymes and other *protein* molecules, it has shown that these molecules hold information about an organism's ancestry. This has made it possible to reconstruct evolutionary events that were previously unknown, and to confirm and adjust the view of events already known. The precision with which these events can be reconstructed is one reason why the evidence from molecular biology is so compelling. Another reason is that *molecular evolution* has shown all living organisms, from bacteria to humans, to be related by descent from common ancestors (see Chapter 3).

A remarkable uniformity exists in the molecular components of organisms—in the nature of the components as well as in the ways in which they are assembled and used. In all bacteria, plants, animals, and humans, the *DNA* comprises a different sequence of the same four component nucleotides, and all the various proteins are synthesized from different combinations and sequences of the same 20 amino acids, although several hundred other amino acids do exist. The genetic code by which the information contained in the *DNA* of the cell *nucleus* is passed on to proteins is virtually everywhere the same. Similar metabolic pathways—sequences of biochemical reactions—are used by the most diverse organisms to produce energy and make up the cell components.

This unity reveals the genetic continuity and common ancestry of all organisms. There is no other rational way to account for their molecular uniformity when numerous alternative structures are equally likely. The genetic code serves as an example. Each particular sequence of three nucleotides in the nuclear *DNA* acts as a pattern for the production of exactly the same amino acid in all organisms. This is no more necessary than it is for a language to use a particular combination of letters to represent a particular object. If it is found that certain sequences of letters—planet, tree, woman—are used with identical meanings in a number of different books, one can be sure that the languages used in those books are of common origin.

Genes and proteins are long molecules that contain information in the sequence of their components in much the same way as sentences of the English language contain information in the sequence of their letters and words. The sequences that make up the genes are passed on from parents to offspring and are identical except for occasional changes introduced by mutations. As an illustration, one may assume that two books are being compared. Both books are 200 pages long and contain the same number of chapters. Closer examination reveals that the two books are identical page for page and word for word, except that an occasional word—say, one in 100—is different. The two books cannot have been written independently; either one has been copied from the other, or both have been copied, directly or indirectly, from the same original book. Similarly, if each component *nucleotide* of *DNA* is represented by one letter, the complete sequence of nucleotides in the *DNA* of a higher organism would require several hundred books of hundreds of pages, with several thousand letters on each page. When the "pages" (or sequences of nucleotides) in these "books" (organisms) are examined one by one, the correspondence in the "letters" (nucleotides) gives unmistakable evidence of common origin.

The two arguments presented above are based on different grounds, although both attest to *evolution*. Using the alphabet *analogy*, the first argument says that languages which use the same dictionary—the same genetic code and the same 20 amino acids—cannot be of independent origin. The second argument, concerning similarity in the sequence of nucleotides in the *DNA* (and thus the sequence of amino acids in the proteins), says that books with very similar texts cannot be of independent origin.

The evidence of *evolution* revealed by molecular biology goes even farther. The degree of similarity in the sequence of nucleotides or of amino acids can be precisely quantified. For example, in humans and chimpanzees the *protein* molecule called cytochrome c, which serves a vital function in respiration within cells, consists of the same 104 amino acids in exactly the same *order*. It differs, however, from the cytochrome c of rhesus monkeys by one amino acid, from that of horses by 11 additional amino acids, and from that of tuna by 21 additional amino acids. The degree of similarity reflects the recency of common ancestry. Thus the inferences from comparative anatomy and other disciplines concerning evolutionary history can be tested in molecular studies of *DNA* and proteins by examining their sequences of nucleotides and amino acids.

Chapter 3

# Life's Origin

## INTRODUCTION

Biologists agree that life on our planet originated spontaneously by natural processes from the same chemicals of which living organisms consist today, mostly carbon, nitrogen, oxygen, and hydrogen, as well as phosphorus, calcium, and others. Does that mean that we know how life began? Not quite. Although there are some good ideas and experiments about the origin of life, there is not as yet general agreement as to how it might have started. What we do know, because the evidence is overwhelming, is that all living organisms on Earth have evolved from a single original form of life (Schrödinger, 1944; Gilbert, 1986; Orgel, 1994).

The diversity of life on Earth is staggering. More than 1 million existing *species* of fungi, plants, and animals have been named and described, and many more remain to be discovered—at least 10 million according to most estimates. In addition there are microscopic organisms, such as eukaryotic protozoa and the much smaller bacteria and archaea, which each amount to more millions of *species* than the fungi, plants, and animals combined. What is impressive is not just the numbers, but also the incredible heterogeneity in size, shape, and way of life: from lowly bacteria less than one-thousandth of a millimeter in diameter to the stately sequoias of California, rising 100 m above the ground and weighing several thousand tons; from microorganisms living in the hot springs of Yellowstone National Park at temperatures near the boiling point of water, some like *Pyrolobus fumarii* able to grow at more than 100°C, to fungi and algae thriving on the ice masses of Antarctica and in saline pools at −23°C; from the strange wormlike creatures discovered in dark ocean depths, thousands of meters below the surface, to spiders and larkspur plants living on Mount Everest more than 6600 m above sea level (Ayala, 2007; Knoll, 2003; Schopf, 1993).

## THE UNITY OF LIFE

The exact number of *species* on Earth is not known, though estimates for fungi, plants, and animals alone range between 10 and 30 million. The colossal wealth of *species* and their stupendous diversity are the outcome of the evolutionary processes, starting from a single life form.

Evolution, Explanation, Ethics, and Aesthetics. http://dx.doi.org/10.1016/B978-0-12-803693-8.00003-7

The molecular components of organisms are remarkably uniform—in the nature of the components as well as the ways in which they are assembled and used. In all organisms, from bacteria and archaea to plants, animals, and humans, the instructions that guide *development* and functioning are encased in the same hereditary material, *DNA* (deoxyribonucleic acid), which provides the instructions for the synthesis of proteins. The thousands of enormously diverse proteins that exist in organisms are synthesized from different combinations, in sequences of variable length, of 20 amino acids, the same in all proteins and all organisms. Yet several hundred other amino acids exist in nature. There is a very long list of features that could possibly take alternative configurations and yet are uniform throughout all life.

Moreover, the complex machinery by which the hereditary information is conveyed from the *nucleus* to the main body of the cell is everywhere the same: the sequence of nucleotides in the *DNA* is transcribed into a complementary sequence of ribonucleic acid (*RNA*; dubbed "*messenger RNA*" or *mRNA*), which is translated into specified sequences of amino acids that make up the proteins and enzymes that carry out all life processes. This translation involves specific *RNA* molecules ("transfer RNA") and *RNA–protein* complexes (ribosomes), universally shared. The genetic dictionary that guides the translation of the *mRNA* sequence into the amino acid sequence of proteins is also universal.

The unity of life—that is, the similarity of components and processes in all organisms—reveals the genetic continuity and common ancestry of all organisms. Consider the genetic code as an example. Each particular sequence of three nucleotides (*codon*) in the nuclear *DNA* acts as a code for exactly the same amino acid in all organisms. For example, in any given *gene* of any organism, the *codon* GCC determines that the amino acid alanine will be incorporated in the *protein* specified by the *gene*, the *codon* GAC determines the incorporation of the amino acid asparagine, and so on. The universal correspondence between the *DNA* language (*codons*) and the *protein* language (amino acids) is *analogous* with two written languages using the same combination of letters for representing the same particular concept or object. If we find that certain sequences of letters—planet, tree, woman—are used with identical meaning in a number of different books, we can be sure that they have a common origin (Knoll, 2003; Schopf, 1999; Orgel, 1994; Maynard Smith and Szathmány, 1995).

## EARLY LIFE

When did life begin? The oldest known rocks on Earth are estimated to be 4.3 billion years old, only a few hundred million years younger than the Sun, which is thought to have formed somewhat earlier than 4.5 billion years ago. Life originated on Earth several hundred million years later, at least 3.4 billion years ago (Bell et al., 2015). Stromatolites are structures formed by columns of cyanobacteria, a type of photosynthetic bacteria. Ancient structures that resemble stromatolites have been discovered in Australia, Canada, and Baja California, Mexico. Some of these stromatolite-like structures have been dated by radiometric methods at approximately 3.5 billion years. Electron microscopy has revealed structures that resemble cyanobacteria in some stromatolites (Schopf, 1993; Knoll and Bargoorn, 1977).

The oldest fossils ever found, dated 3.4 billion years old, were discovered not long ago between cemented sand grains from an ancient beach in Western Australia. Other sources of evidence from different parts of the world indicate various forms of microscopic bacteria-like organisms that are at least 3 billion years old. It is commonly accepted by experts that life originated on Earth during a 300-million-year interval between 3.8 and 3.5 billion years ago[1] (Rivera et al., 1998; Embley and Williams, 2015; Brasier et al., 2015).

## WHAT IS LIFE?

Before we can answer the question "How did life begin?", we need to address a related question, "What is life?". Two essential properties of life are heredity and metabolism. Cells reproduce by dividing, by making copies of themselves. To ensure continuity of life, the daughter cells must inherit the same components that make up the mother cell. These components include instructions about the chemical machinery of the cell: what chemicals to make and how they will operate. The instructions (*DNA*) are themselves encased in chemicals. But the synthesis of the instruction chemicals requires the chemical machinery of the cell. We have a chicken-and-egg problem to get the process started. The instructions detail how the chemical machinery will operate, but the instructions cannot be synthesized *without* the chemical machinery. Think of a document to be copied that tells you how to build a photocopying machine, but you need the photocopier to do the copying. Better yet, think of electronic computers: you need the hardware (the computer) to carry out the instructions and the software to provide the instructions on how to make the computer. In life, the software includes the instructions (*DNA*) as to how to make the computer (the cell's machinery).

The information-carrying constituents are *DNA* and *RNA* molecules. *DNA* and *RNA* are strings made up of four kinds of chemical components, represented by the letters A, C, G, and T in *DNA*, and by A, C, G, and U in *RNA*; that is, U replaces T in *RNA*. Hereditary information is contained in long sequences of the four letters, in the same way that semantic information is contained in sequences of the letters of the alphabet. The machinery that carries out the chemical reactions (called "metabolism") consists of enzymes, which are proteins able to catalyze chemical reactions with great accuracy and at speeds much faster than any human-made machine. Proteins are made up of long strings of 20 different kinds of components, the amino acids.

A cell consists of thousands and thousands of components, including many thousands of different kinds of enzymes, carrying out with tremendous efficiency and precision many thousands of chemical reactions, a great number of them in precisely determined sequences. Graphic representations of cellular processes show them as extremely complex networks, resembling a map of the transportation system in the United States. There are interstate and state highways, plus a grid of additional roads, trails, driveways, and access roads; oil wells, refineries, gas stations, and factories; cars, trucks, motorcycles, and other vehicles; rivers, waterways, harbors, and all sorts of boats; and airports and airplanes.

Consider, now, the question: how did the transportation network of the United States get started? We might think of simple foot trails, followed later by roads, paved or otherwise, suitable for carts and wagons. But it would be difficult to ascertain where the first trails were fashioned, and which towns or villages they connected. The first trails in North America were made a few thousand years ago. In contrast, life originated on Earth a few thousand *million* years ago. So determining the origins of the trails of life is all the more difficult.

## LIFE IN THE LABORATORY?

If we want to know how life first started, we need to identify the primitive constituents that made up the simplest forms of life. From the start we encounter, as pointed out, the chicken-and-egg problem. We need information molecules, such as *DNA* or *RNA*, which convey from mother cell to daughter cells the information as to what enzymes are to be synthesized, how and when to synthesize them, and how they would interact. There is one additional complication: *DNA*, *RNA*, and proteins cannot do their work without fatty lipids, which make up the cell membranes within which they are contained. But the *DNA* or *RNA* molecules need to be synthesized themselves, and for that we need metabolism—the chemical machinery of living processes. Returning to the computer metaphor, in life the software has the information as to how the computer is to be built, but the software (heredity molecules) cannot be read without the computer (metabolic machinery). Once you have computers in existence, there is no problem. The problem is how to get the first computer.

In 1953 Stanley Miller, a graduate student in chemistry at the University of Chicago, simulated in a tabletop glass apparatus the conditions that might have occurred on our planet soon after its birth. He used some inorganic chemicals that may have been present on the early Earth, such as ammonia, methane, and hydrogen, and added water vapor and electric discharges to simulate lightning, as that may also have occurred at that time. After one week, amino acids, including many found in modern proteins, as well as other compounds, such as urea, naturally occurring only in organisms, had formed in the 5 L glass flask in which the experiment was conducted (Miller, 1953; see also Schopf, 1999; Knoll, 2003; Gilbert, 1986). Miller had thus demonstrated that organic compounds could be formed without the mediation of enzymes. Later experiments, under conditions more similar to those of the primitive Earth, have confirmed that simple organic compounds can be formed spontaneously. This possibility is now taken for granted, because of a multitude of experiments, but also because simple organic molecules have been found in meteorites falling on our planet, in comets, and even in interstellar gas clouds (Orgel, 1994; Maynard Smith and Szathmány, 1995).

The question remains of how these basic building blocks got put together into more complex molecules, such as enzymes and *DNA*, and into living cells. One favored scenario is that once the Earth cooled enough to allow oceans to form, something like the processes observed by Miller and others resulted in a broth of organic molecules (a "primordial soup"), which given enough time (many millions of years were available!) produced by chance combinations of molecules, some of which were more successful than others. At some point a replicating entity would have formed that eventually evolved into life as we know it.

## HEREDITY

The chicken-and-egg problem still remains. Can *DNA* or *RNA* molecules come about, able to direct the synthesis of enzymes that in turn would be able to synthesize *DNA* and *RNA* molecules identical, or at least similar, to the pre-existing ones? Could there be continuity of life through the spontaneous synthesis of hereditary molecules that specify the synthesis of the enzymes to carry out the life processes before there are suitable enzymes? An important advance occurred in the early 1980s when Thomas R. Cech and Sydney Altman independently discovered that some *RNA* molecules can catalyze chemical reactions, including their own synthesis, a discovery for which they received the Nobel Prize in 1989. This discovery contributed importantly to solving the chicken-and-egg problem, in that these *RNA* molecules, dubbed ribozymes, can fill the two roles of heredity and metabolism, roles mostly played by separate molecules—*DNA* and proteins, respectively—in current living organisms (Cech, 1985, 1986, 1987; Altman et al., 2005; Pomeranz Krummel and Altman, 1999a,b). Many scientists now believe that life went through an RNA-dominated phase, called the "*RNA* world," which preceded the current *DNA* world, where biological heredity is prevailingly encased in *DNA* molecules (Gilbert, 1986; Cech, 1993).

The idea that *RNA* could serve a dual function, as a carrier of information and a catalyst that would carry out the instructions, had already been suggested in the 1960s by Francis Crick, who, with James Watson, had discovered in 1953 the double-helix structure of *DNA* (Watson and Crick, 1953; Watson, 1968). The discovery of ribozymes brought attention to Crick's suggestion. The viability of the ribozyme hypothesis is enhanced by consideration of the multiple roles that *RNA* molecules play in modern organisms. While the role of *DNA* is largely limited to carrying out the genetic information, *RNA* is primarily involved in the translation process by which that genetic information is conveyed. As pointed out, different molecules of *mRNA*, transfer *RNA*, and *RNA—protein* complexes are involved in the translation process. Small *RNA* molecules are also involved in *DNA* splicing and *DNA* and *RNA* processing, and there is a large diversity of *RNA* molecules (known as RNA interference or RNA*i*) that is extensively involved in all sorts of *DNA* and *RNA* signaling (Watson and Tooze, 1981).

Once *RNA* molecules were formed that could reproduce by copying themselves, albeit subject to some error (*mutation*) in the synthesis of new *RNA* molecules, *natural selection* would occur, gradually leading to greater molecular complexity and eventually to cells: first simpler cells, as in bacteria, and later more advanced ones, as in animals, plants, and other eukaryotic organisms. Once there were molecules (*RNA*) able to reproduce, some were likely to reproduce more effectively than others. And once there were primitive cells able to reproduce, some of these were likely to reproduce more effectively than other cells. The characteristics of the more effectively reproducing cells would increase in frequency at the expense of the less effectively reproducing ones. It stands to reason that the more effectively reproducing cells would often be those that had more precise heredity and more efficient metabolism. The process of *evolution*, the gigantic project that would give rise to the immense diversity of life forms, had started (Cech, 1987, 1993; Cech and Bass, 1986).

## ORIGINS

The issue arises of how ribozyme *RNA* molecules may have formed spontaneously in the early Earth and how *RNA* molecules would replicate, leading eventually to the *RNA* world. Ribozymes, like other *RNA* molecules, consist of four kinds of nucleotides, represented by the letters A, C, G, and U, as mentioned. They are made up of a limited number of nucleotides, say two dozen or so. But nucleotides are far from simple, containing as they do three molecular components: a ribose sugar, phosphate, and a *nitrogen base*. The *nitrogen base* is the only constituent that varies from one *nucleotide* to another. There are four different kinds of nitrogen bases, which correspond to the A, C, G, and U nucleotides. Scientists have recently shown how the ribose sugar may become spontaneously linked to the nitrogen bases C and U, which was previously the step most difficult to account for.

Modern *RNA* molecules are right handed (called D-enantiomers). Joyce et al. (1984) showed that, in the absence of enzymes, an *RNA* strand would not be readily able to synthesize a complementary strand of itself, because *nucleotide* monomers of different handedness would become incorporated into the *RNA* being synthesized and would bring the replicating *RNA* process to a halt. However, Sczepanski and Joyce (2014) were able to show by a process of several cycles of in vitro *selection* that a D-ribozyme could be isolated that could perform template-directed joining of left handed RNA (L-RNA) molecules at fast rates, consistent with living cell processes. Sczepanski and Joyce (2014) concluded that simple forms of *RNA* nucleic acids could have served as templates for the synthesis and polymerization of *RNA* nucleotides on the prebiotic Earth (Shelke and Piccirilli, 2014; see also Attwater et al., 2013).

A minimal functional unit of life requires molecular information (*DNA* or, in the primitive Earth, *RNA*), metabolism (enzymatic activity), and compartment-forming molecules (lipids making up membranes). All these three molecular components, as a minimum, may thus make up a simple functional cell, which may then reproduce itself. John Sutherland and collaborators (Patel et al., 2015) showed that all three component subsystems can rise simultaneously, starting with just hydrogen cyanide (HCN), hydrogen sulfide ($H_2S$), and ultraviolet (UV) light. These conditions may have obtained on the early Earth. Abundant impacts from comets pelting the early Earth would have produced enough energy to synthesize HCN from hydrogen, carbon, and nitrogen. $H_2S$ is thought to have been common in the early Earth, as well as the UV radiation necessary to catalyze the reactions that would have given rise to the nucleic acids, peptides (proteins), and fatty acids (lipids) to make up primitive cells. According to Sutherland (Patel et al., 2015; see also Service, 2015; Bracher, 2015) it would seem unlikely that the reactions required to generate all three kinds of building blocks—nucleic acids, peptides, and lipids—would have occurred in the same location and at the same time. But rainwater would have simply washed them into a common pool, where the three biomolecular building blocks would be integrated into minimal cells.

So, do we now know how life began? We do not. What we do know is that in the absence of previous life and under conditions that may plausibly have existed on the early Earth, spontaneous chemical processes can give rise to organic compounds, including those that are the fundamental building blocks of life: the nucleic acids that encase heredity and the enzymes that account for metabolism, that is, the proteins that catalyze the chemical reactions that make up all living processes. Returning to the transportation network *analogy*, we now know that early foot trails could be made and how more advanced roads might be developed from them, but we still do not know where those first trails were made or precisely how. What we also know is that life originated on Earth only once; or that if it originated more than once, all diverse original forms of life but one became extinct.

## Endnote

1. Bell et al. (2015) acknowledge that the microfossil record only extends to ~3.5 billion years and the chemofossil record to ~3.8 billion years (Knoll and Bargoorn, 1977; Knoll, 2003; Brasier et al., 2015). They report that carbon isotopic measurements of detrital zircon inclusions from Jack Hills, Western Australia, yield an estimate of 4.10 ± 0.01 billion years. They assert that their analysis is consistent with a biogenic origin and "may be evidence that a terrestrial biosphere had emerged by 4.1 Ga, or ~300 My earlier than previously proposed" (Bell et al., 2015, p. 14,518). They acknowledge, however, that nonbiological processes could also explain their results.

Chapter 4

# LUCA and the Tree of Life

## INTRODUCTION

The Earth is probably the only planet in the solar system that presently has life. There are about 100 billion stars in our galaxy, and many of them have planetary systems. And there are more than 100 billion galaxies in the universe. It may very well be that life exists elsewhere. Some planets, perhaps many of them throughout the universe, may have life if temperature, chemical composition, and other features favorable for life occur—something that seems possible given the immense number of galaxies, stars, and planets. Life occurs on Earth because conditions favorable for life exist on our planet. Given similar conditions and eons of time, life is likely to happen on other planets. Planets with conditions (at least temperature) suitable for life have recently been identified in our galaxy. Life as it may exist elsewhere would likely have features different from those of life on Earth. The basic chemical elements might be different; for example, silicon rather than carbon might combine with hydrogen, oxygen, and other elements to make up the fundamental molecules of life (Woese, 1987; Conway Morris, 1998; Ayala, 2007).

## TAXONOMY

As pointed out in earlier chapters, more than 1 million *species* have been named and described, and many more are known to exist; and the number of extinct *species* is more than 100 times the number of *species* now in existence. The number of different *species* that have lived on Earth from the beginning of life to the present is likely to be in the order of 1 billion. The diversity of the living world is apparent in not only the large number of *species* but also their heterogeneity. Organisms vary enormously in size, way of life, and *habitat*, as well as in structure and form.

Despite their prodigious diversity, organisms have much in common. There are features that all organisms have in common, features that characterize life as such, at least as it exists on our planet, where life as it now exists must have had a single origin. Certain similarities are shared by some organisms, but not by all. These similarities can be used to classify, ie, characterize, some groups of organisms and distinguish them from others. The basic process responsible

Evolution, Explanation, Ethics, and Aesthetics. http://dx.doi.org/10.1016/B978-0-12-803693-8.00004-9

for the hierarchy of similarities among living things is *evolution*—some organisms resemble each other more than others because they are more closely related by lines of descent; they had a more recent common ancestor than they have with organisms that are less similar.

Some *species* resemble each other quite closely, and some groups of organisms have features in common so as to give them a sort of *family* resemblance. Mosquitoes, for example, all look pretty much alike. Although separate groups of insects are rather different from each other when viewed in detail, insects as a whole tend to resemble one another when compared with other distinctive groups of animals, such as worms, clams, or fish. There are many more *species* of insects than of any other comparable group of animals, even though insects are nearly all restricted to the terrestrial *habitat*; only relatively few live in truly aquatic environments. In the oceans, the snails are the group with the most *species*. Organisms seem to live in every conceivable *habitat* on the Earth's surface, from the greatest ocean depths to Himalayan mountain peaks, and from subzero arctic plains to hot springs. A major reason for the great diversity of living things is the great diversity of environmental conditions found on Earth.

Aristotle (384−322 BCE), the great philosopher of Ancient Greece, was also the most important biologist of ancient times. He was the first author to develop a system of classification for animals. As pointed out in Chapter 1, he established two higher categories: blooded animals and bloodless animals. He classified the blooded animals into five groups: mammals, birds, amphibians, reptiles, and fish. He divided the bloodless animals into cephalopods, crustaceans, insects, and a fourth group, testaceans, which included the remaining animals. He also introduced the binomial method of nomenclature: the *genus*, which identified a group of closely related organisms, and their difference, which identified each subgroup within the closely related group. None of the surviving texts by Aristotle deals with what we call botany, but it is commonly accepted that he wrote at least two treatises on plants. Aristotle's and his Greek contemporaries' understanding of plants is known through the writings of Theophrastus, Aristotle's disciple, especially a treatise that Aristotle had dedicated to the morphology, natural history, and therapeutic use of plants.

The Aristotelian method of definition by *genus* and difference dominated medieval and modern systems of classification—not only of organisms but also of other entities, from books to rocks—up to the 18th and 19th centuries, when we meet the two giants, Carl Linnaeus (1707−78) and Charles Darwin (1809−1882). Important intermediate milestones include Porphyry and St. Albert the Great. Porphyry (234−305 CE) introduced the *category* of "*species*" combined with "*genus*" to identify organisms, thus refining the combination of "*genus*" and "difference" used by Aristotle (Ayala, 2014).

St. Albert the Great (1193−1280), in addition to being recognized as a great theologian and the teacher of St. Thomas Aquinas, was undoubtedly the greatest naturalist of the Middle Ages. He wrote seven books on botany

(*De vegetabilibus*) and 26 on zoology (*De animalibus*), in which he gathered the collected knowledge of other authors, even Aristotle, but also included many observations and studies of his own, such as a precise description of the anatomy of the leaves of the plants he investigated and descriptions of their various systems of propagation and reproduction, including sexual reproduction in animals. Albert the Great contributed significantly to the rejection of many superstitions accepted by his contemporaries, paving the way for future European science.

Carl Linnaeus (1707−78) is considered the father of *taxonomy*, and he is certainly that in regard to modern *taxonomy*. Linnaeus adopted a hierarchical system of *taxonomy* that initially identified four levels of categories within each kingdom (*class*, *order*, *genus*, and *species*), and used the system of nomenclature in which each organism is defined by two terms, the *genus* and the *species*, the system already established by Porphyry that can be traced to Aristotle. Linnaeus was not the first to use a hierarchical *taxonomy* or a two-term definition, but because he was generally recognized as an authority on the subject, his hierarchical *taxonomy* and binomial nomenclature became generally accepted by taxonomists and other biologists. In 1753, when he published his *Species Plantarum*, Linnaeus knew of only some 6000 plant *species* but believed that the total could reach as many as 10,000 *species*. In 1758, in the tenth edition of his *Systema Naturae*, Linnaeus included some 4000 *species* of animals and thought that the total could be 10,000, as in the case of plants. As Ernst Mayr (1982) points out, Eberhard August Wilhelm von Zimmermann, a contemporary of Linnaeus, estimated in 1778 that the number of plant *species* could be 150,000 and animal *species* could be 7 million, a much more realistic estimate.

Linnaeus, as an Aristotelian essentialist, believed in the immutability of *species* and was less correct than Aristotle in his classification of animals. As pointed out in Chapter 1, Linnaeus divided animals into only six classes: mammals, birds, reptiles, fish, insects, and worms. The first four classes are similar to Aristotle's, although amphibians were missing (Linnaeus included them with the reptiles), but the last two classes correspond to four of the Aristotelian categories. Jean-Baptiste Lamarck (1744−1829), best known now for his erroneous theory of *evolution*, was an excellent taxonomist and established the distinction between *vertebrate* and invertebrate animals. The latter, defined by a characteristic they lack instead of one they possess, include more than 90% of animal *species*. An example of Lamarck's competence as a taxonomist is his affirmation that barnacles are not molluscs (as other taxonomists believed), but arthropods. His *category* of arthropods included crustaceans and arachnids, although he did not at first include insects among them.

One of the great taxonomists of the 19th century is the German Ernst Haeckel (1834−1919), who in 1866 proposed the kingdom *Protista* to include all unicellular organisms, arguing that it is impossible to distinguish clearly between plants and animals at the level of a single cell. In 1866, Haeckel constructed a tree of life with three principal branches coming from the trunk, representing plants, protists, and animals (Fig. 4.1). Each of the branches in turn is successively divided into other branches, which are each divided again. Among the animals, for example, we find branches representing coelenterates, echinoderms, arthropods, molluscs, and amniotes, each with additional branches. Another version of this tree that was published later (1875) shows the relationship of man to other living organisms (Fig. 4.2).

A significant refinement to the tree was made by Edouard Caton, who in 1937 divided living organisms into two groups, prokaryotes and eukaryotes, depending on the presence or absence of a *nucleus* in the cell. In 1959, Robert Whittaker proposed a tree of five kingdoms (adding fungi) consisting of three levels of organization: prokaryotes (kingdom monera), unicellular eukaryotes (kingdom protista), and multicellular eukaryotes (kingdoms of plants, fungi, and animals) (Doolittle, 2000).

## HOMOLOGY AND ANALOGY

Evolutionary theory offers a causal explanation for the similarities among living beings. Organisms evolve by a process of descent with modification. Changes, and thus differences, gradually accumulate over the generations. If the last common ancestor of two *species* is recent, they will have accumulated few differences. Consequently, similarities in form and function reflect evolutionary proximity, and therefore evolutionary affinities can be inferred from the degrees of similarity. This principle is the scientific foundation for the reconstruction of evolutionary relationships based on comparative analyses of living organisms through anatomical, embryological, biogeographical, and molecular investigations.

One important distinction, particularly in the case of anatomical consideration, is "*homology*" versus "*analogy*." The concept of *homology* was defined in 1843 by the biologist Richard Owen (independently of biological *evolution*, which he did not consider) as "the same organ in different animals under every variety of form and function" (Owen, 1843). Nowadays, *homology* is explained in evolutionary terms. Two characters (such as human arms and dog forelimbs) are *homologous* when the resemblance between them is due to them being inherited from a common ancestor which had limbs basically similar to those now found in its descendants (Chapter 2, Fig. 2.4). *Analogy* applies to similarities that originated independently in different lineages because they serve similar functions. For instance, the wings of bats, birds, and butterflies are *analogous*. These structures were not inherited from a common ancestor with wings, but evolved separately in each lineage as an *adaptation* to flight.

The degree of detail in the resemblance provides a practical way to distinguish between *homology* and *analogy*. *Homology* involves detailed similarity

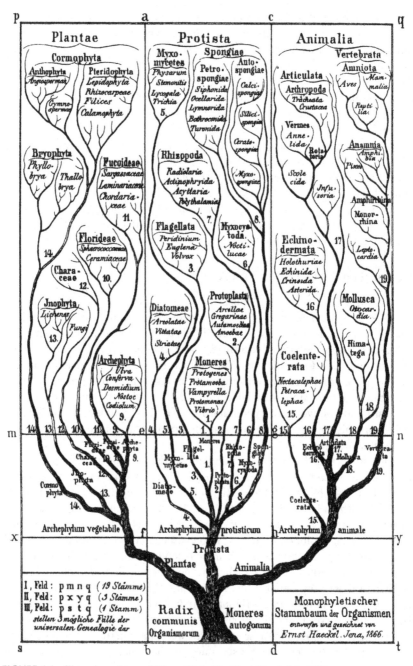

**FIGURE 4.1**  The universal tree of life, published by Ernst Haeckel in 1866.

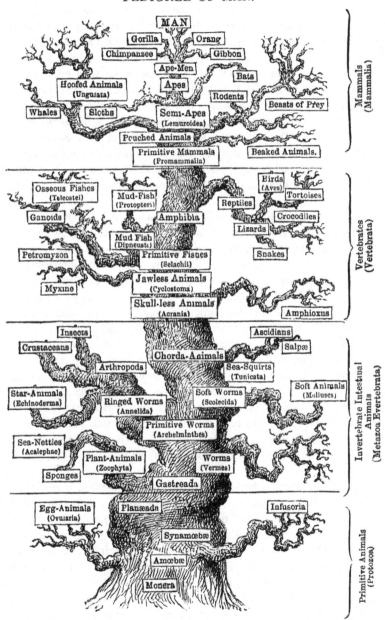

PEDIGREE OF MAN.

**FIGURE 4.2**    The tree of life showing humans at the top, by Ernst Haeckel in 1875.

(as is the case with each of the bones and muscles of human arms and dog forelegs). *Analogy* involves similarities in the global configuration (the wings of butterflies, like those of eagles and bats, are wide thin surfaces) but not in the details of structure and organization. The parts of an organ in different animals may be *analogous* in some respect but *homologous* in other respects. The wings of bats and birds are *analogous* with respect to their function of flying, because wings evolved independently. But they are *homologous* with respect to their underlying bone structure. Birds and mammals inherited it from their common reptilian ancestor. The classification of organisms and the reconstruction of evolutionary history depend, of course, on the degrees of *homology* between different organisms. The greater the degree of *homology* between two *species*, the more closely related they will be—that is, the more recent their common ancestor. Anatomy, embryology, physiology, and other features of organisms can all be used to evaluate degrees of *homology* and thus degrees of common ancestry (Mayr, 1982).

In addition to traditional criteria, accumulated through the experience of evolutionists during the 19th and early 20th centuries, two new theories of classification known as "*phenetics*," or numerical *taxonomy*, and "*cladistics*" emerged in the 1950s. *Phenetics* proceeds by formulating numerical algorithms, known as phenograms, in which each character can take one of two states: "present" or "absent" (they can be morphological characters, such as the thumb, or an amino acid in a particular *protein*, such as valine at position six in hemoglobin beta, or any other trait). Each character receives a zero (if it is absent) or a one (if it is present) for each *species* (or higher-ranking categories, such as genera, families, or classes). The degree of phenetic affinity among different taxa is determined by the number of ones in the strings of zeroes and ones. This measure does not necessarily reflect evolutionary affinity: it only indicates the extent to which two organisms are similar in form. Indeed, *phenetics* seeks to avoid any theoretical underpinnings (such as *evolution*). It does not seek to address the reasons behind the resemblances (Michener and Sokal, 1957; Sneath, 1957; Cain and Harrison, 1958; Sokal and Sneath, 1963).

*Cladistics* is currently the most extensively used method in evolutionary *taxonomy*. *Cladistics* is a method developed by the German zoologist Willi Hennig (1950, 1966). It starts from the requisite that *species* (or other taxa: genera, families, etc.) be classified according to their phylogenetic (evolutionary history) relationship, rather than based on their degree of morphological or phenetic similarity. The graphical representation of phylogenetic relationships in a *cladogram* is a branching diagram where one branch splits into two whenever one *species* (or other *taxon*) splits into two *species* (or other taxa). Phylogenies are inferred by comparing primitive and derived traits of current *species* with each other and with those of *fossil species*, or by comparing with each other *fossil species* living at different times.

## MOLECULAR BIOLOGY

Molecular biology provides the most detailed and precise evidence available for determining the *evolution* of organisms. In the second half of the 20th century, *population* genetics and evolutionary genetics became very active disciplines which eventually incorporated molecular biology, a new discipline emerged from the 1953 discovery by James Watson and Francis Crick of the molecular structure of *DNA* (deoxyribonucleic acid), the hereditary chemical contained in the chromosomes of every cell *nucleus* (Watson and Crick, 1953). The genetic information is encoded within the sequence of nucleotides, the elementary components that make up the chainlike *DNA* molecules. This information determines the sequence of amino acid building blocks of *protein* molecules, which include structural proteins as well as the numerous enzymes that carry out the organism's fundamental life processes. Genetic information could first be investigated by examining the sequences of amino acids in the proteins, and eventually by examining directly the sequences of the nucleotides that make up each *DNA* (Nei, 1987).

In the mid-1960s laboratory techniques, such as electrophoresis and selective assay of enzymes became available for the rapid and inexpensive study of differences among enzymes and other proteins. These techniques made possible the pursuit of evolutionary issues, such as quantifying genetic variation in natural populations (which variation sets bounds on the evolutionary potential of a *population*) and determining the amount of genetic change that occurs during the formation of new *species*. Comparisons of the amino acid sequences of corresponding proteins in different *species* provided precise measures of the divergence among *species* evolved from common ancestors, a considerable improvement over the typically much more qualitative evaluations obtained by comparative anatomy and other evolutionary subdisciplines (Ayala et al., 1970, 1971; Ayala and Powell, 1972).

The laboratory techniques of *DNA cloning* and sequencing, developed much more recently, have provided a new and powerful means of investigating *evolution* at the molecular level. The fruits of this technology began to accumulate during the 1980s following the *development* of automated DNA-sequencing machines and the invention of the polymerase chain reaction, a simple and inexpensive technique that obtains, in a few hours, billions or trillions of copies of a specific *DNA* sequence or *gene* (Li, 1997; Graur and Li, 2000). Major research efforts, such as the Human Genome Project further improved the technology for obtaining long *DNA* sequences rapidly and inexpensively. By the first few years of the 21st century, the full *DNA* sequence—ie, the full genetic complement, or *genome*—had been obtained for more than 20 higher organisms, including human beings. A draft of the chimpanzee *genome* was published in 2005 (Li and Saunders, 2005), followed by the orangutan *genome* in 2011 and the gorilla *genome* in 2012. The complete *genome* sequences of hundreds of organisms, from bacteria to all sorts of animals as well as viruses, are currently available to investigators (Cavalli-Sforza and Feldman, 2005; Ayala, 2015).

Molecular biology has made it possible to reconstruct evolutionary events that were previously unknown, and to confirm or adjust the view of events already known. The precision with which these events can be reconstructed is one reason why the evidence from molecular biology is so compelling. Another reason is that *molecular evolution* has shown all living organisms, from bacteria to humans, to be related by descent from common ancestors (Fig. 4.3).

Nucleic acids (*DNA* and *RNA*—ribonucleic acid) and proteins are "macromolecules," long linear molecules made up of sequences of units—nucleotides in the case of nucleic acids, amino acids in the case of proteins—which retain considerable amounts of evolutionary information. Comparing two macromolecules establishes the number of their units that are different. Because *evolution* usually occurs by changing one unit at a time, the number of differences is an indication of the recency of common ancestry.

Macromolecular studies have four notable advantages over comparative anatomy and the other classical disciplines used to classify organisms: quantification, universality, multiplicity, and versatility (Ayala, 2015). Macromolecular changes are readily quantifiable. The number of units that are different is readily established when the sequence of units is known for a given macromolecule in different organisms. The second advantage is that comparisons can be made even between very different sorts of organisms. There is very little that comparative anatomy can say when organisms as diverse as yeasts, pine trees, and human beings are compared, but there are *homologous* macromolecules that

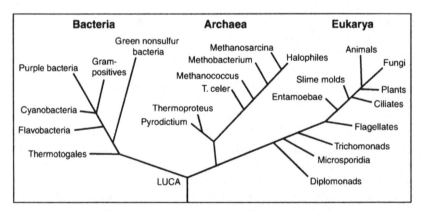

**FIGURE 4.3**   The universal tree of life, reconstructed with ribosomal ribonucleic acid (rRNA) genes. The last universal common ancestor (LUCA) is at the bottom. Branches represent different kinds of organisms. There are three major groups of organisms: bacteria, archaea, and eukaryotes. Bacteria, archaea, and most eukaryotes are microscopic. Plants, animals, and fungi are multicellular (macroscopic) branches of eukaryotes. *Adapted from Woese, C.R., 2000. Interpreting the universal phylogenetic tree. Proceedings of the National Academy of Sciences of the United States of America 97 (15), 8392–8396.*

can be compared in all three. The third advantage is multiplicity. Each organism possesses thousands of genes and proteins, which all reflect the same evolutionary history. If the investigation of one particular *gene* or *protein* does not resolve the evolutionary relationship of a set of *species*, additional genes and proteins can be investigated until the matter has been settled.

The fourth advantage of *molecular evolution* is versatility: implicit in their multiplicity is that genes and proteins evolve at very different rates. Genes and proteins that evolve at fast rates, such as the proteins knows as fibrinopeptides, can be used to investigate the evolutionary relationships among closely related *species*, such as primates, while slow-evolving genes and proteins can be investigated to determine the evolutionary relationships among *species* more remotely related (Fig. 4.4). Thus Carl Woese (2000) was able to establish the main evolutionary relationships among all sorts of organisms, from bacteria to archea and eukaryotes, with the analysis of only ribosomal ribonucleotide genes (Fig. 4.3).

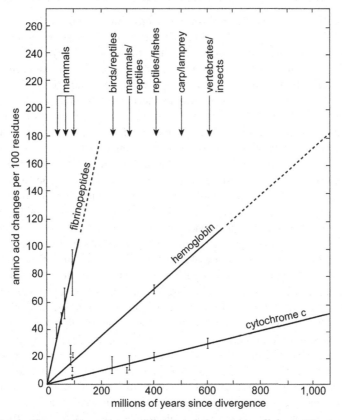

**FIGURE 4.4**  Three proteins with very different evolutionary rates: fibrinopeptides (very fast), hemoglobin (intermediate), and cytochrome *c* (slow).

Yet a *phylogeny* obtained with the analysis of one single *gene*, or one single *protein*, no matter how informative it may be, is unsatisfactory. The tree of 20 organisms obtained with cytochrome *c* (see Fig. 4.7 later; Fitch and Margoliash, 1967) enormously impressed the scientific community, even though it was far from accurate. Similarly, it has become apparent in recent years that the universal tree of life (Fig. 4.3) is also inaccurate in important ways. Significantly, the early splits among bacteria, archea, and eukaryotes are no longer accepted as represented in Fig. 4.3. Rivera and Lake (2004) provided evidence of *gene* fusion between two diverse *prokaryote* genomes in the origin of the *eukaryote genome*. As they assert it, "the tree of life is actually a ring of life" (Rivera and Lake, 2004, p. 152). Evidence of *gene* flow between lineages of the bacterial branch and the eukaryotes had been evidenced earlier. The three-domain universal evolutionary tree has been challenged (Embley and Williams, 2015).

## CLADOGENESIS AND ANAGENESIS: EVOLUTIONARY TREES

Informational macromolecules provide information not only about the branching of lineages from common ancestors (*cladogenesis*) but also about the amount of genetic change that has occurred in any given lineage (anagenesis). It might seem at first that quantifying anagenesis for proteins and nucleic acids would be impossible, because it would require comparison of molecules from organisms that lived in the past with those from living organisms. Organisms of the past are sometimes preserved as fossils, but their *DNA* and proteins have largely disintegrated. Nevertheless, comparisons between living *species* give information about anagenesis.

As a concrete example, consider the *protein* cytochrome *c*, involved in cell respiration (Fitch and Margoliash, 1967; Ayala, 2007). The sequence of amino acids in this *protein* is known for many organisms, from bacteria and yeasts to insects and humans; in animals cytochrome *c* consists of 104 amino acids. When the amino acid sequences of humans and rhesus monkeys are compared, they are found to be different at position 66 (isoleucine in humans, threonine in rhesus monkeys) but identical at the other 103 positions. When humans are compared with horses, 12 amino acid differences are found, but when horses are compared with rhesus monkeys there are only 11 amino acid differences. Even without knowing anything else about the evolutionary history of mammals, one would conclude that the lineages of humans and rhesus monkeys diverged from each other much more recently than they diverged from the horse lineage. Moreover, it can be concluded that the amino acid difference between humans and rhesus monkeys must have occurred in the human lineage after its separation from the rhesus monkey lineage, given that horses and rhesus monkeys are identical at position 66 (both have threonine), while humans differ from both (Figs. 4.5 and 4.6).

| Human | G – D – V – E – K – G – K – K – I – F – I – M – K – C – S – Q – C – |
| Rhesus monkey | • • • • • • • • • • • • • • • • • |
| Horse | • • • • • • • • • • V Q • • A • • |

H – Y – V – E – K – G – G – K – H – K – Y – G – P – N – L – H – G – L – F – G – R – K – T –
• • • • • • • • • • • • • • • • • • • • • • •
• • • • • • • • • • • • • • • • • • • • • • •

G – Q – A – P – G – Y – S – Y – T – A – A – N – K – N – K – G – I – I – W – G – E – D – T –
• • • • • • • • • • • • • • • • • T • • • • •
• • • • • F T • • D • • • • • • • T • K • E •

L – M – E – Y – L – E – N – P – K – K – Y – I – P – G – T – K– M – I – F – V – G – I – K –
• • • • • • • • • • • • • • • • • • • • • •
• • • • • • • • • • • • • • • • • • A • • •

K – K – E – E – R – A – D – L – I – A – Y – L – K – K – A – Y – N – E
• • • • • • • • • • • • • • • • • •
• • T • • E • • • • • • • • • • •

FIGURE 4.5 The 104 amino acids in the cytochrome c of humans are shown on top (using standard one-letter representations for each amino acid). At one position rhesus monkeys have threonine, while humans have isoleucine. Humans differ from horses by 12 amino acids; monkey and horse differ by 11 amino acids. Dots indicate amino acids identical to those of human cytochrome c. *After Fitch, W.M., Margoliash, E., 1967. Science 155, 279–284.*

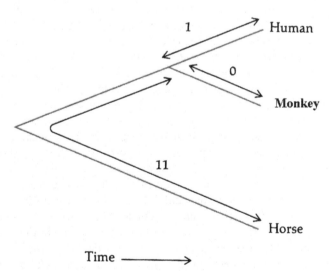

FIGURE 4.6 Evolutionary tree of human, rhesus monkey, and horse, based on their cytochrome c. The one difference between human and monkey (Fig. 4.5) is due to a change in the human lineage after it diverged from the monkey lineage. This conclusion is reached because monkey and horse (as well as other animals) have the same amino acid (T) at this position, while the human is different (amino acid I).

Evolutionary trees are models that seek to reconstruct the evolutionary history of taxa—ie, *species* or other groups of organisms, such as genera, families, or orders. The trees embrace two kinds of information related to evolutionary change, *cladogenesis* and anagenesis. Figs. 4.5 and 4.6 can be used to illustrate both kinds. The branching relationships of the tree reflect the relative relationships of ancestry, or *cladogenesis*. Thus, in the right side of Fig. 4.6, humans and rhesus monkeys are seen to be more closely related to each other than either is to the horse. Stated another way, this tree shows that the last ancestor common to all three *species* lived in a more remote past than the last ancestor common to humans and monkeys.

Evolutionary trees may also indicate the changes that have occurred along each lineage, or anagenesis. Thus, in the *evolution* of cytochrome *c* since the last common ancestor of humans and rhesus monkeys (again, the right side of Fig. 4.6), one amino acid changed in the lineage going to humans but none in the lineage going to rhesus monkeys.

There are several methods for constructing evolutionary trees. Some were developed for interpreting morphological data, others for interpreting molecular data; some can be used with either kind of data. The main methods currently in use are called distance, *parsimony*, and maximum likelihood (Fitch and Margoliash, 1967; Nei, 1987).

## Distance Methods

A "distance" is the number of differences between two taxa. The differences are measured with respect to certain traits (ie, morphological data) or certain macromolecules (primarily the sequence of amino acids in proteins or the sequence of nucleotides in *DNA* or *RNA*). The tree illustrated in Fig. 4.6 was obtained by taking into account the distance, or number of amino acid differences, between three organisms with respect to a particular *protein*. The amino acid sequence of a *protein* contains more information than is reflected in the number of amino acid differences. This is because in some cases the replacement of one amino acid by another requires no more than one *nucleotide* substitution in the *DNA* that codes for the *protein*, whereas in other cases it requires at least two *nucleotide* changes. Table 4.1 shows the minimum number of *nucleotide* differences in the genes of 20 separate *species* that are necessary to account for the amino acid differences in their cytochrome *c*. An evolutionary tree based on the data in Table 4.1, showing the minimum numbers of *nucleotide* changes in each branch, is illustrated in Fig. 4.7.

The relationships between *species* as shown in Fig. 4.7 correspond fairly well to the relationships determined from other sources, such as the *fossil* record, but not in all cases. According to the figure, chickens are less closely related to ducks and pigeons than to penguins, and humans and monkeys diverged from the other mammals before the marsupial kangaroo separated from the nonprimate placentals. These particular relationships are known to be

**TABLE 4.1** Minimum Number of Nucleotide Differences in Genes Coding for Cytochrome c in 20 Different Organisms

| Organism | 1 | 2 | 3 | 4 | 5 | 6 | 7 | 8 | 9 | 10 | 11 | 12 | 13 | 14 | 15 | 16 | 17 | 18 | 19 | 20 |
|---|---|---|---|---|---|---|---|---|---|---|---|---|---|---|---|---|---|---|---|---|
| 1. Human | — | 1 | 13 | 17 | 16 | 13 | 12 | 12 | 17 | 16 | 18 | 18 | 19 | 20 | 31 | 33 | 36 | 63 | 56 | 66 |
| 2. Monkey | | | 12 | 16 | 15 | 12 | 11 | 13 | 16 | 15 | 17 | 17 | 18 | 21 | 32 | 32 | 35 | 62 | 57 | 65 |
| 3. Dog | | | | 10 | 8 | 4 | 6 | 7 | 12 | 12 | 14 | 14 | 13 | 30 | 29 | 24 | 28 | 64 | 61 | 66 |
| 4. Horse | | | | | 1 | 5 | 11 | 11 | 16 | 16 | 16 | 17 | 16 | 32 | 27 | 24 | 33 | 64 | 60 | 68 |
| 5. Donkey | | | | | | 4 | 10 | 12 | 15 | 15 | 15 | 16 | 15 | 31 | 26 | 25 | 32 | 64 | 59 | 67 |
| 6. Pig | | | | | | | 6 | 7 | 13 | 13 | 13 | 14 | 13 | 30 | 25 | 26 | 31 | 64 | 59 | 67 |
| 7. Rabbit | | | | | | | | 7 | 10 | 8 | 11 | 11 | 11 | 25 | 26 | 23 | 29 | 62 | 59 | 67 |
| 8. Kangaroo | | | | | | | | | 14 | 14 | 15 | 13 | 14 | 30 | 27 | 26 | 31 | 66 | 58 | 68 |
| 9. Duck | | | | | | | | | | 3 | 3 | 3 | 7 | 24 | 26 | 25 | 29 | 61 | 62 | 66 |
| 10. Pigeon | | | | | | | | | | | 4 | 4 | 8 | 24 | 27 | 26 | 30 | 59 | 62 | 66 |
| 11. Chicken | | | | | | | | | | | | 2 | 8 | 28 | 26 | 26 | 31 | 61 | 62 | 66 |
| 12. Penguin | | | | | | | | | | | | | 8 | 28 | 27 | 28 | 30 | 62 | 61 | 65 |
| 13. Turtle | | | | | | | | | | | | | | 30 | 27 | 30 | 33 | 65 | 64 | 67 |
| 14. Snake | | | | | | | | | | | | | | | 38 | 40 | 41 | 61 | 61 | 69 |
| 15. Tuna | | | | | | | | | | | | | | | | 34 | 41 | 72 | 66 | 69 |
| 16. Screwworm | | | | | | | | | | | | | | | | | 16 | 58 | 63 | 65 |
| 17. Moth | | | | | | | | | | | | | | | | | | 59 | 60 | 61 |
| 18. *Neurospora* (mold) | | | | | | | | | | | | | | | | | | | 57 | 61 |
| 19. *Saccharomyces* (yeast) | | | | | | | | | | | | | | | | | | | | 41 |
| 20. *Candida* (yeast) | | | | | | | | | | | | | | | | | | | | — |

From Fitch, W.M., Margoliash, E., 1967. Science 155, 281.

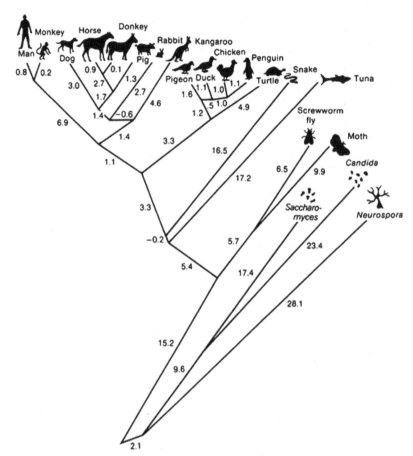

**FIGURE 4.7** Evolutionary history of 20 species, based on the cytochrome c amino acid sequence. The common ancestor (at the bottom) of yeast and humans lived more than 1 billion years ago. The numbers along the branches estimate the nucleotide substitutions occurring in the span of evolution represented by each branch. Although fractional (or negative) numbers of nucleotide substitutions cannot occur, the numbers along the branches are those that best fit the data. More detailed studies would make it possible to determine the exact number of changes along each branch. *After Fitch, W.M., Margoliash, E., 1967. Science 155, 279–284.*

erroneous, but the power of the method is apparent in that a single *protein* yields a fairly accurate reconstruction of the evolutionary history of 20 organisms that started to diverge more than 1 billion years ago. Moreover, as pointed out earlier, molecular biology provides "multiplicity," One can investigate two or more genes or proteins, even up to thousands, so there is enough confidence that the combined *phylogeny* obtained is the correct one.

Morphological data can also be used for constructing distance trees. The first step is to obtain a distance matrix, such as that making up the *nucleotide* differences table, but based on a set of morphological comparisons between *species* or other taxa. For example, in some insects one can measure body length, wing length, wing width, number and length of wing veins, and other traits. The most common procedure to transform a distance matrix into a *phylogeny* is called cluster analysis. The distance matrix is scanned for the smallest distance element, and the two taxa involved (say, A and B) are joined at an internal *node*, or branching point. The matrix is scanned again for the next smallest distance, and the two new taxa (say, C and D) are clustered. The procedure is continued until all taxa have been joined. When a distance involves a *taxon* that is already part of a previous cluster (say, E and A), the average distance is obtained between the new *taxon* and the preexisting cluster (say, the average distance between E to A and E to B). This simple procedure, which can also be used with molecular data, assumes that the rate of *evolution* is uniform along all branches.

Other distance methods (including the one used to construct the tree in Fig. 4.7) relax the condition of uniform rate and allow for unequal rates of *evolution* along the branches. One of the most extensively used methods of this kind is called neighbor joining. The method starts, as before, by identifying the smallest distance in the matrix and linking the two taxa involved. The next step is to remove these two taxa and calculate a new matrix in which their distances from other taxa are replaced by the distance between the *node* linking the two taxa and all other taxa. The smallest distance in this new matrix is used for making the next connection, which will be between two other taxa or between the previous *node* and a new *taxon*. The procedure is repeated until all taxa have been connected with one another by intervening nodes.

## Maximum Parsimony Methods

*Maximum parsimony* methods seek to reconstruct the tree that requires the fewest (ie, most parsimonious) number of changes summed along all branches. This is a reasonable assumption, because it usually will be the most likely. But *evolution* may not necessarily have occurred following a minimum path, because the same change instead may have occurred independently along different branches, and some changes may have involved intermediate steps. Consider three *species*—C, D, and E. If C and D differ by two amino acids in a certain *protein* and either one differs by three amino acids from E, *parsimony* will lead to a tree with the structure shown in Fig. 4.8. It may be the case, however, that in a certain position at which C and D both have amino acid g while E has h, the ancestral amino acid was g. Amino acid g did not change in the lineage going to C but changed to h in a lineage going to the ancestor of D and E and then changed again, back to g, in the lineage going to D. The correct *phylogeny* would lead then from the common ancestor of all three *species* to C in one branch (in which no amino acid changes occurred), and to the last

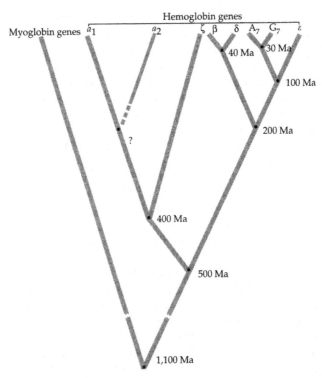

**FIGURE 4.8** Evolutionary history of the globin genes. The dots indicate points at which ancestral genes duplicated, giving rise to new gene lineages. The approximate times when these duplications occurred are indicated in millions of years ago. The time when the duplication of $\alpha_1$ and $\alpha_2$ occurred is uncertain.

common ancestor of D and E in the other branch (in which g changed to h), with one additional change (from h to g) occurring in the lineage from this ancestor to E.

Not all evolutionary changes, even those that involve a single step, may be equally probable. For example, among the four *nucleotide* bases in *DNA*, cytosine (C) and thymine (T) are members of a *family* of related molecules called pyrimidines; likewise, adenine (A) and guanine (G) belong to a *family* of molecules called purines. A change within a *DNA* sequence from one pyrimidine to another (C $\rightleftarrows$ T) or from one purine to another (A $\rightleftarrows$ G), called a transition, is more likely to occur than a change from a purine to a pyrimidine or the converse (G or A $\rightleftarrows$ C or T), called a transversion. *Parsimony* methods take into account different probabilities of occurrence if they are known.

Maximum *parsimony* methods are related to *cladistics*. The critical feature in *cladistics* is the identification of derived shared traits, called synapomorphic traits. A synapomorphic trait is shared by some taxa but not others because the former inherited it from a common ancestor that acquired the trait after its lineage separated from the lineages going to the other taxa. In the *evolution* of carnivores, for example, domestic cats, tigers, and leopards are clustered together because they possess retractable claws, a trait acquired after their common ancestor branched off from the lineage leading to dogs, wolves, and coyotes. It is important to ascertain that the shared traits are *homologous* rather than *analogous*. For example, mammals and birds, but not lizards, have a four-chambered heart. Yet birds are more closely related to lizards than to mammals; the four-chambered heart evolved independently in the bird and mammal lineages by parallel *evolution*.

## Maximum Likelihood Methods

Maximum likelihood methods seek to identify the most likely tree given the available data. They require that an evolutionary model be identified which would make it possible to estimate the probability of each possible individual change. For example, as mentioned in the preceding subsection, transitions are more likely than transversions among *DNA* nucleotides, but a particular probability must be assigned to each. All possible trees are considered. The probabilities for each individual change are multiplied for each tree. The best tree is the one with the highest probability (or maximum likelihood) among all possible trees.

Maximum likelihood methods are computationally expensive when the number of taxa is large, because the number of possible trees (for each of which the probability must be calculated) grows factorially with the number of taxa. With 10 taxa there are about 3.6 million possible trees; with 20 taxa the number of possible trees is about two followed by 18 zeros ($2 \times 10^{18}$). Even with powerful computers, maximum likelihood methods can be prohibitive if the number of taxa is large. Heuristic methods exist in which only a subsample of all possible trees is examined and thus an exhaustive search is avoided.

## EVALUATION OF EVOLUTIONARY TREES

The statistical degree of confidence of a tree can be estimated for distance and maximum likelihood trees. The most common method is called bootstrapping. It consists of taking samples of the data by removing at least one data point at random and then constructing a tree for the new dataset. This random sampling process is repeated hundreds or thousands of times. The bootstrap value for each *node* is defined by the percentage of cases in which all *species* derived from that *node* appear together in the trees. Bootstrap values above 90% are regarded as statistically strongly reliable; those below 70% are considered unreliable.

Cytochrome *c* consists of only 104 amino acids, encoded by 312 nucleotides. Nevertheless, this short *protein* stores enormous evolutionary information, which made possible a fairly good approximation, shown in Fig. 4.7, to the evolutionary history of 20 very diverse *species* over a period longer than 1 billion years. But cytochrome *c* is a slowly evolving *protein*. Widely different *species* have in common a large proportion of the amino acids in their cytochrome *c*, which allows the study of genetic differences between organisms only remotely related. For the same reason, however, comparing cytochrome *c* molecules cannot determine evolutionary relationships between closely related *species*. For example, the amino acid sequence of cytochrome *c* in humans and chimpanzees is identical, although they diverged about 6 million years ago; between humans and rhesus monkeys, which diverged from their common ancestor 35–40 million years ago, it differs by only one amino acid replacement.

Molecular biology provides versatility in the construction of evolutionary trees, as pointed out earlier. Proteins that evolve more rapidly than cytochrome *c* can be studied to establish phylogenetic relationships between closely related *species*. Some proteins evolve very fast; the fibrinopeptides—small proteins involved in the blood-clotting process—are suitable for reconstructing the *phylogeny* of recently evolved *species*, such as the primates or other groups of closely related mammals. Other proteins evolve at intermediate rates; the hemoglobins, for example, can be used for reconstructing evolutionary history over a fairly broad range of time (see Fig. 4.8; Ayala, 1985, 2007).

## THE MOLECULAR CLOCK

One conspicuous attribute of *molecular evolution* is that differences between *homologous* molecules can readily be quantified and expressed as, for example, proportions of nucleotides or amino acids that have changed. Rates of evolutionary change can therefore be more precisely established with respect to *DNA* or proteins than with respect to phenotypic traits of form and function. Studies of *molecular evolution* rates have led to the proposition that macromolecules may serve as evolutionary clocks (Ayala, 1982, 1986; Kimura, 1983).

It was first observed in the 1960s that the numbers of amino acid differences between *homologous* proteins of any two given *species* seem to be nearly proportional to the time of their divergence from a common ancestor. If the rate of *evolution* of a *protein* or *gene* were approximately the same in the evolutionary lineages leading to different *species*, proteins and *DNA* sequences would provide a *molecular clock* of *evolution*. The sequences could then be used to reconstruct not only the sequence of branching events of a *phylogeny* but also the time when the various events occurred.

Consider, for example, Fig. 4.7 depicting a 20-organism *phylogeny*. If the substitution of nucleotides in the *gene* coding for cytochrome *c* occurred at a constant rate through time, one could determine the relative time elapsed along any branch of the *phylogeny* simply by examining the number of *nucleotide* substitutions along that branch. To determine the actual time elapsed, one would need only to calibrate the clock by reference to an outside source, such as the *fossil* record, that would provide the actual geological time elapsed in at least one specific lineage, such as the time when vertebrates diverged from insects.

The molecular evolutionary clock, of course, is not expected to be a metronomic clock, like a watch or other timepiece that measures time exactly, but a stochastic clock, like radioactive decay. In a stochastic clock the probability of a certain amount of change is constant (for example, a given quantity of atoms of radium-226 is expected, through decay, to be reduced by half in 1620 years), although some variation occurs in the actual amount of change. Over fairly long periods of time a stochastic clock is quite accurate. The enormous potential of the molecular evolutionary clock lies in the fact that each *gene* or *protein* is a separate clock. Each clock "ticks" at a different rate—the rate of *evolution* characteristic of a particular *gene* or *protein*—but each of the thousands and thousands of genes or proteins provides an independent measure of the same evolutionary events.

Evolutionists have found that the amount of variation observed in the *evolution* of *DNA* and proteins is greater than is expected from a stochastic clock—in other words, the clock is erratic. The discrepancies in evolutionary rates along different lineages are not excessively large, however. So, it is possible, in principle, to time phylogenetic events with as much accuracy as may be desired, but more genes or proteins (about two to four times as many) must be examined than would be required if the clock was stochastically constant. The average rates obtained for several proteins taken together become a fairly precise clock, particularly when many *species* are studied and the evolutionary events involve long time periods (in the *order* of 50 million years or longer).

This conclusion is illustrated in Fig. 4.9, which plots the cumulative number of *nucleotide* changes in seven proteins against the dates of divergence of 17 *species* of mammals (16 pairings) as determined from the *fossil* record. The overall rate of *nucleotide* substitution is fairly uniform. Some *primate species* (the pairs represented by the points below the line at the lower left of Fig. 4.9) appear to have evolved at a slower rate than the average for the rest of the *species*. This anomaly occurs because the more recent the divergence of any two *species*, the more likely it is that the changes observed will depart from the average evolutionary rate. As the length of time increases, periods of rapid and slow *evolution* in any lineage are mostly likely to cancel one another out.

Evolutionists have discovered, however, that molecular time estimates tend to be systematically older than estimates based on other methods and, indeed, to be older than the actual dates. This is a consequence of the statistical properties of molecular estimates, which are asymmetrically distributed. Because of chance, the number of molecular differences between two *species* may be larger or smaller than expected. But overestimation errors are

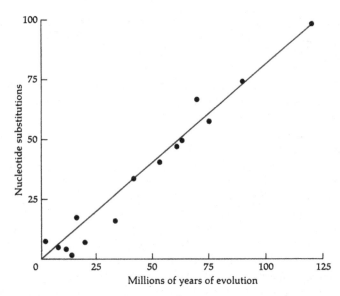

**FIGURE 4.9** The molecular clock of evolution. The numbers of nucleotide substitutions for seven proteins in 17 species of mammals have been estimated for each comparison between pairs of species whose ancestors diverged at the time indicated in the abscissa. Each dot represents the number of substitutions for the seven proteins added up. The line has been drawn from the origin to the outermost point and corresponds to a rate of 0.41 nucleotide substitutions per million years for all seven proteins combined. The proteins are cytochrome c, fibrinopeptides A and B, hemoglobins α and β, myoglobin, and insulin c-peptide. *From Ayala, F.J., 1985. The theory of evolution: recent successes and challenges. In: McMullin, E. (Ed.), Evolution and Creation. Notre Dame Press, Notre Dame, Indiana, pp. 59–90.*

unbounded, whereas underestimation errors are bounded, since they cannot be smaller than zero. Consequently, a graph of a typical distribution (a "normal distribution") of estimates of the age when two *species* diverged, gathered from a number of different genes, is likely to skew from the normal bell shape, with a large number of estimates of younger age clustered together at one end and a long "tail" of older-age estimates trailing away toward the other end. The average of the estimated times will thus consistently overestimate the true date. The overestimation bias becomes greater when the rate of *molecular evolution* is slower, when the sequences used are shorter, and when the time becomes increasingly remote.

## THE NEUTRALITY THEORY OF MOLECULAR EVOLUTION

In the late 1960s, it was proposed that at the molecular level most evolutionary changes are selectively "neutral," meaning that they are due to "*genetic drift*" (random equally probable changes) rather than to *natural selection* (which favors one evolutionary change over another). *Nucleotide* and amino acid substitutions appear in a *population* by *mutation*. If alternative alleles (alternative *DNA* sequences) have identical *fitness*—if they are identically able to perform their function—changes in allelic frequency from generation to generation will occur only by *genetic drift*. Rates of allelic substitution will be stochastically constant—that is, they will occur with a constant probability for a given *gene* or *protein*. This constant rate is the *mutation* rate for neutral alleles (Kimura, 1983; Ayala, 1977, 1986).

According to the neutrality theory, a large proportion of all possible mutants at any *gene locus* are harmful to their carriers. These mutants are eliminated by *natural selection*, just as standard evolutionary theory postulates. The neutrality theory also agrees that morphological, behavioral, and ecological traits evolve under the control of *natural selection*. What is distinctive in the theory is the claim that at each *gene locus* there are several favorable mutants equivalent to one another with respect to *adaptation*, so they are not subject to *natural selection* among themselves. Which of these mutants increases or decreases in frequency in one or another *species* is purely a matter of chance, the result of random *genetic drift* over time.

Neutral alleles are those that differ so little in *fitness* that their frequencies change by random *drift* rather than by *natural selection*. This definition is formally stated as $4N_e s < 1$, where $N_e$ is the effective size of the *population* (the number of breeding individuals) and $s$ is the selective coefficient that measures the difference in *fitness* between the alleles.

Assume that $k$ is the rate of substitution of neutral alleles per unit time in the course of *evolution*. The time units can be years or generations. In a random-mating *population* with $N$ (simplified representation of $N_e$) diploid individuals, $k = 2Nux$, where $u$ is the neutral *mutation* rate per *gamete* per unit time (time measured in the same units as for $k$) and $x$ is the probability of ultimate fixation of a neutral *mutant*. The derivation of this equation is straightforward: there are $2Nu$ mutants per time unit, each with a probability $x$ of becoming fixed. In a *population* of $N$ diploid individuals there are $2N$ genes at each *locus*, all of them, if they are neutral, with an identical probability, $x = 1/(2N)$, of becoming fixed. If this value of $x$ is substituted in the equation $k = 2Nux$, the result is $k = u$. In terms of the theory, then, the rate of substitution of neutral alleles is precisely the rate at which the neutral alleles arise by *mutation*, independently of the number of individuals in the *population* or any other factors.

If the neutrality theory of *molecular evolution* were strictly correct, it would provide a theoretical foundation for the hypothesis of the molecular evolutionary clock, since the rate of neutral *mutation* would be expected to remain constant through evolutionary time and in different lineages. The number of amino acid or *nucleotide* differences between *species* would, therefore, simply reflect the time elapsed since they shared their last common ancestor.

Evolutionists debate whether the neutrality theory is valid. Tests of the *molecular clock* hypothesis indicate that the variations in the rates of *molecular evolution* are substantially larger than would be expected according to the neutrality theory. Other tests have revealed substantial discrepancies between the amount of genetic *polymorphism* found in populations of a given *species* and the amount predicted by the theory. But defenders of the theory argue that these discrepancies can be assimilated by modifying the theory somewhat—by assuming, for example, that alleles are not strictly neutral but their differences in *selective value* are quite small. Be that as it may, the neutrality theory provides a "null hypothesis" for measuring *molecular evolution*.

Chapter 5

# Three Grand Challenges of Human Biology

## INTRODUCTION

Human biology in the 21st century faces three great research frontiers: *ontogenetic* decoding, the brain–mind puzzle, and the ape-to-human transformation. By *ontogenetic* decoding I refer to the problem of how the unidimensional genetic information encoded in the *DNA* (deoxyribonucleic acid) of a single cell becomes transformed into a four-dimensional being, the individual that grows, matures, and dies. Cancer, disease, and aging are epiphenomena of *ontogenetic* decoding. By the brain–mind puzzle I refer to the interdependent questions of how the physicochemical signals that reach our sense organs become transformed into perceptions, feelings, ideas, critical arguments, aesthetic emotions, and ethical values; and how, out of this diversity of experiences, there emerges a unitary reality, the mind or self. Free will and language, social and political institutions, technology, and art are all epiphenomena of the human mind. By the ape-to-human transformation I refer to the mystery of how a particular ape lineage became a hominid lineage from which emerged, after only a few million years, humans able to think and love, who have developed complex societies, and uphold ethical, aesthetic, and other values. The human *genome* differs little from the chimp *genome*.

I refer to these three issues as the egg-to-adult transformation, the brain-to-mind transformation, and the ape-to-human transformation. The egg-to-adult transformation is essentially similar, and similarly mysterious, in humans and other mammals. The brain-to-mind transformation and the ape-to-human transformation are distinctively human; they define the *humanum*, that which makes us specifically human. No other issues in human *evolution* are of greater consequence for understanding ourselves and our place in nature.

Erect posture and a large brain are two of the most significant anatomical traits that distinguish us from nonhuman primates. But humans are also different from chimpanzees and other animals, and, no less importantly, in their behavior, both as individuals and as socially. Distinctive human behavioral attributes include tool making and technology; abstract thinking,

Evolution, Explanation, Ethics, and Aesthetics. http://dx.doi.org/10.1016/B978-0-12-803693-8.00005-0

categorizing, and reasoning; symbolic (creative) language; self-awareness and death-awareness; science, literature, and art; legal codes, *ethics*, and religion; and complex social organization and political institutions. These traits may all be said to be components of human culture, a distinctively human mode of *adaptation* to the environment that is far more versatile and successful than the biological mode.

Cultural *adaptation* is more effective than biological *adaptation* because its innovations are directed, rather than random or undirected mutations; it can be transmitted "horizontally" to any other individuals, related or not, rather than only "vertically" to descendants; and cultural heredity is Lamarckian, rather than Mendelian, so acquired characteristics can be inherited.

## LIFE TO HUMAN

The oldest known *fossil* remains of living organisms are dated somewhat earlier than 3500 million years ago (Mya), just a few hundred million years after the Earth had cooled. The organisms were microscopic individual cells, but already had considerable complexity of organization and elaborate biochemical machinery to carry on the functions of life. We do not know when life started, but it likely was at least 100 million years earlier.

There are several hypotheses about how life was first started, but none is sufficiently well supported by evidence and thus none of them is accepted by all scientists (see Chapter 3). But the fact that it took "only" one or a few hundred million years from the formation of the Earth to the appearance of the first celled organisms suggests that life in some form is likely to appear in any planet that has water and a few other elements (notably, in our planet, carbon, nitrogen, phosphorus, and sulfur). The temperature must also be "right" within a certain range, as is the case for planet Earth, owing to the 150 million kilometers that separate it from the Sun, so water can exist in liquid phase (rather than only as either ice or vapor, if the temperature is too low or too high).

There are three large groups of organisms on Earth: eukaryotes, bacteria, and archaea (Chapter 4). The eukaryotes include animals, plants, and fungi. Eukaryotes are organisms that have their genetic material enclosed in a special capsule, or *organelle*, called the *nucleus*. Humans are eukaryotes. Animals, plants, and fungi are the only organisms that we can directly experience with our senses, and thus they were the only organisms whose existence was known to humans up to three centuries ago, when magnifying glasses and microscopes became available. Yet they account for only a fraction of the total diversity of the eukaryotes. The other eukaryotes are all microscopic. Some cause well-known diseases, such as *Plasmodium,* which causes malaria, or *Entamoeba*, which causes severe intestinal maladies.

A second group of organisms are the bacteria. Humans have known of the existence of bacteria for more than a century. We associate them with diseases, but bacteria perform many useful functions, including the incorporation from the atmosphere of the nitrogen that animals and plants need but are not able to get directly from the atmosphere (where it is very abundant, about 75% of the total; the rest is mostly oxygen). Also, bacteria are responsible for the decomposition of dead matter, a process that is essential in the maintenance of the cycle of life and death, because it makes again available, for new organisms, valuable components that had been incorporated into the now dead organisms. Millions of bacteria, belonging to hundreds of *species*, are associated with each human being. Without bacteria we could not perform many essential functions, including the digestion of food.

The genetic diversity and number of *species* of bacteria are at least as large as in the eukaryotes. There are many more kinds of bacteria than there are kinds of animals, plants, and fungi combined. And they are so abundant that their total weight (their "biomass") is at least as great as (and probably much greater than) all plants, fungi, and animals combined, even though individually they are so much smaller. This is a humbling thought. We see ourselves, the human *species*, as the summit of life and we are the most numerous of all large animals; and we see animals and plants as the *dominant* forms of life on Earth. However, modern biology teaches as that, numerically as well as in biomass, the nearly 2 million known *species* of animals (including humans) amount only to a very small fraction of life on Earth. From the perspective of numbers and biomass, the bacteria alone count much more than we do.

There is another group of microscopic organisms, the archaea, likely to be about as numerous in *species* and individuals as the eukaryotes or the bacteria. The existence of the archaea is a very recent discovery of molecular (modern) biology. Because these organisms do not directly interact much with us, biologists were not aware of their existence. Four decades ago scientists only knew a few *species*, such as some that exist in the hot springs of Yellowstone National Park in the United States and in other volcanic hot springs, where they thrive at temperatures approaching the boiling point of water. Biologists thought that these were some unusual forms of bacteria. Now we know they belong to a very diverse and numerous group of organisms, abundant in the top water layers of the seas and oceans. A bucket of seawater studied with the modern techniques of molecular biology may yield tens or hundreds of archaea *species*.

The number of living *species* on Earth is estimated to be between 10 and 30 million, but some biologists think that there may be as many as 100 million *species* if bacteria and archeaea are included. Animals represent a small fraction of all *species* now living. More than 99% of all animal *species* that lived in the past have become extinct without issue. This is most likely true for all other kinds of organisms as well. Thus the total number of *species* that have existed since the beginning of the Earth is more than 1 billion. We humans are but one of them.

Humans are animals, but a very distinct and unique kind of animal. Our anatomical differences include bipedal gait and enormous brains. But we are notably different also, and more importantly, in our individual and social behaviors, and in the products of those behaviors. With the advent of humankind, biological *evolution* transcended itself and ushered in *cultural evolution*, a more rapid and effective mode of *evolution* than the biological mode. Products of *cultural evolution* are the distinctive human features mentioned earlier. They include science and technology; complex social and political institutions; religious and ethical traditions; language, literature, and art; and radio and electronic communication.

## HUMAN ORIGINS

Our closest biological relatives are the chimpanzees, who are more closely related to us than they are to the gorillas, and much more than to the orangutans. (The chimpanzees include two *species* closely related to one another, but both equally related to humans, *Pan troglodytes* or common chimpanzee, and *P. paniscus* or bonobo.) The *hominin* lineage going to modern humans diverged from the chimpanzee lineage about 7 Mya and evolved exclusively in the African continent until the *emergence* of *Homo erectus*, somewhat before 1.8 Mya (Cela-Conde and Ayala, 2007, 2016; Coppens, 1991, 1994). The first known hominins are the recently discovered *Sahelanthropus tchadensis* (dated 6.0−7.0 Mya; Brunet et al., 2002, 2005; Brunet, 2010; Wolpoff et al., 2002), *Orrorin tugenensis* (dated 5.8−6.1 Mya; Senut et al., 2001), and *Ardipithecus ramidus* (dated 5.2−5.8 Mya; Haile-Selassie, 2001; Haile-Selassie et al., 2004, 2012). They were bipedal when on the ground, but retained tree-climbing abilities. It is not certain that they all are in the direct line of descent to modern humans, *Homo sapiens*; rather, some may represent side branches of the *hominin* lineage, after its divergence from the chimpanzee lineage. *Australopithecus anamensis*, dated 3.9−4.2 Mya, was habitually bipedal and has been placed in the line of descent to *Australopithecus afarensis*, *Kenyanthropus platyops*, *Homo habilis*, *H. erectus*, and *H. sapiens*. Other hominins, not in the direct line of descent to modern humans, are *Australopithecus africanus*, *Paranthropus aethiopicus*, *Paranthropus boisei*, and *Paranthropus robustus*, who lived in Africa at various times between 1.0 and 3.0 Mya, a period when three or four now-extinct *hominin species* lived contemporaneously in the African continent (see Cela-Conde and Ayala, 2007, 2016, for extensive reviews of *hominin evolution*).

The first intercontinental wanderer among our ancestors was *Homo erectus*. Shortly after its *emergence* in tropical or subtropical eastern Africa, *H. erectus* dispersed to other continents of the Old World. *Fossil* remains of *H. erectus* are known from Africa, Indonesia (Java), China, the Middle East, and Europe. *H. erectus* fossils from Java have been dated to $1.81 \pm 0.04$ and $1.66 \pm 0.04$ Mya, and from Georgia to between 1.6 and 1.8 Mya. Anatomically distinctive *H. erectus* fossils have been found in Spain and Italy, deposited about 800,000 years ago, the oldest known in Western Europe.

*Fossil* remains indicate that Neanderthal hominins (*Homo neanderthalensis*), with brains as large as those of *H. sapiens*, appeared in Europe, the Middle East, and further east into what is today Iran, starting around 400,000 years ago (400 kya) and persisted until 40 kya. The Neanderthals were thought to be ancestral to anatomically modern humans, but now we know that modern humans appeared in Africa at least 100 kya, long before the disappearance of the Neanderthals, who never seem to have inhabited the African continent. Moreover, in caves in the Middle East, fossils of modern humans have been found dated nearly 100 kya and Neanderthals dated at 60 and 70 kya, followed again by modern humans dated at 40 kya. It is unclear whether the two forms repeatedly replaced one another by migration from other regions, or whether they coexisted in the same areas. Recent genetic evidence indicates that interbreeding between *sapiens* and *neanderthalensis* occasionally occurred, so as much as 2% of our *genome* may be of Neanderthal origin (Pääbo, 2014; Cela-Conde and Ayala, 2007, 2016; Sankararaman et al., 2012).

The origin of anatomically modern humans is controversial. Some anthropologists argue that the transition from *Homo erectus* to archaic *H. sapiens* and later to anatomically modern humans occurred consonantly in various parts of the Old World. This is the so-called "multiregional model." Other scientists argue instead that modern humans first arose in Africa between 150 kya and 100 kya, and from there spread throughout the world, replacing elsewhere the preexisting populations of *H. erectus* or archaic *H. sapiens*, and eventually *H. neanderthalensis*. This is called the *"Out of Africa"* hypothesis, and is now favored by most anthropologists. Genetic and molecular evidence shows greater difference between African and nonAfrican populations than between all nonAfrican human populations. This pattern of differentiation endorses the hypothesis that the origin of anatomically modern humans was in Africa, whence modern humans expanded to the rest of the world starting about 100 kya. It is not possible, however, to exclude completely a partial participation of archaic *H. sapiens* from the Old World in the origin of modern humans (Wolpoff et al., 1988, 2001, 2002; Adcock et al., 2001; Abi-Rached et al., 2011). In any case, genetic analysis supports the occurrence of at least two, not just one, major migrations *out of Africa*, well after the original range expansion of *H. erectus* (Templeton, 2002, 2005).

It was earlier stated that *H. sapiens*, our *species*, is only one of more than 1 billion *species* that have lived on Earth since the beginning of the planet. From that perspective, humans are but a speck on our planet. This is also the case from the perspective of time. The hominins diverged from the apes about 7.0 Mya, and modern humans came into existence about 100−150 kya. Yet life has existed on Earth for more than 3500 Mya.

It is difficult to think in millions of years. So let me transform the time line of *evolution* into a one-year scale, so that life arises in our planet on January 1 at zero hours, and it is now midnight on December 31. In this one-year scale, for the first eight months there is only microscopic life; the first animals appear around September 1, and are marine animals. The land is colonized around December 1; the primates originate on December 26; the hominins separate from the chimpanzee lineage on December 31 at noon; and modern humans arise on that last day of the year at 23 h 45 min. We have been around for a total of 15 min. That also is a humbling thought. But I hasten to add that even though we are "but a reed," we are a "*thinking* reed," as Blaise Pascal (1669) famously put it, and to this we shall presently return.

## THE HUMAN GENOME SEQUENCE

Biological heredity is based on the transmission of genetic information from parents to offspring, in humans very much the same as in other animals. The genetic information is encoded in the linear sequence of *DNA*'s four *nucleotide* components (the "letters" of the genetic alphabet, represented by A, C, G, and T) in a similar fashion as semantic information is encoded in the sequence of letters of a written text. The *DNA* is compactly packaged in the chromosomes inside the *nucleus* of each cell. Humans have two sets of 23 chromosomes, receiving one set from each parent. The total number of *DNA* letters in each set of chromosomes is about 3000 million. The Human Genome Project, first started in 1989, has deciphered the sequence of the 3000 million letters in the human *genome* (the human *genome* sequence varies among individuals).

The King James Bible contains fewer than 3 million letters, punctuation marks, and spaces. Writing down the *DNA* sequence of one human *genome* demands 1000 volumes of the size of the Bible. The human *genome* sequence is, of course, not printed in books but stored in electronic form, in computers where fragments of information can be retrieved by investigators. But if a printout is wanted, 1000 volumes will be needed just for one human *genome*.

The two genomes (*chromosome* sets) of each individual are different from one another, and from the genomes of any other human being (with the trivial exception of identical twins, who share the same two sets, since identical twins develop from one single fertilized human egg). Therefore, printing the complete *genome* information for just one individual would demand 2000 volumes, 1000 for each of the two *chromosome* sets. Surely, again, there are more economic ways of presenting the information in the second set than listing the complete letter sequence; for example, by indicating the position of each variant letter in the second set relative to the first set. The number of variant letters between one individual's two sets is about 3 million, about one in 1000.

The Human Genome Project of the United States is funded through two agencies, the National Institutes of Health and the Department of Energy. (A private enterprise, Celera Genomics, started in the United States somewhat later, but joined the government-sponsored project in achieving, largely independently, similar results.) The goal set at the start was to obtain the complete sequence of one human *genome* in 15 years at an approximate cost of US$3 billion, coincidentally about one dollar per *DNA* letter. A draft of the *genome* sequence was completed ahead of schedule in 2001 (International Human Genome Sequencing Consortium, 2001). In 2003 the Human Genome Project was finished. The sequence had become known with as much precision as wanted. The analysis of the *DNA* sequences *chromosome* by *chromosome* continued over the following years. Results of this detailed analysis were published on June 1, 2006, by the Nature Publishing Group in a special supplement entitled *Nature Collections: Human Genome.*

Proponents of the project used inflated rhetoric to extol its anticipated achievements. The project was called the "Holy Grail" of biology, and would meet the Biblical injunction of "know thyself." Nobel Laureate Walter Gilbert said about a computer disk encoding an individual's *DNA* sequence information, "this is you."[1] The Nobel Laureate and first director of the project, James Watson, asserted that "our fate is in our genes."[2] Daniel Koshland, editor at the time of *Science,* proclaimed that with knowledge of the *genome* sequence, "we may be able to prevent the damage" caused by violent behavior.[3] Has the Human Genome Project accomplished any of these lofty objectives? Has knowledge of the human *genome* sequence accomplished the anticipated promise of curing human diseases? Certainly not.

## BEYOND THE HUMAN GENOME

As stated in the introduction, human biology faces three great research frontiers: the egg-to-adult transformation, the brain-to-mind transformation, and the ape-to-human transformation. Knowing the *DNA* sequence of human beings is of great use as a database to biologists and health scientists. But such knowledge about the human *genome* does not by itself contribute much to the solution of any of these three conundrums, or to the solution of any other fundamental biological problem.[4]

## ONTOGENETIC DECODING

The instructions that guide the *ontogenetic* process, or the egg-to-adult transformation, are carried in the hereditary material. The theory of biological heredity was formulated by the Augustinian monk Gregor Mendel in 1866, but it became generally known by biologists only in 1900: genetic information is contained in discrete factors, or genes, which exist in pairs, one received from each parent. The next step toward understanding the nature of genes was completed during the first quarter of the 20th century. It was established that genes are parts of the chromosomes, filamentous bodies present in the *nucleus* of the cell, and are linearly arranged along the chromosomes. It took another quarter-century to determine the chemical composition of genes—*DNA. DNA* consists of four kinds of chemical components (nucleotides) organized in long, double-helical structures. As pointed out earlier, the genetic information is contained in the linear sequence of the nucleotides, very much in the same way as the semantic information of an English sentence is conveyed by the particular sequence of the 26 letters of the alphabet.

The first important step toward understanding how the genetic information is decoded came in 1941 when George W. Beadle and Edward L. Tatum demonstrated that genes determine the synthesis of enzymes—the catalysts that control chemical reactions in living beings. It became known later that a series of three consecutive nucleotides in a *gene* codes one amino acid (the components that make up enzymes and other proteins). This relationship accounts for the precise linear correspondence between a particular sequence of coding nucleotides and the sequence of the amino acids that make up the encoded *protein*.

Chemical reactions in organisms must occur in an orderly manner, differently in different cells of a multicellular organism; organisms must have ways of switching genes on and off. The first control system was discovered in 1961 by François Jacob and Jacques Monod for a *gene* determining the synthesis of an *enzyme* that digests sugar in the bacterium *Escherichia coli*. The *gene* is turned on and off by a system of several switches consisting of short *DNA* sequences adjacent to the coding part of the *gene*. (The coding sequence of a *gene* is the part that determines the sequence of amino acids in the encoded *protein*.) The switches acting on a given *gene* are activated or deactivated by feedback loops that involve molecules synthesized by other genes. Various *gene* control mechanisms were soon discovered, in bacteria and other microorganisms. Two elements are typically present: feedback loops and short *DNA* sequences acting as switches. The feedback loops ensure that the presence of a substance in the cell induces the synthesis of the *enzyme* required to digest it, and that an excess of the *enzyme* in the cell represses its own synthesis. (For example, the *gene* encoding a sugar-digesting *enzyme* in *E. coli* is turned on or off by the presence or absence of the sugar to be digested.)

The investigation of *gene* control mechanisms in mammals (and other complex organisms) became possible in the mid-1970s with the *development* of recombinant *DNA* techniques. This technology made it feasible to isolate single genes (and other *DNA* sequences) and multiply or "clone" them to obtain the quantities necessary for ascertaining their *nucleotide* sequence. One unanticipated discovery was that most genes come in pieces: the coding sequence of a *gene* is divided into several fragments separated one from the next by noncoding *DNA* segments. In addition to the alternating succession of coding and noncoding segments, mammalian genes contain short control sequences, like those in bacteria but typically more numerous and complex, that act as control switches and signal where the coding sequence begins.

Much remains to be discovered about the control mechanisms of human and other mammalian genes. The daunting speed at which molecular biology is advancing makes it reasonable to anticipate that the main prototypes of mammalian *gene* control systems will soon be fully unraveled: I am thinking on a scale of one or two decades. But understanding the control mechanisms of individual genes will be only the first major step toward solving the mystery of *ontogenetic* decoding. The second major step will be solving the puzzle of differentiation.

A human being consists of 1 trillion cells of some 200 different kinds, all derived by sequential division from the fertilized egg, a single cell 0.1 mm in diameter. The first few cell divisions yield a spherical mass of amorphous cells. Successive divisions are accompanied by the appearance of folds and ridges in the mass of cells and, later on, the variety of tissues, organs, and limbs characteristic of a human individual. The full complement of genes duplicates with each cell division, so that two complete genomes are present in every cell. Experiments with other animals (and some with humans) indicate that all the genes in any cell have the potential of becoming activated. (The sheep Dolly was conceived using the *DNA* extracted from a cell in an adult sheep.) Yet different sets of genes are active in different cells. This must be so for cells to differentiate: a nerve cell, a muscle cell, and a skin cell are vastly different in size, configuration, and function. The differential activity of genes must continue after differentiation, because different cells fulfill different functions, which are controlled by different genes.

The information that controls cell and organ differentiation is ultimately contained in the *DNA* sequence, but mostly in very short segments of it. What sort of sequences are these controlling elements, where are they located, and how are they decoded? In mammals, insects, and other complex organisms, there are control circuits that operate at higher levels than the control mechanisms that activate and deactivate individual genes. These higher-level circuits (such as the so-called *homeobox* genes) act on sets rather than individual genes. The details of how these sets are controlled, how many control systems there are, and how they interact, as well as many other related questions, need to be resolved to elucidate the egg-to-adult transformation. The advances of knowledge over the last two decades have been impressive. Experiments with stem cells are likely to provide important additional knowledge as scientists ascertain how stem cells become brain cells in one case, muscle cells in another, and how some cells become the heart and others the liver.

The benefits that the elucidation of *ontogenetic* decoding will bring to humankind are enormous. This knowledge will make possible the understanding of the modes of action of complex genetic diseases, including cancer, and therefore their cure. It will also bring an understanding of the process of aging, the unforgiving disease that kills all those who have won the battle against other infirmities.

Cancer is an anomaly of *ontogenetic* decoding: cells proliferate although the welfare of the organism demands otherwise. Individual genes (oncogenes) have been identified that are involved in the causation of particular forms of cancer. But whether or not a cell will turn out cancerous depends on the interaction of the oncogenes with other genes and with the internal and external environment of the cell. Aging is also a failure of the process of *ontogenetic* decoding: cells fail to carry out the functions imprinted in their genetic code script or are no longer able to proliferate and replace dead cells.

In 1985 healthcare expenditures in the United States were $425 billion; in 2004 they surpassed $1 trillion. Most of these expenditures go for supportive *therapy* and technological fixes seeking to compensate for the debilitating effects of diseases that we do not know how to prevent or truly cure. By contrast, those diseases whose causation is understood—tuberculosis, syphilis, smallpox, and viral childhood diseases, for example—can now be treated with relatively little cost and the best of results.[5] A mere 3% of the nation's total healthcare expenditures is devoted to basic research. Doubling or tripling this percentage would result in only a modest rise in total expenditures, but would yield large savings in the near future as cancer, degenerative diseases, and other debilitating infirmities become preventable or curable, and thus no longer require the expensive and ultimately ineffectual *therapy* now in practice.

## THE BRAIN–MIND PUZZLE

The brain is the most complex and most distinctive human organ. It consists of 30 billion nerve cells, or neurons (in addition to glial cells, equally or even more numerous), each connected to many others through two kinds of cell extensions, known as the axon and the dendrites. From the evolutionary point of view, the animal brain is a powerful biological *adaptation*; it allows the organism to obtain and process information about environmental conditions and then to adapt to them. This ability has been carried to the limit in humans, in which the extravagant hypertrophy of the brain makes possible abstract thinking, free will, language, social and political institutions, art, and technology. By these means, humankind has ushered in a new mode of *adaptation* far more powerful than the biological mode: *adaptation* by culture.

The most rudimentary ability to gather and process information about the environment is found in certain single-celled microorganisms. The protozoan *Paramecium* swims apparently at random, ingesting the bacteria it encounters, but when it meets unsuitable acidity or salinity it checks its advance and starts in a new direction. The single-celled alga *Euglena* not only avoids unsuitable environments but seeks suitable ones by orienting itself according to the direction of light, which it perceives through a light-sensitive spot in the cell. Plants have not progressed much further. Except for those with tendrils that twist around any solid object and the few "carnivorous" plants that react to touch, they mostly react only to gradients of light, gravity, and moisture.

In animals the ability to secure and process environmental information is mediated by the nervous system. The simplest nervous systems are found in corals and jellyfish; they lack coordination between different parts of their bodies, so any one part is able to react only when it is directly stimulated. Sea urchins and starfish possess a nerve ring and radial nerve cords that coordinate stimuli coming from different parts, hence they respond with direct and unified actions of the whole body. They have no brain, however, and seem unable to learn from experience. Planarian flatworms have about the most rudimentary brain known; their central nervous system and brain processes coordinate information gathered by the sensory cells. These animals are capable of simple learning and hence of variable responses to repeatedly encountered stimuli. Insects and their relatives have much more advanced brains; they obtain precise chemical, acoustic, visual, and tactile signals from the environment and process them, making possible complex behaviors, particularly in their search for food and their *selection* of mates.

Vertebrates—animals with backbones—are able to obtain and process much more complicated signals and to respond to the environment more variably than insects or any other invertebrates. The *vertebrate* brain contains an enormous number of associative neurons arranged in complex patterns. In vertebrates the ability to react to environmental information is correlated with an increase in the relative size of the cerebral hemispheres and the neopallium, an organ involved in associating and coordinating signals from all receptors and brain centers. In mammals, the neopallium has expanded and become the cerebral cortex. Humans have a very large brain relative to their body size, and a cerebral cortex that is disproportionately large and complex even for their brain size. Abstract thinking, symbolic language, complex social organization, values, and *ethics* are manifestations of the wondrous capacity of the human brain to gather information about the external world, integrate that information, and react flexibly to what is perceived.

With the advanced *development* of the human brain, biological *evolution* has transcended itself, opening up a new mode of *evolution*: *adaptation* by technological manipulation of the environment. Organisms adapt to the environment by means of *natural selection*, by changing their genetic constitution over the generations to suit the demands of the environment. Humans, and humans alone to any significant extent, have developed the capacity to adapt to hostile environments by modifying the environments according to the needs of their genes. The discovery of fire, the manufacture of clothing, and the building of shelter have allowed humans to spread from the warm tropical and subtropical regions of the Old World, to which we are biologically adapted, to almost the whole Earth; it was not necessary for wandering humans to wait until genes would evolve providing anatomical protection by means of fur or hair. Nor are humans biding their time in expectation of wings or gills; we have conquered the air and seas with artfully designed contrivances, airplanes and ships. It is the human brain (the human mind) that has made humankind the most successful living *species* by most meaningful standards.

There are not enough bits of information in the complete *DNA* sequence of a human *genome* to specify the trillions of connections among the 30 billion neurons of the human brain. Accordingly, the genetic instructions must be organized in control circuits operating at different hierarchical levels, as described earlier, so an instruction at one level is carried through many channels at a lower level in the hierarchy of control circuits. The *development* of the human brain is indeed one particularly intriguing component of the egg-to-adult transformation.

Within the last few decades neurobiology has developed into one of the most exciting biological disciplines. An increased commitment of financial and human resources has brought an unprecedented rate of discovery. A great deal has been learned about how light, sound, temperature, resistance, and chemical impressions received in our sense organs trigger the release of chemical transmitters and electric potential differences that carry the signals through the nerves to the brain and elsewhere in the body. Much has also been learned about how neural channels for information transmission become reinforced by use or may be replaced after damage; about which neurons or groups of neurons are committed to processing information derived from a particular organ or environmental location; and about many other matters. But, for all this *progress*, neurobiology remains an infant discipline, at a stage of theoretical *development* comparable perhaps to that of genetics at the beginning of the 20th century. Those things that count most remain shrouded in mystery: how physical phenomena become mental experiences (the feelings and sensations, called "qualia" by philosophers, that contribute the elements of consciousness), and how out of the diversity of these experiences emerges the mind, a reality with unitary properties, such as free will and the awareness of self, that persist through an individual's life.

These mysteries are not unfathomable; rather, they are puzzles that the human mind can solve with the methods of science and illuminate with philosophical analysis and reflection. Over the next half-century or so many of these puzzles will be solved. We will then be well on our way toward answering the injunction: "Know thyself."

## THE APE-TO-HUMAN TRANSFORMATION

Knowing the human *DNA* sequence is only a first step toward understanding the genetic make-up of a human being. Think of the 1000 Bible-sized volumes. We now know the orderly sequence of the 3 billion letters, but this sequence does not provide an understanding of human beings any more than we would understand the contents of 1000 Bible-sized volumes written in an extraterrestrial language, of which we only know the alphabet, just because we would be able to decipher their letter sequence.

Human beings are not *gene* machines. The expression of genes in mammals takes place in interaction with the environment, in patterns that are complex and all but impossible to predict in the details—and it is in the details that the self resides. In humans, the "environment" takes a new dimension, which becomes the *dominant* one. Humans manipulate the natural environment so that it fits the needs of their biological make-up; for example, using clothing and housing to live in cold climates. Moreover, the products of human technology, art, science, political institutions, and the like become *dominant* features of human environments. As mentioned earlier, a distinctive characteristic of human *evolution* is *adaptation* by means of "culture," which may be understood as the set of nonstrictly biological human activities and creations.

Two conspicuous features of human anatomy are erect posture and a large brain. We are the only *vertebrate species* with a bipedal gait and erect posture. Birds are bipedal, but their backbone stands horizontal rather than vertical (penguins are a minor exception); kangaroos are mostly bipedal, but without proper erect posture or bipedal gait. Brain size is generally proportional to body size; relative to body mass, humans have the largest (and most complex) brain. The chimpanzee's brain weighs less than a pound; a gorilla's slightly more. The human male adult brain has a volume of 1400 cubic centimeters (cc), about three pounds in weight.

In earlier decades evolutionists raised the question whether bipedal gait or large brain came first, or whether they evolved consonantly. The issue is now resolved. Our *hominin* ancestors had a bipedal gait from at least 4 Mya, but their brain was still small, no more than 450 cc and a pound in weight, until about 2 Mya. Brain size started to increase notably with our *Homo habilis* ancestors, who had a brain about 650 cc and also became tool makers (hence the name *habilis*), and who lived for a few hundred thousand years, starting about 2.5 Mya. Their immediate descendants were *H. erectus*, with adult brains reaching up to 1200 cc in size. (I use the name *H. erectus* in the broad sense that encompasses a fairly diverse group of ancestors and their relatives which current paleoanthropologists classify in several *species*, including *Homo ergaster*, *Homo antecessor*, and *Homo heidelbergensis*.) Our *species*, *H. sapiens*, has a brain of 1300−1400 cc, about three times as large as that of the early hominins. Our brain is not only much larger than that of chimpanzees or gorillas, but also much more complex. The cerebral cortex, where the higher cognitive functions are processed, is in humans disproportionally much greater than the rest of the brain when compared to apes (Cela-Conde and Ayala, 2007, 2016).

## BIOLOGICAL EVOLUTION VERSUS CULTURAL EVOLUTION

Culture has an individual and a social component. It includes ideas, habits, dispositions, preferences, values, and beliefs of each individual. It also includes the public outcomes of human intellectual activity: technology; humanistic and scientific knowledge; language, literature, music, and art; codes of law and social and political institutions; ethical codes and religious systems; and other creations of the human mind. The individual and social components of culture correspond to the World 2 and World 3 of the eminent philosopher Karl Popper.

The difference between the two becomes apparent when we consider that the extinction of humankind on Earth would eliminate World 2, while World 3 could survive in part or on the whole, and could be assimilated by humans or humanoids coming from a different planet: technology, works of art, books, and electronic information could persist beyond the last humans (Popper, 1974). The advent of culture brought with it *cultural evolution*, a superorganic mode of *evolution* superimposed on the organic mode, which has, in the last few millennia, become the *dominant* mode of human *evolution*.

There are in humankind two kinds of heredity, the biological and the cultural; they may also be called organic and superorganic, or endosomatic and exosomatic systems of heredity. Biological inheritance in humans is very much like that in any other sexually reproducing organism: it is based on the transmission of genetic information encoded in *DNA* from one generation to the next by means of the sex cells.

Cultural inheritance, in contrast, is based on transmission of information by a teaching–learning process, which is in principle independent of biological parentage. Culture is transmitted by instruction and learning, by example and imitation, through books, newspapers, radio, television, and motion pictures, through works of art, and by any other means of communication. Culture is acquired by every person from parents, relatives, and neighbors, and from the whole human environment (Dobzhansky, 1962; Ehrlich, 2000; Ehrlich and Ehrlich, 2008; Cavalli-Sforza and Feldman, 1981; Boyd and Richerson, 1985; Richerson and Boyd, 2005).

Cultural inheritance makes possible for humans what no other organism can accomplish—the cumulative transmission of experience from generation to generation. Animals can learn from experience, but they do not transmit their experiences, their "discoveries" (at least not to any large extent) to the following generations. Animals have individual memory, but do not have a "social memory." Humans, on the other hand, have developed culture because they can transmit cumulatively their experiences from generation to generation. Some cultural transmission has been identified in chimpanzees and orangutan populations, but the "cultures" developed by these apes amount to trivial rudiments when compared to human cultures (Whiten et al., 1999, 2005; Whiten, 2005).

Cultural inheritance makes possible *cultural evolution*; that is, the *evolution* of knowledge, social structures, ethical systems, religion, and all other components of human culture. Cultural inheritance makes possible a new mode of *adaptation* to the environment that is not available to nonhuman organisms—*adaptation* by means of culture.

Organisms in general adapt to the environment by means of *natural selection*, by changing over generations their genetic constitution to suit the demands of the environment. But humans, and humans alone, can also adapt by changing the environment to suit the needs of their genes. (Some animals build nests and also modify their environment in other ways, but the manipulation of the environment by any nonhuman *species* is trivial compared to humankind, even in the case of the apes.)

For the last few millennia humans have been adapting the environments to their genes more often than their genes to the environments. To extend its geographical *habitat*, or to survive in a changing environment, a *population* of organisms must become adapted, through slow accumulation of genetic variants sorted out by *natural selection*, to the new climatic conditions, different sources of food, different competitors, and so on. The discovery of fire and the use of shelter and clothing allowed humans to spread from the warm tropical and subtropical regions of the Old World to the whole Earth, except for the frozen wastes of Antarctica, without the anatomical *development* of fur or hair. Humans did not wait for genetic mutants promoting wing *development*; they have conquered the air in a somewhat more efficient and versatile way by building flying machines. People travel the rivers and the seas without gills or fins. The exploration of outer space started without waiting for mutations providing humans with the ability to breathe under low oxygen pressures or to function in the absence of gravity; astronauts carry their own oxygen and specially equipped pressure suits. From their obscure beginnings in Africa, humans have become the most widespread and abundant *species* of mammal on Earth. It was the appearance of culture as a superorganic form of *adaptation* that made humankind the most successful animal *species*.

Whenever a need arises, humans can directly pursue the appropriate cultural "mutations," that is, *design* changes to meet the challenge. These changes are the discoveries and inventions that pervade human life. The invention and use of fire, the construction of bridges and skyscrapers, the telephone, and the internet are examples of technological cultural mutations; science, art, political institutions, codes of *ethics*, and religious systems also are cultural mutations. In contrast, biological *adaptation* depends on the accidental availability of a favorable *mutation*, or a combination of several mutations, at the time and place where the need arises.

Cultural heredity and biological heredity also drastically differ in their mode of transmission, with important consequences in the speed with which a favorable *adaptation* spreads. Biological heredity is transmitted only vertically, from parents to their offspring, while cultural heredity spreads "horizontally" as well as vertically, as noted earlier. A favorable genetic *mutation* newly arisen in an individual can be transmitted to a sizable part of the human *species* only through innumerable generations. However, a new scientific discovery or technical innovation can be transmitted to the whole of humankind, potentially at least, in less than one generation. Witness the worldwide spread of cellular phones or the internet in less than a decade, or of the personal computer in less than a quarter-century.

Biological heredity is Mendelian because only the genes received from one's own parents are transmitted to the progeny. (The presence in an individual of newly acquired *gene* variations by spontaneous *mutation* does not materially challenge this statement.) But acquired characteristics—that is, the inventions, technological developments, and any kind of learning or experience acquired throughout an individual's life—can all be transmitted to other humans, whether or not they are direct descendants of the individual. Cultural heredity is Lamarckian in this sense, because "acquired characteristics," and not only inherited ones, can be transmitted to others.

The draft *DNA* sequence of the chimpanzee *genome* was published on September 1, 2005.[6] In the *genome* regions shared by humans and chimpanzees, the two *species* are 99% identical. The differences appear to be very small or quite large, depending on how one chooses to look at them: 1% of the total seems very little, but it amounts to a difference of 30 million *DNA* letters out of the 3 billion in each *genome*. Of the enzymes and other proteins encoded by the genes, 29% are identical in both *species*. Out of the 100 to several hundred amino acids that make up each *protein*, the 71% of nonidentical proteins differ by only two amino acids on average. The two genomes are about 96% identical if one takes into account *DNA* stretches found in one *species* but not the other. That is, a large amount of genetic material, about 3% or some 90 million *DNA* letters, has been inserted or deleted since humans and chimps initiated their separate evolutionary ways, 7 Mya. Most of this *DNA* does not seem to contain genes coding for proteins.

Comparison of the two genomes provides insights into the rate of *evolution* of particular genes in the two *species*. One significant finding is that genes active in the brain have changed more in the human lineage than in the chimp lineage. Also significant is that the fastest evolving human genes are those coding for "transcription factors." These are "switch" proteins, which control the expression of other genes—that is, when they are turned on and off. On the whole, 585 genes have been identified as evolving faster in humans, including genes involved in resistance to malaria and tuberculosis. (It might be mentioned that malaria is a much more severe disease for humans than for chimps.) Genes located in the Y *chromosome* (the *chromosome* that determines maleness; females have two X chromosomes, while males have one X and one Y *chromosome*, the Y being much smaller than the X) have been much better protected by *natural selection* in the human than in the chimpanzee lineage, where several genes have incorporated disabling mutations that make the genes nonfunctional. There are several regions of the human *genome* that seem to contain beneficial genes that have rapidly evolved within the last 250,000 years. One region contains the *FOXP2 gene*, which had earlier been discovered to be involved in the *evolution* of speech (Pääbo, 2014).

Extended comparisons of the human and chimp genomes and experimental exploration of the functions associated with significant genes will surely advance considerably our understanding, over the next decade or two, of what it is that accounts for the *humanum*, what makes us distinctively human. Surely also full understanding will only come from the joint solution of the three conundrums I have identified. The distinctive features that make us human begin early in *development*, well before birth, as the linear information encoded in the *genome* gradually becomes expressed in a four-dimensional individual. In an important sense, the most distinctive human features are those expressed in the brain, those that account for the human mind and for human identity. It is human intelligence that makes possible human culture.

Some Christian believers will say that the fundamental difference between humans and apes is that we have a soul, created by God, which the apes do not have. This is a religious or theological answer that will be satisfying for many believers, but it is not *scientifically* satisfactory. What I mean is that, soul or no soul, scientists still want to learn how the anatomical and behavioral differences between humans and apes come to emerge from genetic differences between them. Surely, believers in the soul would not, I hope, believe that there are no biological correlates that account for the ape-to-human differences. That is what scientists seek to understand: what are the genetic and other features that distinguish our *species* from apes and other animals. Consider, by *analogy*, a human individual. People of faith may believe it is the soul infused by God that accounts for what each person is. But surely this does not deny that each individual develops from a fertilized egg in the mother's womb and later by multiple cell divisions. Nor will we want to ignore the genetic and other features that distinguish one person from another.

Religious believers might also assert that the soul accounts for the mind, thus providing the answer to the brain-to-mind puzzle. Once again, however, scientists want to understand the biological (as well as the chemical and electrical) correlates that account for mental experiences.[7] As I have asserted, I do not believe that the mysteries of the mind are unfathomable. On the contrary, I am convinced that they will be solved with the methods of science and will provide valuable religious and theological insights for people of faith. As biological understanding advances, there will surely be much left for philosophical reflection, as well as plenty of issues with great theological significance. Rather, scientific knowledge may provide a basis for theological insights.[8]

## Endnotes

1. Cited by D. Nelkin and M. S. Lindee, *The DNA Mystique. The Gene as a Cultural Icon,* W.H. Freeman, New York, 1995, p. 7.
2. Quoted in Leon Jaroff, "The Gene Hunt," *Time*, March 20, 1989, pp. 62–67.
3. D. Koshland, "Elephants, Monstrosities and the Law," *Science* 25 (February 4, 1992), p. 777.
4. I am not challenging here that the Human Genome Project has many public health applications or that the deciphering of the genomes of other *species* is of great consequence in healthcare, agriculture, animal husbandry, and industry. The question is how much it can contribute to solve the three fundamental problems faced by human biology which are expounded in this chapter.

5. This statement is overly optimistic, and it may be outright erroneous if the phrase "understood causation" is not precisely construed. Malaria and AIDS are two diseases whose causation is understood at a number of levels, yet we fail to treat them "with relatively little cost and the best results." In any case, one can anticipate that increased knowledge of the etiology of these diseases may lead to successful *development* of effective vaccines and curative drugs.

6. *Nature*, volume 437, September 1, 2005; see also *Science*, volume 309, September 2, 2005.

7. Some biologists, philosophers, and other people wonder whether dogs, whales, or monkeys might not have a sense of self, whether they are self-aware just as we humans are. They assert that we do not know for certain, because we cannot know what is in the mind of nonhuman animals. This is of course correct, but I am persuaded that only humans have self-awareness on the basis of the following argument. Being self-aware implies being aware of one's own finitude, that our self will come to an end when we die. That is, self-awareness implies death-awareness. Death-awareness, in turn, calls for ceremonial burial of the dead. We treat other dead humans with respect, because we want to be so treated when we die. Humans are the only animals that ceremonially bury their dead. We are, I conclude, the only animals that are self-aware. This conclusion is, of course, congenial to people of faith who believe in the existence of the soul.

8. The philosopher and mathematician René Descartes (1596–1650) proposed that the soul influences the body by acting through the pineal gland of the brain. This fanciful suggestion leaves unresolved the issues that I have raised: what are the physiological correlates of the mental experiences? We know, for example, that memories are stored in the brain. However, our understanding of the brain states that encompass our mental experiences, or vice versa (ie, understanding how our mental experiences impact our brain states), remains in its infancy.

# Part II

# Explanation

Chapter 6

# Design Without Designer

*It is also frequently asked what our belief must be about the form and shape of heaven according to Sacred Scripture. Many scholars engage in lengthy discussions on these matters... Such subjects are of no profit for those who seek beatitude, and, what is worse, they take up very precious time that ought to be given to what is spiritually beneficial. What concern is it of mine whether heaven is like a sphere and the earth is enclosed by it and suspended in the middle of the universe?... In the matter of the shape of heaven the sacred writers... did not wish to tech men these facts that would be of no avail for their salvation.*
St. Augustine, The Literal Meaning of Genesis Book 2, Chapter 9.

*New knowledge has led us to realize that the theory of evolution is no longer a mere hypothesis. It is indeed remarkable that this theory has been progressively accepted by researchers, following a series of discoveries in various fields of knowledge. The convergence, neither sought nor fabricated, of the results of work that was conducted independently is in itself a significant argument in favor of this theory.*
Pope John Paul II, Address to the Pontifical Academy of Sciences<br>October 22, 1996.

## INTRODUCTION

I vividly remember the day in 1971 when it was announced in New York that the Metropolitan Museum of Art had acquired at auction in Christie's London headquarters the painting "Juan de Pareja" by Diego Velázquez (Fig. 6.1). The Metropolitan had paid the staggering sum of $5,544,000, more than had ever before been paid for a painting, no matter how illustrious the artist or distinguished the work. Thomas Hoving, the Metropolitan's director at the time, later referred to "Juan de Pareja" as "the most important painting in world history" (Hoving, 1993, p. 254). This is hyperbole, but there can be little doubt that the painting is one of the finest portraits crafted by the great Spanish master. In the summer of 1649 Velázquez had brought his Moorish servant with him from Madrid to Rome, and painted him while getting ready to portray Pope Innocent X. "Juan de Pareja" is distinctively Velázquez's in the long but firm brush stroke, the clarity of execution, and the color palette, with the body executed in browns textured with brilliant reds and blacks, and set

Evolution, Explanation, Ethics, and Aesthetics. http://dx.doi.org/10.1016/B978-0-12-803693-8.00006-2

**FIGURE 6.1** "Juan de Pareja" by Diego Velázquez.

dramatically off from the face by a large "white" shirt collar extending from neck to shoulder. Like other Velázquez portraits of this period, form is created by color. Velázquez is said to have made a point of sending Pareja to visit his friends carrying the portrait so as to astonish them with its vivid likeness. In the spring of 1650 the painting was exhibited to great critical acclaim in Rome's Pantheon, and this success may have been a reason for Velázquez's election to the Roman Academy later that year.[1]

In distant northern India a few years later, an unknown Muslim craftsman forged the Dagger of Aurangzeb (Fig. 6.2), one of the most exquisite treasures in the great collection of Indian decorative arts held by the Los Angeles County Museum of Art. The hilt, in the form of a horse's head and neck, is crafted from green jade highlighted in dark orange. The blade, made of damascened steel, is shaped in the curvilinear "khanjar" style and exhibits floral ornaments and an inscription inlaid in gold with the date 1660/61. The Taj Mahal had just been built at Agra by the Mogul emperor Shah Jahan as a mausoleum to honor his dead wife, Mumtaz Mahal. The Dagger of Aurangzeb was designed as a weapon, but also as a decorative object; its author was a refined artist, not just a craftsman.

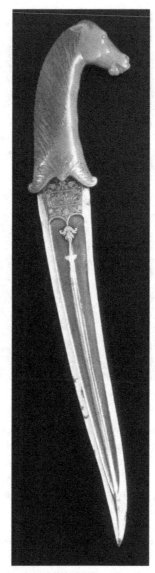

**FIGURE 6.2**   The Dagger of Aurangzeb, Los Angeles County Museum of Art.

## THE ARGUMENT FROM DESIGN

It would be ridiculously misplaced to suggest that the Dagger of Aurangzeb was a product of chance rather than *design*. It would be equally ridiculous to suggest that Velázquez executed "Juan de Pareja" without any preconceived plan or purpose. The exquisite *design* of organisms and their marvelous adaptations were similarly explained, up to the mid-19th century, as an outcome of the plan and purpose of a Creator. God had created the birds and bees, the fish and corals, the trees in the forest, and, best of all, humans. God had given us eyes so that we might see, and He had provided fish with gills to breathe in water. Scientists, philosophers, and theologians argued that the complex and exquisite *design* of organisms manifests indeed the existence of an all-wise Creator. Wherever there is *design*, there is a designer; the existence of a watch evinces the existence of a watchmaker.

The English theologian William Paley in his *Natural Theology* (1802), for example, elaborated the "argument from design" as a forceful demonstration of the existence of the Creator. The functional *design* of the human eye, argued Paley, provides conclusive evidence of an all-wise Creator. It would be absurd to suppose, he wrote, that the human eye by mere chance "should have consisted, first, of a series of transparent lenses... secondly of a black cloth or canvas spread out behind these lenses so as to receive the image formed by pencils of light transmitted through them, and placed at the precise geometrical distance at which, and at which alone, a distinct image could be formed... thirdly of a large nerve communicating between this membrane and the brain." The *Bridgewater Treatises*, published between 1833 and 1840, were written by eminent scientists and philosophers to set forth "the Power, Wisdom, and Goodness of God as manifested in the Creation." Thus the fanciful structure and mechanisms of the human hand were cited as incontrovertible evidence that the hand had been designed by the same omniscient Power that had created the world (see Chapter 16).

The exquisite *design* of Velázquez's "Juan de Pareja" proclaims that it was created by a gifted artist following a preconceived plan. Nor is the Dagger of Aurangzeb an accident of nature, a random assemblage of materials, but rather it is the product of *design*. Similarly, the structures, organs, and behaviors of living beings are directly organized to serve certain purposes or functions. The functional *design* of organisms and their features would therefore seem to argue for the existence of a Designer.

Darwin accepted that organisms are "designed" for certain purposes—that is, they are functionally organized. Organisms are adapted to certain ways of life and their parts are adapted to perform certain functions. Fish are adapted to live in water, kidneys are designed to regulate the composition of blood, and the human hand is made for grasping. But Darwin went on to provide a natural explanation of the *design*. The seemingly purposeful aspects of living beings could now be explained, like the phenomena of the inanimate world, by the methods of science, as the result of natural laws manifested in natural processes.

## EMERGENCE OF MODERN SCIENCE

Darwin is deservedly given credit for the theory of biological *evolution*, because he accumulated evidence demonstrating that organisms evolve and discovered the process, *natural selection*, by which they evolve their functional organization. But, as pointed out in Chapter 1, the import of *Origin of Species* is that he completed the Copernican Revolution, initiated three centuries earlier, and thereby radically changed our conception of the universe and the place of humankind in it. Darwin completed the Copernican Revolution by drawing out for biology the ultimate conclusion of the notion of nature as a lawful system of matter in motion that human reason can explain without recourse to extranatural agencies. The adaptations and diversity of organisms, the origin of novel and highly organized forms, and the origin of humankind itself could now be explained by an orderly process of change governed by natural laws.

The advances of physical science had driven humankind's conception of the universe to a split-personality state of affairs, which persisted well into the mid-19th century. Scientific explanations, derived from natural laws, dominated the world of nonliving matter, on the Earth as well as in the heavens. Supernatural explanations, depending on the unfathomable deeds of the Creator, accounted for the origin and configuration of living creatures—the most diversified, complex, and interesting realities of the world. It was Darwin's genius to resolve this conceptual schizophrenia.

The conundrum faced by Darwin can hardly be overestimated. The strength of the *argument from design* to demonstrate the role of the Creator was easily seen. Wherever there is function or *design*, we look for its author. Paley had belabored this argument with great skill and profusion of detail. It was Darwin's greatest accomplishment to show that the complex organization and functionality of living beings can be explained as the result of a natural process, *natural selection*, without any need to resort to a Creator or other external agent. The origin and *adaptation* of organisms in their profusion and wondrous variations were thus brought into the realm of science.

## DARWIN'S DISCOVERY

The central argument of the theory of *natural selection* is summarized by Darwin in *The Origin of Species* as follows:

> As more individuals are produced than can possibly survive, there must in every case be a struggle for existence, either one individual with another of the same species, or with the individuals of distinct species, or with the physical

*conditions of life (Chapter III, p. 63). Can it, then, be thought improbable, seeing that variations useful to man have undoubtedly occurred, that other variations useful in some way to each being in the great and complex battle of life, should sometimes occur in the course of thousands of generations? If such do occur, can we doubt (remembering that more individual are born than can possibly survive) that individuals having any advantage, however slight, over others, would have the best chance of surviving and of procreating their kind? On the other hand, we may feel sure that any variation in the least degree injurious would be rigidly destroyed. This preservation of favorable variations and the rejection of injurious variations, I call Natural Selection (Chapter IV, pp. 80–81).*

Darwin's argument addresses the same issues that had concerned Christian theologians over the centuries: how to account for the adaptive configuration of organisms, the obvious "design" of their parts to fulfill certain functions. Darwin argues that hereditary adaptive variations ("variations useful in some way to each being") occasionally appear, and that these are likely to increase the reproductive chances of their carriers. The success of pigeon fanciers and animal breeders clearly evinces the occasional occurrence of useful hereditary variations. Over the generations favorable variations will be preserved, multiplied, and conjoined; injurious ones will be eliminated. In one place, Darwin adds: "I can see no limit to this power [natural selection] in slowly and beautifully *adapting* each form to the most complex relations of life." *Natural selection* was proposed by Darwin primarily to account for the adaptive organization, or "design," of living beings; it is a process that preserves and promotes *adaptation*. Evolutionary change through time and evolutionary diversification (multiplication of species) are not directly promoted by *natural selection* (hence the so-called evolutionary *stasis* emphasized by the theory of punctuated equilibrium), but they often ensue as byproducts of *natural selection* fostering *adaptation*.

There is a possible reading of Darwin's *Origin of Species* that sees it, first and foremost, as a sustained effort to solve Paley's problem within a scientific explanatory framework. It is, indeed, how Darwin's masterpiece may be interpreted (see Chapter 1). Organisms exhibit *design*, but it is not "intelligent design," imposed by God as a supreme engineer, but the result of *natural selection* promoting the *adaptation* of organisms to their environments. Organisms exhibit complexity, but it is not *irreducible complexity* emerging all of a sudden in its current elaboration, but has arisen gradually and cumulatively, step by step, promoted by the adaptive success of individuals with incrementally more complex elaborations.

## NATURAL SELECTION AS A "DESIGN" PROCESS

The modern understanding of the principle of *natural selection* is formulated in genetic and statistical terms as differential reproduction. *Natural selection* implies that some genes and genetic combinations are transmitted to the following generations with a higher probability than their alternatives. Such genetic units will become more common in subsequent generations and their alternatives less common. *Natural selection* is a statistical bias in the relative rate of reproduction of alternative genetic units.

*Natural selection* does not operate in the manner of the Greek philosopher Empedocles' unacceptable hypothesis, acting on randomly formed organisms, allowing the functional ones to survive while the great majority die. *Natural selection* does not operate, either, as a sieve that retains the rarely arising useful genes and lets go the more frequently arising harmful mutants; or at least, not only in this way. *Natural selection* acts in the filtering way of a sieve, but it is much more than a purely negative process, for it is able to generate *novelty* by increasing the probability of otherwise extremely improbable genetic combinations. *Natural selection* is thus a creative process. It does not create the entities upon which it operates, but it produces adaptive (functional) genetic combinations that could not have existed otherwise.

The creative role of *natural selection* must not be understood in the sense of the absolute *creation* that traditional Christian theology predicates of the divine act by which the universe was brought into being *ex nihilo*, or in the manner of *creation* in which they assume God, the supreme engineer, had created the adaptations of organisms. *Natural selection* may rather be compared to a painter who creates a picture by mixing and distributing pigments in various ways over the canvas. The canvas and pigments are not created by the artist, but the paining is. It is inconceivable that a random combination of the pigments might result in the orderly whole that is the final work of art, say Leonardo's "Mona Lisa." In the same way, the combination of genetic units that carries the hereditary information responsible for the formation of the *vertebrate* eye could have never been produced by a random process such as *mutation*—not even if we allow for the 3 billion years plus during which life has existed on Earth. The complicated anatomy of the eye, like the exact functioning of the kidney, is the result of a nonrandom process—*natural selection*.

How *natural selection*, a purely material process, can generate *novelty* in the form of accumulated hereditary information may be illustrated by the following example. Some strains of the colon bacterium *Escherichia coli*, to be able to reproduce in a culture medium, require a certain substance, the amino acid histidine, to be provided in the medium. When a few such bacteria are added to 10 cubic centimeters of liquid culture medium, they multiply rapidly and produce 20–30 billion bacteria in a few hours. Spontaneous mutations to streptomycin resistance occur in normal (ie, sensitive) bacteria at rates of the order of one in 100 million ($1 \times 10^{-8}$) cells. In the bacterial culture we expect 200–300 bacteria to be resistant to streptomycin due to spontaneous *mutation*. If a proper concentration of the antibiotic is added to the culture, only the resistant cells survive. The 200–300 surviving bacteria will start reproducing, however, and allowing one to two days for the necessary number of cell divisions, 20 billion or so bacteria are produced, all resistant to streptomycin. Among cells requiring histidine as a growth factor, spontaneous mutants able to reproduce in the absence of histidine arise at rates of about four in 100 million ($4 \times 10^{-8}$) bacteria. The streptomycin-resistant cells may now be transferred to a culture with streptomycin but with no histidine. Most of them will not be able to reproduce, but about 1000 will and will start reproducing until the available medium is saturated.

*Natural selection* has produced in two steps bacterial cells resistant to streptomycin and not requiring histidine for growth. The probability of the two mutational events happening in the same bacterium is about four in 10 million billion ($1 \times 10^{-8} \times 4 \times 10^{-8} = 4 \times 10^{-16}$) cells. An event of such low probability is unlikely to occur even in a large laboratory culture of bacterial cells. With *natural selection*, cells having both properties are the common result.

Critics have sometimes alleged as evidence against Darwin's theory of *evolution* examples showing that random processes cannot yield meaningful, organized outcomes. It is thus pointed out that a series of monkeys randomly striking letters on a typewriter would never write *The Origin of Species*, even if we allow millions of years and many generations of monkeys pounding at typewriters.

This criticism would be valid if *evolution* depended only on random processes. But *natural selection* is a nonrandom process that promotes *adaptation* by selecting combinations that "make sense," ie, that are useful to the organisms. The *analogy* of the monkeys would be more appropriate if a process existed by which, first, meaningful words were chosen every time they are produced by the typewriter; and then we have other typewriters with previously selected words rather than just letters in the keys, and again there is a process to select meaningful sentences every time they are generated by this second typewriter. If every time words such as "the," "origin," "species," and so on appeared in the first kind of typewriter they each became a key in the second kind of typewriter, meaningful sentences would occasionally be produced by this second typewriter. If such sentences became incorporated into keys of a third type of typewriter, in which meaningful paragraphs were selected whenever they appeared, it is clear that pages and even chapters "making sense" would eventually be produced. The end-product would be an irreducibly complex text.

We need not carry this too far, since the *analogy* is not fully satisfactory, but the point is clear. *Evolution* is not the outcome of purely random processes, but rather there is a "selecting" process which picks up adaptive combinations because these reproduce more effectively and thus become established in populations. These adaptive combinations constitute, in turn, new levels of organization upon which the *mutation* (random) plus *selection* (nonrandom or directional) process again operates. The complexity of organization of animals and plants is "irreducible" to simpler components in one or very few steps, but not through the millions and millions of generations and the multiplicity of steps and levels made possible by eons of time.

The critical point is that *evolution* by *natural selection* is an incremental process, operating over eons of time and yielding organisms better able to survive and reproduce than others, which typically differ from one another at any one time only in small ways; for example, the difference between having or lacking an *enzyme* able to catalyze the synthesis of the amino acid histidine. Notice also that increased complexity is not a necessary outcome of *natural selection*, although such increases occur from time to time, so that, although rare, they are very conspicuous over eons of time. Increased complexity is not a necessary consequence of *evolution* by *natural selection*, but rather emerges occasionally as a matter of statistical bias. The longest-living organisms on Earth are the microscopic bacteria, which have continuously existed on our planet for 3.5 billion years, and yet those now living exhibit no greater complexity than their first ancestors. More complex organisms came about much later, without the elimination of their simpler relatives. For example, the primates appeared on Earth some 50 million years ago and our *species, Homo sapiens,* came about between one hundred and two hundred thousand years ago (Chapter 5).

## MATHEMATICAL CHALLENGES

On April 25 and 26, 1966, the Wistar Institute of Philadelphia in the United States held a symposium on "Mathematical Challenges to the Neo-Darwinian Interpretation of Evolution" (Moorhead and Kaplan, 1967) in which some mathematicians and physical scientists, skippered by Dr Murray Eden of Massachusetts Institute of Technology, held the neo-Darwinian account of *evolution* to be wanting because of the enormous improbability that, say, a particular *protein* consisting of 146 residues would arise by chance combination of the 20 possible amino acids. The critics failed to understand that all events in the real world have infinitesimal a priori probabilities, and *natural selection* is not a random process (Ayala, 1992b).

The first point was well made by botanist Conway Zirkle of the University of Pennsylvania during the discussion of Eden's paper in the symposium's first session:

> *Mr. Chairman, I wish merely to indulge in a little improbability, one that is at least as great as that cited by Dr. Eden. If we can assume, I think quite reasonably, that our parents were heterozygous for about 10,000 loci, we can see how slight the chances are that any one of us would have been born instead of some nonexisting brother or sister... The chances against any one of us having been born is [sic] practically infinite; and this forces me to accept a solipsism and to assume that this room is empty, except for myself, of course, and that the only existence any of you have is in my imagination.*
>
> Moorhead and Kaplan, 1967, p. 19.

This comment elicited no response. Apparently neither the force of Zirkle's argument nor his irony was appreciated.

An additional ironical way of making the same point as Conway Zirkle's is that if you happen to be reading this paragraph, you may consider stopping since you are performing an impossibly improbable act. There are $7 \times 10^9$ people in the world, $2.5 \times 10^{11}$ paragraphs in the world libraries, $10^{1061}$ possible letter combinations in a paragraph with 750 characters, and $8 \times 10^{-15}$ min have elapsed since the origin of the universe. Your reading this paragraph at this moment could not possibly occur because it has a probability of $10^{-1098}$. Of course, you are reading it and thus showing my argument's fallacy, just as Samuel Johnson kicked the cat to prove its existence against Bishop Berkeley's idealism.

The second point, the notion of the nonrandomness of *natural selection*, apparently remained impenetrable to Eden and others. In the symposium's final session Eden expressed his willingness to accept that evolutionary events with low probabilities can be measured in the laboratory: "I am told that nowadays a *mutation* with a frequency of about $10^{-9}$ can be detected. This obviously requires an awfully large number of Petri dishes." But he went on to say that it is impossible, now or "at any time in the future," to study experimentally events with probabilities of $10^{-14}$ to $10^{-16}$, "unless you presumably agar-plate a great part of the surface of the earth" (Moorhead and Kaplan, 1967, p. 99).

The notion that *natural selection* can yield extremely improbable outcomes by stepwise accumulation of moderately improbable events was, however, discerned by mathematician Stanislaw M. Ulam of the Los Alamos Laboratories:

> *A mathematical treatment of evolution… must include the mechanism of the advantages that single mutations bring about and the process of how these advantages, no matter how slight, serve to sieve out parts of the population, which then get additional advantages. It is the process of selection which might produce the more complicated organisms that exist today.*
>
> Moorhead and Kaplan, 1967, p. 21.

What was to come at the symposium on mathematical challenges could have been anticipated from a position paper by Eden ("Inadequacies of Neo-Darwinian Evolution as a Scientific Theory") distributed to participants in advance. The first paragraph reads as follows:

> *During the course of development of neo-Darwinian evolution as a theory, a variety of suggested universal postulates with empirical content have been invalidated. For example, the postulate that environmental influences on parents cannot affect offspring was invalidated by the discovery of induced mutations. In like manner, the notions that genes alone govern inheritance or that no morphological changes in a phenotype will propagate in its descendants have*

*also been experimentally contradicted. In consequence the theory has been modified to the point that virtually every formulation of the principles of evolution is a tautology.*

Moorhead and Kaplan, 1967, p. 109.

Like this paragraph, the rest of the paper expounds bad biology, bad *epistemology*, and bad logic.

As illustrated by the bacterial example, *natural selection* produces combinations of genes that would otherwise be highly improbable because *natural selection* proceeds stepwise. The *vertebrate* eye did not appear suddenly in all its present perfection. Its formation required the appropriate integration of many genetic units, and thus the eye could not have resulted from random processes alone, nor did it come about suddenly or in a few steps. The ancestors of today's vertebrates had for more than half a billion years some kind of organs sensitive to light. Perception of light, and later vision, were important for these organisms' survival and reproductive success. Accordingly, *natural selection* favored genes and *gene* combinations increasing the functional efficiency of the eye. Such genetic units gradually accumulated, eventually leading to the highly complex and efficient *vertebrate* eye.

*Natural selection* can account for the rise and spread of genetic constitutions, and therefore of types of organisms, that would never have existed under the uncontrolled action of random *mutation*. In this sense, *natural selection* is a creative process, although it does not create the raw materials—the genes—upon which it acts. A common objection posed to the account I have sketched of how *natural selection* gives rise to otherwise improbable features is that some postulated transitions, for example, from a leg to a wing, cannot be adaptive. The answer to this kind of objection is well known to evolutionists. For example, there are rodents, primates, and other living animals that exhibit modified legs used for both running and gliding. The *fossil* record famously includes the reptile Archaeopteryx and many other intermediates showing limbs incipiently transformed into wings endowed with feathers. One other example is described later in this chapter, namely the transition involving bones that make up the lower jaw of reptiles but later evolved into bones now found in the mammalian ear. What possible function could a bone have, either in the mandible or in the ear, during the intermediate stages?

## DESIGN AND CHANCE

There is an important respect in which an artist makes a poor *analogy* for *natural selection*. A painter has a preconception of what he wants to paint and will consciously modify the painting so that it represents what he wants. *Natural selection* has no foresight, nor does it operate according to some preconceived plan; rather it is a purely natural process resulting from the interacting properties of physicochemical and biological entities. *Natural selection* is simply a consequence of the differential multiplication of living beings, as pointed out. It has some appearance of purposefulness because it is conditioned by the environment: which organisms reproduce more effectively depends on what variations they possess that are useful in the place and at the time when the organisms live.

*Natural selection* does not anticipate the environments of the future; drastic environmental changes may be insuperable to organisms that were previously thriving. *Species* extinction is the common outcome of the evolutionary process. The *species* existing today represent the balance between the origin of new *species* and their eventual extinction. More than 99% of all *species* that have ever lived on Earth have become extinct without issue. There may have been more than 1 billion *species*; the available inventory of living *species* has identified and described fewer than 2 million out of some 10 million estimated to be now in existence.

The team of typing monkeys is also a bad *analogy* of *evolution* by *natural selection*, because it assumes that there is "somebody" who selects letter combinations and word combinations that make sense. In *evolution* there is no one selecting adaptive combinations. These select themselves because they multiply more effectively than less adaptive ones.

There is a sense in which the *analogy* of the typing monkeys is better than the *analogy* of the artist, at least if we assume that no particular statement was to be obtained from the monkeys' typing endeavors, but just any statements making sense. *Natural selection* does not strive to produce predetermined kinds of organisms, but only organisms that are adapted to their present environments. Which characteristics will be selected depends on which variations happen to be present at a given time in a given place. This in turn depends on the random process of *mutation*, as well as on the previous history of the organisms (ie, on the genetic make-up they have as a consequence of their previous evolution). *Natural selection* is an opportunistic process. The variables determining in what direction it will go are the environment, the preexisting constitution of the organisms, and the randomly arising mutations.

Thus *adaptation* to a given environment may occur in a variety of different ways. An example may be taken from the adaptations of plant life to the desert climate. The fundamental *adaptation* is to the condition of dryness, which involves the danger of desiccation. During a major part of the year, sometimes for several years in succession, there is no rain. Plants have accomplished the urgent necessity of saving water in different ways. Cacti have transformed their leaves into spines, having made their stems into barrels storing a reserve of water; photosynthesis is performed on the surface of the stem instead of in the leaves. Other plants have no leaves during the dry season, but after it rains they burst into leaves and flowers and produce seeds. Ephemeral plants germinate from seeds, grow, flower, and produce seeds—all within the space of the few weeks while rainwater is available; the rest of the year the seeds lie quiescent in the soil.

The opportunistic character of *natural selection* is also well evidenced by the phenomenon of adaptive radiation. The *evolution* of *Drosophila* flies in Hawaii is a relatively recent adaptive radiation. There are about 1500 *Drosophila species* in the world. Approximately 500 of them have evolved in the Hawaiian archipelago, which has a small land area, about one twenty-fifth the size of California. Moreover, the morphological, ecological, and behavioral diversity of Hawaiian *Drosophila* exceeds that of *Drosophila* in the rest of the world. There are more than 1000 *species* of land snails in Hawaii, all of which have evolved in the archipelago. There are 72 bird *species*, all but one of which exist nowhere else.

Why should have such explosive *evolution* have occurred in Hawaii? The overabundance of *Drosophila* flies there contrasts with the absence of many other insects. The ancestors of Hawaiian *Drosophila* reached the archipelago before other groups of insects did, and thus found a multitude of unexploited opportunities for living. They responded by a rapid adaptive radiation; although they are all derived from a single colonizing *species*, they adapted to the diversity of opportunities available in diverse places or at different times by developing appropriate adaptations, which range broadly from one to another *species*. The geographic remoteness of the Hawaiian archipelago seems, in any case, a more reasonable explanation for these explosions of diversity of a few kinds of organisms than assuming an inordinate preference on the part of the Creator for providing the archipelago with numerous *Drosophila* but not with other insects, or a peculiar distaste for creating land mammals in Hawaii, since none existed there until introduced by humans.

The process of *natural selection* can explain the adaptive organization of organisms, as well as their diversity and *evolution*, as a consequence of their *adaptation* to the multifarious and ever-changing conditions of life. The *fossil* record shows that life has evolved in a haphazard fashion. The radiations, expansions, relays of one form by another, occasional but irregular trends, and the ever-present extinctions are best explained by *natural selection* of organisms subject to the vagaries of genetic *mutation* and environmental challenge. The scientific account of these events does not necessitate recourse to a preordained plan, whether imprinted from without by an omniscient and all-powerful Designer or resulting from some immanent force driving the process toward definite outcomes. Biological *evolution* differs from a painting or an artifact in that it is not the outcome of preconceived *design*.

*Natural selection* accounts for the "design" of organisms, because adaptive variations tend to increase the probability of survival and reproduction of their carriers at the expense of maladaptive, or less adaptive, variations. The arguments of Paley against the incredible improbability of chance accounts of the adaptations of organisms are well taken as far as they go. But neither Paley nor any other author before Darwin was able to discern that there is a natural process (namely, natural selection) that is not random, but rather is oriented and able to generate *order* or "create." The traits that organisms acquire in their evolutionary histories are not fortuitous but determined by their functional utility to the organisms, designed as it were to serve their life needs.

Chance is, nevertheless, an integral part of the evolutionary process. The mutations that yield the hereditary variations available to *natural selection* arise at random, independently of whether they are beneficial or harmful to their carriers. But this random process (as well as others that come to play in the great theater of life) is counteracted by *natural selection*, which preserves what is useful and eliminates the harmful. Without hereditary *mutation, evolution* could not happen because there would be no variations that could be differentially conveyed from one generation to another. But without *natural selection*, the *mutation* process would yield disorganization and extinction because most mutations are disadvantageous. *Mutation* and *selection* have jointly driven the marvelous process that, starting from microscopic organisms, has yielded orchids, birds, and humans.

The theory of *evolution* conveys chance and necessity jointly linked in the stuff of life, randomness and determinism interlocked in a natural process that has spawned the most complex, diverse, and beautiful entities in the universe: the organisms that populate the Earth, including humans who think and love, are endowed with free will and creative powers, and are able to analyze the process of *evolution* itself that brought them into existence. This is Darwin's fundamental discovery, that there is a process that is creative though not conscious. And this is the conceptual revolution that Darwin completed: that everything in nature, including the "design" of living organisms, can be accounted for as the result of natural processes governed by natural laws. This is nothing if not a fundamental vision that has forever changed how humankind perceives itself and its place in the universe.

## EVIDENCE OF EVOLUTION

As pointed out in Chapter 2, the biological disciplines provide overwhelming evidence that organisms are related by common descent with modification: paleontology, comparative anatomy, *biogeography*, embryology, biochemistry, molecular genetics, and others. The idea first emerged from the observations of graded changes in the succession of *fossil* remains found in a sequence of layered rocks, as well as numerous remains of kinds of organisms no longer in existence. The layers have a cumulative thickness of tens of kilometers that represent up to 3.5 billion years of geological time. The general sequence of fossils from bottom upward in layered rocks had been recognized before Darwin proposed that the succession of biological forms strongly implied *evolution*. The farther back into the past one looked, the less the fossils resembled recent forms, the more the various lineages merged, and the broader the implications of a common ancestry for organisms presently quite diverse, such as fish, reptiles, and mammals.

Although gaps in the paleontological record remain, many have been filled by the research of paleontologists since Darwin's time. Millions of *fossil* organisms found in well-dated rock sequences represent a succession of forms through time and manifest many evolutionary transitions. Microbial life of the simplest type (ie, procaryotes, which are cells whose genetic matter is not bound by a nuclear membrane) was already in existence more than 3 billion years ago. The oldest evidence of more complex organisms (ie, eukaryotic cells with their genetic matter enclosed in a chamber known as the nucleus) has been discovered in flinty rocks approximately 1.4 billion years old. More advanced forms, like algae, fungi, higher plants, and a great variety of animals, have been found only in younger geological strata.

The sequence of observed forms and the fact that all (except the procaryotes) are constructed from the same basic cellular type strongly imply that all these major categories of life (including animals, plants, algae, and fungi) have a common ancestry in the first eukaryotic cells. Moreover, there have been so many discoveries of intermediate forms between fish and amphibians, between amphibians and reptiles, between reptiles and mammals, and so on that it is often difficult to identify categorically along the line when the transition occurs from one to another particular *genus* or, more generally, from one to another kind of organism. Nearly all fossils can be regarded as intermediates in some sense; they are life forms that come between ancestral forms that preceded them and those that followed.

Inferences about common descent derived from paleontology have been reinforced by comparative anatomy. The skeletons of humans, dogs, whales, and bats are strikingly similar, despite the different ways of life led by these animals and the diversity of environments in which they have flourished. The correspondence, bone by bone, can be observed in every part of the body, including the limbs: a person writes, a dog runs, a whale swims, and a bat flies with structures built of the same bones organized in the same pattern. Structures that manifest great similarity in their composition and configuration are called "homologous," and are best explained by common descent from a kind of organism that already exhibited the same composition and configuration, but modifications followed that made the structures suitable to the way of life of the descendants. Comparative anatomists investigate such homologies, not only in bone structure but in other parts of the body as well, working out degrees of relationships from degrees of similarity.

The mammalian ear and jaw offer an example in which paleontology and comparative anatomy combine to show common ancestry through transitional stages. The lower jaws of mammals contain only one bone, whereas those of reptiles have several. The additional bones in the reptile jaw are homologous with bones now found in the mammalian ear. What function could these bones have had, either in the mandible or in the ear, during intermediate stages? Paleontologists have discovered two transitional forms of mammal-like reptiles (Therapsida) with a double jaw-joint—one joint composed of the bones that persist in the mammalian jaw, the other consisting of the quadrate and articular bones that eventually became the hammer and anvil of the mammalian ear. The complex structure of the jaw of the Therapsida made possible the gradual *evolution* of some of its bones into a different function, while the remainder retained the jaw function. Similar examples are numerous.

Other biological disciplines that manifest biological *evolution* include embryology and *biogeography*, already known in Darwin's time, as well as more recently developed disciplines such as biochemistry, genetics, and comparative ethology. It is not my intention here to review the evidence coming from these biological disciplines, because it is readily available in numerous textbooks and treatises and has been discussed, if only briefly, in previous chapters. However, I do want to reiterate that the most encompassing as well as detailed evidence comes from molecular biology, a discipline that emerged in the second half of the 20th century (see Chapter 4). It is the most encompassing evidence because the most diverse kinds of organisms can all be compared in many different respects at once, from the lowly bacteria and the microscopic protozoa to the multicellular plants, fungi, and animals visible to the human eye. Molecular biology is remarkable in that organisms encompass thousands of genes and proteins, each one of which can be evaluated as an independent test of the evolutionary relationships among any particular organisms. Moreover, the evidence can readily be quantified. The possibility exists today of determining the evolutionary history of any group of organisms with as much detail as wanted. Only the limitations of human or other resources stand in the way of reconstructing the grand panorama of the *evolution* of all life, from the microscopic creatures of 3.5 billion years ago to the microorganisms, animals, and plants of today.

The proteins and nucleic acids that are essential to the make-up of all organisms are informational macromolecules that retain a record of their evolutionary history. The evolutionary information is contained in the linear sequence of their component elements in much the same way as semantic information is contained in the sequence of letters of an English sentence. This evolutionary information is so detailed that it not only makes it possible to reconstruct the phylogenetic topology, or evolutionary relationships of parentage among organisms, but also opens up the possibility of timing the events in that history, even those that occurred in the remote past of life on Earth. The information is quantifiable because the number of units that differ between organisms is readily established when the sequences of the component units are obtained for a given *protein* or *gene*. There is very little that comparative anatomy can say about the relative similarity of organisms as diverse as yeasts, pine trees, and human beings, but there are homologous macromolecules that can be compared among all three.

Nucleic acids (such as *DNA*—deoxyribonucleic acid—which embodies the hereditary information) and proteins are linear molecules made up of units, called nucleotides in the case of nucleic acids, amino acids in the case of proteins. *Evolution* typically occurs by the substitution of some of these units gradually, one at a time, so that the number of differences between two organisms is an indication of the recency of their common ancestry, in a similar way as the distance between two cars reflects how long they have been traveling in opposite directions.

Some antievolutionists argue that the theory of *evolution* is only that, a theory and not a fact. Science relies on observation, replication, and experimentation, but, they say, nobody has seen the origin of life or the *evolution* of *species*, nor have these events been replicated in the laboratory or by experiment.

This argument ignores that when scientists talk about the theory of *evolution*, they use the word "theory" differently from its use in ordinary language. In everyday English a theory is an imperfect fact, as in "I have a theory as to why there are so many religious people in the United States." In science, however, a theory is based on and incorporates a body of knowledge. According to the theory of *evolution*, organisms are related by common descent. There is a multiplicity of *species* because organisms change from generation to generation, and different lineages change in different ways. *Species* that share a recent ancestor are, therefore, more similar than those with more remote ancestors. Thus humans and chimpanzees are, in configuration and genetic make-up, more similar to each other than they are to baboons or elephants. That *evolution* has occurred is, in ordinary language, a fact.

How is this factual claim compatible with the accepted view that science relies on observation, replication, and experimentation since nobody has observed the *evolution* of *species*, much less replicated it by experiment? What scientists observe are not the concepts or general conclusions of theories, but their consequences. Copernicus's heliocentric theory affirms that the Earth revolves around the Sun. Nobody has observed this phenomenon, but we accept it because of numerous confirmations of its predicted consequences. We accept that matter is made of atoms, even though nobody has seen them, because of corroborating observations and experiments in physics and chemistry. The same applies with the theory of *evolution*. For example, the claim that humans and chimpanzees are more closely related to each other than they are to baboons leads to the prediction that the *DNA* is more similar between humans and chimps than between chimps and baboons. To test this prediction, scientists select a particular *gene*, examine its *DNA* structure in each *species*, and thus corroborate the inference. Experiments of this kind are replicated in a variety of ways to gain further confidence in the conclusion. This holds for myriad predictions and inferences between all sorts of organisms.

## DEFECTS, DEFICIENCIES, AND DYSFUNCTIONS

Traditional Christian philosophies and theologians, as well as recent proponents of *intelligent design*, seek to account for the imperfections of organisms either by attributing them to the "Fall," that is, the original sin associated with the narrative of *creation* in the Bible's book of Genesis, or by asserting that the Creator, or an unnamed Designer, may not have intended to achieve perfection. This is how a recent proponent of *intelligent design* responds to the critics who point out the imperfections of organisms:

> *The most basic problem is that the argument [against intelligent design] demands perfection at all. Clearly, designers who have the ability to make better designs do not necessarily do so... I do not give my children the best, fanciest toys because I don't want to spoil them, and because I want them to learn the value of a dollar. The argument from imperfection overlooks the possibility that the designer might have multiple motives, with engineering excellence oftentimes relegated to a secondary role... Another problem with the argument from imperfection is that it critically depends on psychoanalysis of the unidentified designer. Yet the reasons that a designer would or would not do anything are virtually impossible to know unless the designer tells you specifically what those reasons are.*[2]

So God may have had His reasons for not designing organisms as perfectly as they could have been.

A problem with this explanation is that it destroys *intelligent design* as a scientific hypothesis, because it provides it with an empirically impenetrable shield.[3] If we cannot reject *intelligent design* because the designer may have reasons that we could not possibly ascertain, there would seem to be no way to test *intelligent design* by drawing out predictions logically derived from the hypothesis that are expected to be observed in the world of experience. *Intelligent design* as an explanation for the adaptations of organisms could be (natural) theology, as Paley would have it, but, whatever it is, it is not a scientific hypothesis (see Chapter 16).

I would argue, moreover, that is not good theology either, because it leads to conclusions about the nature of the designer quite different from the omniscience, omnipotence, and omnibenevolence that Paley had inferred as the attributes of the Creator. It is not only those organisms and their parts that are less than perfect, but also that deficiencies and dysfunctions are pervasive, evidencing defective *design*. Consider the human jaw. We have too many teeth for the jaw's size, so wisdom teeth need to be removed and orthodontists make a decent living straightening the others. Would we want to blame God for such defective design? A human engineer could have done better. *Evolution* gives a good account of this imperfection. Brain size increased over time in our ancestors, and the remodeling of the *skull* to fit the larger brain entailed a reduction of the jaw. *Evolution* responds to the organisms' needs through *natural selection*, not by optimal *design* but by tinkering, as it were, by slowly modifying existing structures. Consider now the birth canal of women, much too narrow for easy passage of the infant's head, so that thousands upon thousands of babies die during delivery. Surely we do not want to blame God for this defective *design* or for the children's deaths. Science makes it understandable, a consequence of the evolutionary enlargement of our brain. Females of other animals do not experience this difficulty. Theologians in the past struggled with the issue of dysfunction because they thought it had to be attributed to God's *design*. Science, much to the relief of many theologians, provides an explanation that convincingly attributes defects, deformities, and dysfunctions to natural causes.

One more example: why are our arms and our legs, which are used for such different functions, made of the same materials, the same bones, muscles, and nerves, all arranged in the same overall pattern? *Evolution* makes sense of the anomaly. Our remote ancestors' forelimbs were legs. After our ancestors became bipedal and started using their forelimbs for functions other than walking, these became gradually modified, but retained their original composition and arrangement. Engineers start with raw materials and a *design* suited for a particular purpose; *evolution* can only modify what is already there. An engineer who designed cars and airplanes, or wings and wheels, using the same materials arranged in a similar pattern would surely be fired.

Examples of deficiencies and dysfunctions in all sorts of organisms can be endlessly multiplied, reflecting the opportunistic, tinkering character of *natural selection* rather than *intelligent design*. The world of organisms also abounds in characteristics that might be called "oddities," as well as those that have been characterized as "cruelties," an apposite qualifier if the cruel behaviors were designed outcomes of a being holding on to human or higher standards of *morality*. But the cruelties of biological nature are only meta-phorical cruelties when applied to the outcomes of *natural selection*. Examples of cruelty involve not only the familiar predators tearing apart their prey (say, a small monkey held alive by a chimpanzee biting large flesh morsels from the screaming monkey), or parasites destroying the functional organs of their hosts, but also, and very abundantly, between organisms of the same *species*, even between individuals of different sexes in association with their mating. A well-known example is the female praying mantis devouring the male after coitus is completed. Less familiar is that, if she gets the opportunity, the fe-male will eat the head of the male *before* mating, which thrashes the headless male mantis into spasms of sexual frenzy that allow the female to connect his genitalia with hers (Lawrence, 1992; Elgar, 1992). In some midges (tiny flies), the female captures the male as if he were any other prey and with the tip of her proboscis she injects into his head her spittle, which starts digesting the male's innards that are then sucked by the female; partly protected from digestion are the relatively intact male organs that break off inside the female and fertilize her (Downes, 1978). Male cannibalism is known in dozens of *species*, particularly spiders and scorpions. Diverse sorts of oddities associated with mating behavior are described in a delightful, accurate, and documented book by Olivia Judson (2002).

The defective *design* of organisms could be attributed to the gods of the Ancient Greeks, Romans, and Egyptians, who fought with one another, made blunders, and were clumsy in their endeavors. But it is not compatible with special action by the omniscient and omnipotent God of Judaism, Christianity, and Islam. With a somewhat more strident tone, the distinguished American philosopher of biology David Hull has made the same point: "What kind of God can one infer from the sort of phenomena epitomized by the *species* on Darwin's Galapagos Islands? The evolutionary process is rife with happenstance, contingency, incredible waste, death, pain and horror... Whatever the God implied by evolutionary theory and the data of *natural selection* may be like, he is not the Protestant God of waste not, want not. He is also not the loving God who cares about his productions. He is not even the awful God pictured in the Book of Job. The God of the Galapagos is careless, wasteful, indifferent, and almost diabolical. He is certainly not the sort of God to whom anyone would be inclined to pray" (Hull, 1992).

## POWERS AND LIMITS OF SCIENCE

This final comment should not be necessary, but probably is, owing to the hubris of some scientists and the pusillanimity of some believers. Science is a wondrously successful way of knowing. Science seeks explanations of the natural world by formulating explanations based on observation and experimentation that are subject to the possibility of rejection or *corroboration* by cycles upon cycles of additional observations and experimentation. A scientific explanation is tested by ascertaining whether or not predictions about the world of experience derived from the explanation agree with what is later observed.

Science as a mode of inquiry into the nature of the universe has been successful and of great consequence. Witness the proliferation of academic science departments in universities and other research institutions, the enormous budgets that the body politic and the private sector willingly commit to scientific research, and its economic impact. The Office of Management and the Budget of the US government has estimated that 50% of all economic growth in the United States since the Second World War can directly be attributed to scientific knowledge and technical advances. The technology derived from scientific knowledge pervades our lives: the high-rise buildings of our cities, thruways and long-span bridges, rockets that take men to the Moon, telephones that provide instant communication across continents, computers that perform complex calculations in millionths of a second, vaccines and drugs that keep bacterial parasites at bay, and *gene* therapies that replace *DNA* in defective cells. All these remarkable achievements bear witness to the validity of the scientific knowledge from which they originated.

Scientific knowledge is also remarkable in the way it emerges by way of consensus and agreement among scientists, and in the way new knowledge builds upon past accomplishments rather than starting anew with each generation or each new practitioner. Certainly scientists disagree with each other on many matters; but these are issues not yet settled, and the points of disagreement generally do not bring into question previous knowledge. Modern scientists do not challenge that atoms exist, or that there is a universe with a myriad stars, or that heredity is encased in *DNA*.

I want to add something that seems rather obvious to me: science is a way of knowing, but it is not the only way. Knowledge also derives from other sources, such as common sense, artistic and religious experience, and philosophical reflection. The validity of the knowledge acquired by nonscientific modes of inquiry can be simply established by pointing out that science (in the modern sense of the word) dawned in the 16th century, but humankind had for centuries built cities and roads, brought forth political institutions and sophisticated codes of law, advanced profound philosophies and value systems, and created magnificent plastic art, music, and literature. We thus learn about ourselves and the world in which we live, and we also benefit from products of this nonscientific knowledge. We learn about the human predicament by reading Shakespeare's "King Lear," looking at a Rembrandt self-portrait, and listening to Tchaikovsky's "Symphonie Pathetique" or Elton John's "Candle in the Wind." The crops we harvest and the animals we husband emerged millennia before science's dawn from practices set down by farmers in the Middle East, Andean Sierras, and Mayan plateaus.

It is not my intention to belabor the extraordinary fruits of nonscientific modes of inquiry. But I have set forth the view that nothing in the world of nature escapes the scientific mode of knowledge, and that we owe this universality to Darwin's revolution. Here, I wish simply to state that successful as it is, and universally encompassing as its subject is, a scientific view of the world is hopelessly incomplete. There are matters of value, meaning, and purpose that are outside science's scope. Even when we have a satisfying scientific understanding of a natural object or process, we are still missing matters that may well be thought by many to be of equal or greater import. Scientific knowledge may enrich aesthetic and moral perceptions, and illuminate the significance of life and the world, but these are matters outside science's realm.

## Endnotes

1. From the perspective of the year 2016, $5,544,000 does not seem a "staggering" sum, as thought in 1971 by Thomas Hoving and others. Nowadays, as much as $100 million and even more have been paid at auction for a single painting. Just as I am reviewing this chapter in January 2016, the Museum of Modern Art in New York is holding a "Picasso Sculpture" show highlighting Picasso's spectacular "Bust of a Woman (Marie-Thérèse)," courtesy of Larry Gagosian, who says he bought the sculpture in May 2015 for $105.8 million (Sunday Review, *New York Times*, January 17, 2016, p. 8).

2. Michael J. Behe, *Darwin's Black Box. The Biochemical Challenge to Evolution.* Touchstone, Simon & Schuster, New York, 1996, p. 223.
3. Robert T. Pennock (ed.), *Intelligent Design Creationism and Its Critics. Philosophical, Theological, and Scientific Perspectives*, MIT Press, Cambridge, MA, 2001, p. 249. The implications of this point with respect to the teaching of *evolution* in schools have been drawn in the public arena. In *The Washington Times*, March 21, 2002, US Senator Edward Kennedy, who had publicly supported the teaching of alternate scientific theories when there is diversity of opinion among scientists, wrote that "intelligent *design* is not a genuine scientific theory and, therefore, has no place in the curriculum of our nation's public school science classes."

Chapter 7

# Adaptation and Novelty: Teleological Explanations in Evolutionary Biology

## INTRODUCTION

In *The Origin of the Species* Darwin accumulated an impressive number of observations supporting the evolutionary origin of living organisms. Moreover, and most importantly, he provided a causal explanation of evolutionary processes—the theory of natural selection. The principle of natural selection makes it possible to give a natural explanation of the design of organisms and their adaptation to their environments. Darwin recognized, and accepted without reservation, that organisms are adapted to their environments, and that their parts are adapted to the functions they serve. Penguins are adapted to live in the cold, the wings of birds are made to fly, and the eye is made to see. Darwin accepted the facts of adaptation, and then provided a natural explanation for the facts. One of his greatest accomplishments was to bring the teleological aspects of nature into the realm of science. He substituted a scientific teleology for a theological one. The teleology of nature could now be explained, at least in principle, as the result of natural laws manifested in natural processes, without recourse to an external Creator or to spiritual or nonmaterial forces. At that point biology came to maturity as a science.

The concept of teleology is held in general disrepute in modern science. More frequently than not, it is considered to be a mark of superstition, or at least a vestige of the nonempirical, a prioristic approach to natural phenomena characteristic of the prescientific era. The main reason for this discredit is that the notion of teleology is equated with the belief that future events—the goals or end-products of processes—are active agents in their own realization. In evolutionary biology, teleological explanations are understood to imply the belief that there is a planning agent external to the world, or a force immanent to the organisms, directing the evolutionary process toward the production of specified kinds of organisms. The nature and diversity of organisms are, then, explained teleologically as the goals or ends-in-view intended from the beginning by the Creator, or implicit in the nature of the first organisms.

Evolution, Explanation, Ethics, and Aesthetics. http://dx.doi.org/10.1016/B978-0-12-803693-8.00007-4

Biological evolution can, however, be explained without recourse to a Creator or a planning agent external to the organisms themselves. There is no evidence either of any vital force or of immanent energy directing the evolutionary process toward the production of specified kinds of organisms. The evidence of the fossil record is against any directing force, external or immanent, leading the evolutionary process toward specified goals. Teleology in the stated sense is, then, appropriately rejected in biology as a category of explanation.

In this chapter, I discuss the role of teleological explanations in biology. I show that these explanations constitute patterns of explanation that apply to organisms while they do not apply to any other kind of objects in the natural world. I further claim that, although teleological explanations are compatible with causal accounts, they cannot be reformulated in nonteleological form without loss of explanatory content. Consequently, I conclude that teleological explanations cannot be dispensed with in biology, and are therefore distinctive of biology as a natural science. There are a variety of biological phenomena to which different categories of teleological explanations can be applied.

## TELEOLOGY

Teleology (from the Greek *telos* = end) is "the use of design or purpose as an explanation of natural phenomena" (*Webster's Collegiate Dictionary*, tenth ed., 1994). An object or a behavior is said to be teleological or telic when it gives evidence of design or appears to be directed toward certain ends. The behavior of human beings is often teleological. A person who buys an airplane ticket, reads a book, or cultivates the earth is trying to achieve a certain end: getting to a given city, acquiring knowledge, or producing food. Objects and machines made by people also are usually teleological: a knife is made for cutting, a clock is made for telling time, a thermostat is made to regulate temperature. Features of organisms are teleological as well: a bird's wings are *for* flying, eyes are *for* seeing, kidneys are constituted *for* regulating the composition of the blood. The features of organisms that may be said to be teleological are those that can be identified as adaptations, whether they are structures like a wing or a hand, or organs like a heart or a kidney, or behaviors like a wolf hunting a rabbit or the courtship displays of a peacock. Adaptations are features of organisms that have come about by natural selection because they serve certain functions and thus increase the reproductive success of their carriers.

Biologists make statements such as the following (*emphasis* added).

- "[E]cological compatibility is one of the most important species characteristics. *In order to* survive, each species must be supreme master in its own niche" (Mayr, 1963, p. 69).

- "It is folly or ignorance to deny that the *purpose* of nests is to protect the relatively helpless young of birds and mammals... The *purpose* of teeth... is mastication; of eyes to see, and of ears to hear" (Medawar and Medawar, 1983, p. 256).
- "Any biological mechanism produces at least one effect that can properly be called its *goal*, vision for the eye or reproduction and dispersal for the apple... Thus I would say that reproduction and dispersal are the *goals*, or *purposes* of apples and that the apple is a means or mechanism by which such *goals* are realized by apple trees" (Williams, 1966, pp. 8−9).
- "Generation by generation, step by step, the *designs* of all the diverse organisms alive today—from redwoods and manta rays to humans and yeast—were permuted out of the original, very simple, single-celled ancestor through an immensely long sequence of successive modifications" (Tooby and Cosmides, 1992, p. 52).
- "[T]he *design* of eyes reflects the properties of light, objects, and surfaces; the *design* of milk reflects the dietary requirements of infants; the *design* of claws reflects things such as the properties of prey animals, the strength of predator limbs, and the task of capture and dismemberment" (Tooby and Cosmides, 1992, p. 68).
- "Fig wasps don't transport pollen for food. They deliberately take it on board, using special pollen-carrying pockets, *for the sole purpose* of fertilizing figs (which benefits the wasps only in a more indirect way)" (Dawkins, 1996, p. 302).
- "Our mouth, throat, and larynx... were originally '*designed*' for swallowing food and breathing. They were modified *so that* we could produce sounds that were easy to understand" (Lieberman, 1998, p. 20).

These statements refer in one way or other to the functional organization of organisms and their constituent parts that come about by natural selection, as Darwin saw it. No similar statements are found in the writings of physical scientists. Inanimate objects and processes (other than those created by men) are not teleological because they are not directed toward specific ends and did not come into existence as a consequence of the purposes they serve. The configuration of a sodium chloride molecule depends on the structure of sodium and chlorine, but it makes no sense to say that this structure is made up so as to serve a certain end (although we can use it as food, that is not why it has its distinctive properties). The shape of a mountain is the result of certain geological processes that can be used for climbing or skiing, but it did not come about to serve those purposes. The motion of the Earth around the Sun results from the laws of gravity, but it does not exist to satisfy certain ends or goals, such as creating the seasons, which humans can use for their own purposes. When we use sodium chloride as food, a mountain for skiing, and take advantage of the seasons, we are not fulfilling functions that explain why these objects or phenomena came into existence or why they have certain configurations. On the other hand, a knife and a car exist and have particular configurations precisely to serve the ends of cutting and transportation. Similarly, the wings of birds came about precisely because they permitted flying, which is reproductively advantageous to the birds. The mating display of peacocks came about because it increases the chances of mating and thus of leaving progeny.

As stated in Chapter 1 and elsewhere, Darwin addressed the problem of explaining the adaptive nature of organisms. Darwin argues that adaptive variations ("variations useful in some way to each being") must occasionally appear, and that these are likely to increase the reproductive chances of their carriers. Over the generations favorable variations will be preserved and injurious ones will be eliminated. In one of the few places where he uses the term "adaptation" or its cognates in *Origin of Species*, Darwin adds: "I can see no limit to this power [natural selection] in slowly and beautifully *adapting* each form to the most complex relations of life." Natural selection was proposed by Darwin primarily to account for the adaptive organization of living beings; it is a process that promotes or maintains adaptation, and fits the features of organisms and their behaviors to the characteristics of the environment.

## TELEOLOGICAL EXPLANATIONS: A DEFINITION

The previous comments point out the essential characteristics of telic phenomena, ie, phenomena whose existence and configuration can be explained teleologically. I propose the following definition. *Teleological explanations account for the existence of a certain feature in a system by demonstrating the feature's contribution to a specific property or state of the system. Teleological explanations require the feature or behavior to contribute to the existence or maintenance of a certain state or property of the system. Moreover, and this is the essential component of the concept, the contribution must be the reason why the feature or behavior exists at all.*

The configuration of a molecule of sodium chloride contributes to its property of tasting salty and therefore to its use as food, not vice versa; the potential use of sodium chloride as food is not the reason why it has a particular molecular configuration or tastes salty. The motion of the Earth around the Sun is the reason why seasons exist; the succession of the seasons is not the purpose or reason why the Earth moves about the Sun. On the other hand, the sharpness of a knife can be explained teleologically because the knife has been created precisely to serve the purpose of cutting. Motorcars and their particular configurations exist because they serve transportation, and thus can be explained teleologically. (Not all features of a car contribute to efficient transportation—some are added for aesthetic or other reasons. But as long as a feature is added because it exhibits certain properties—like appeal to the aesthetic preferences of potential customers—it may be explained teleologically. Nevertheless, there may be features in a car, a knife, or any other human-made object that need not be explained teleologically. That knives have handles may be explained teleologically, but the fact that a particular handle is made of pine rather than oak might simply be due to the availability of materials. Similarly, as pointed out below, not all features of organisms have teleological explanations.)

Many features and behaviors of organisms meet the requirements of teleological explanation. The hand of humans, the wings of birds, the structure and behavior of kidneys, and the mating displays of peacocks are examples already given. In general, as pointed out above, those features and behaviors that are considered adaptations are explained teleologically. This is simply because adaptations are features that come about by natural selection. Among alternative genetic variants that may arise by mutation or recombination, the ones that become established in a population are those that contribute more to the reproductive success of their carriers. The effects on reproductive success are usually mediated by some function or property. Wings and hands acquired their present configuration through long-term accumulation of genetic variants adaptive to their carriers. An alternative feature may be due to a single gene mutation, eg, the presence of normal hemoglobin rather than hemoglobin S in humans. One amino acid substitution in the beta chain in humans results in hemoglobin molecules that are less efficient for oxygen transport. The general occurrence in human populations of normal rather than S hemoglobins is explained teleologically by the contribution of hemoglobin to effective oxygen transport and thus to reproductive success. The difference between peppered-gray and melanic moths is due to one or only a few genes. The replacement of gray moths by melanics in polluted regions is explained teleologically by the fact that melanism decreased the probability of predation in such regions. The predominance of peppered forms in nonpolluted regions is similarly explained.

Not all features of organisms need to be explained teleologically, since not all come about as a direct result of natural selection. Some features may become established by genetic drift, by chance association with adaptive traits, or in general by processes other than natural selection. Proponents of the neutrality theory of protein evolution argue that many alternative protein variants are adaptively equivalent. Most evolutionists would admit that at least in certain cases the selective differences between alternative protein variants must be virtually nil, particularly when population size is very small. The presence in a population of one amino acid sequence rather than another adaptively equivalent to the first would not then be explained teleologically. Needless to say, there should also be amino acid sequences that would not be adaptive. The presence of an adaptive protein rather than a nonadaptive one would be explained teleologically; but the presence of one protein rather than another among those adaptively equivalent would not require a teleological explanation—it could be the result of chance.

## TELEOLOGICAL FEATURES AND BEHAVIORS IN ORGANISMS

There are three kinds of biological phenomena where teleological explanations are pertinent, although the distinction between the categories need not always be clearly delimited, and it is also possible to subdivide or reformulate them in a different or more prolific array. These three classes of teleological phenomena are established according to the mode of relationship between the structure or process and the property or end-state that accounts for its presence. Other classifications of teleological phenomena are possible according to other principles of distinction, including some suggested below.

1. Behaviors such that the end-state or goal is consciously anticipated by the agent. This is purposeful activity which, if it is understood in a strict sense, probably occurs only in humans. With a lesser degree of intentionality, behaviors initiated to reach a goal also occur in other animals. I am acting teleologically when I buy an airplane ticket to fly to Mexico City. A cheetah hunting a gazelle gives at least the appearance of purposeful behavior. We may notice that according to those who believe in "special" creation, the existence of organisms and their adaptations is the result of the consciously intended activity of a Creator seeking to create each kind specifically. Biologists recognize purposeful activity in the living world, at least in humans; but the existence of the living world, including humans, need not be explained as the result of purposeful behavior, the actions of a designer or Creator. A very common reason why biologists do not use the term "teleology" is that they believe it necessarily implies that function and design must be attributed to an external agent; ie, that the design features of organisms have been created by God. I take up this matter below. It is in any case amusing to read statements of denial of teleology in articles and books pervaded with teleological language and explanations. One is reminded that "a rose by any other name is still a rose." Informally attributed to one or another distinguished evolutionist (in particular to the notorious J.B.S. Haldane) is the witticism: "Teleology is like a mistress. A man does not want to be seen in her company, but he cannot do without her."

2. Self-regulating of teleonomic systems, when there exists a mechanism that enables the system to reach or maintain a specific property in spite of environmental fluctuations. The regulation of body temperature in mammals is a teleological mechanism of this kind. In general, the homeostatic reactions of organisms belong to this category of teleological phenomena. Biologists usually distinguish between two types of homeostasis— *physiological* and *developmental*—although intermediate and additional types do exist. For instance, the persistence of a genetic polymorphism (two or more alleles with stable frequencies through time) in a population

due to heterosis (advantage of individuals who inherit a different allele from each parent) may be considered a homeostatic mechanism acting at the population level. One example is the presence of the "normal" and the S forms of hemoglobin in human populations severely infected with malaria. The S form protects against malaria and the "normal" variant prevents falciform anemia.

Physiological homeostatic reactions enable the organism to maintain a certain physiological steady state in spite of environmental shocks. The regulation of the composition of the blood by the kidneys and the hypertrophy of muscle in case of strenuous use are examples of this type of homeostasis. Developmental homeostasis refers to the regulation of the different paths that an organism may follow in its progression from zygote to adult. The development of a chicken from an egg is a typical example of developmental homeostasis. The process can be influenced by the environment in various ways, but the characteristics of the adult individual and its pattern of development, at least within a certain range, are largely predetermined in the fertilized egg.

Self-regulating systems or servomechanisms built by humans belong to this second category of teleological phenomena. A simple example of a servomechanism is a thermostat unit that maintains a specified room temperature by turning on and off the source for heating or cooling. Self-regulating mechanisms of this kind, living or human-made, are controlled by feedback information. Robots programmed to perform certain functions are additional examples.

3. Organs, limbs, and other features anatomically and physiologically constituted to perform a certain function. The human hand is made for grasping, and the eye for vision. Tools and human-made machines are teleological in this third sense. A watch, for instance, is made to tell time, and a faucet to draw water. The distinction between the previous category and this category of teleological systems is sometimes blurred. Thus the human eye is able to regulate itself within a certain range according to the conditions of brightness and distance so as to perform its function more effectively.

The adaptations of organisms—whether organs, homeostatic mechanisms, or patterns of behavior—are explained teleologically as a consequence of natural selection, because their existence is ultimately accounted for in terms of their contribution to the reproductive fitness of the organisms. A feature of an organism that increases its reproductive fitness will be selectively favored. Once it has arisen by genetic mutation or recombination, given enough generations it will extend to the whole species.

Patterns of behavior, such as the migratory habits of certain birds or the web spinning of spiders, have developed because they favored the reproductive success of their possessors in the environments where the organisms lived. Similarly, natural selection can account for the existence of homeostatic mechanisms. Some living processes can operate only within a certain range of conditions. If the environmental conditions oscillate frequently beyond the functional range of the process, natural selection will favor self-regulating mechanisms that maintain the system within the functional range. In humans, death results if the body temperature is allowed to rise or fall by more than a few degrees above or below normal. Body temperature is regulated by the dissipation of heat in warm environments through perspiration and dilation of the blood vessels in the skin; in cool weather the loss of heat is minimized and additional heat is produced by increased activity and shivering. Finally, the adaptation of an organ or feature to its function is also explained teleologically by natural selection, in that the existence of the organ or feature is accounted for in terms of the contribution it makes to the reproductive success of its carriers. The vertebrate eye arose because genetic mutations responsible for its development occurred and were gradually combined in progressively more efficient patterns, the successive changes increasing the reproductive fitness of their possessors in the environments in which they lived.

## PROXIMATE VERSUS ULTIMATE TELEOLOGY

We now further explore the role of teleological explanations in biology by distinguishing three additional categories within each of which we can identify alternative types of teleology and, consequently, teleological explanations: proximate (or particular) versus ultimate (or general); natural versus artificial; and bounded or determinate versus unbounded or indeterminate teleology.

There are in all organisms two levels of teleology that may be labeled *proximate* (or *particular*) and *ultimate* (or *general*). Proximate or particular teleology exists for every specific feature determined by natural selection in every animal or plant. The existence of the feature is thus explained in terms of the function or property that it serves, which function or property can be said to be the particular or proximate end of the feature. Thus seeing is a particular, specific, or proximate end served by eyes, and flying is a particular, specific, or proximate end served by wings. There is also an ultimate goal to which all features contribute or have contributed in the past—reproductive success. The general or ultimate end to which all features and their functions contribute is increased reproductive efficiency. The presence of the functions themselves—and therefore of the features that serve them—is ultimately explained by their contribution to the reproductive fitness of the organisms in which they exist. It is in this sense that the ultimate source of teleological explanation in biology is the principle of natural selection.

Because of the reasoning just advanced, I have suggested that natural selection can be said to be a teleological process in a *causal sense*, namely as a distinctive process, occurring only in the living world, which accounts for the adaptive features of organisms (Ayala, 1968, 1970). I could have said instead that natural selection is a *teleology-inducing* process, intending to convey the same idea. But these terms must not be misunderstood. Natural selection is not an entity or an agent, and thus it is not a cause in the usual sense. Nor does natural selection result in predetermined or preconceived features or organisms, as I further expound. To reiterate the point, natural selection is not an entity but a purely material or natural process governed by the laws of physics and chemistry, and other natural laws. To designate it as a "teleological process" would be to convey exclusively the meaning that natural selection results in the production and preservation of end-directed organs and behaviors, when the functions these serve contribute to the reproductive effectiveness of the organisms.

In any case, the process of natural selection is not at all teleological in a different sense. Natural selection is not in any way directed toward the production of specific kinds of organisms or organisms having certain specific properties. The overall process of evolution cannot be said to be teleological in the sense of proceeding toward certain specified goals, preconceived or not. Natural selection is nothing more than the outcome of differential reproduction. The final result of natural selection for any species may be extinction, as shown by the fossil record, if the species fails to cope with environmental challenges.

I have argued that the presence of organs, processes, and patterns of behavior can be explained teleologically by exhibiting their contribution to the reproductive fitness of the organisms in which they occur. This does not imply that reproductive fitness is a consciously intended goal. Such intent must be denied, except in the case of the voluntary behavior of humans. In teleological explanations the end-state or goal is not to be understood as the efficient cause of the object or process that it explains. The end-state is causally posterior, the outcome of a process, not its cause.

## NATURAL VERSUS ARTIFICIAL TELEOLOGY

We have already noted several kinds of biological phenomena, such as the existence of wings or the functioning of the kidneys, that call for teleological accounts, and that teleological accounts also apply to purposeful behavior and human-made objects. It will be helpful to characterize some differentiating features of these categories of teleological entities, particularly the biological in general and the distinctively human.

Human actions are *purposeful* when an end-state or goal is consciously intended by an agent. Thus a person mowing a lawn is acting teleologically in the purposeful sense; a lion hunting deer and a bird building a nest manifest at least the appearance of purposeful behavior. A knife, a car, and a thermostat are objects or systems intended for (and produced by) humans. Actions or objects resulting from human purposeful behavior may be said to exhibit *artificial* teleology. Their teleological features have come about because they were consciously intended for and by some agent.

Systems with teleological features that are not due to the purposeful action of an agent but result from natural process may be said to exhibit *natural* teleology. The wings of birds have a natural teleology; they serve an end, flying, but their configuration is not due to the conscious design of any agent. The development of an egg into a chicken is a teleological process also of an internal or natural kind, since it comes about as a natural process, both in terms of its proximate causation, the concatenation of events by which the egg develops into a chicken, and its remote causation, the evolutionary process by which chickens and their developmental processes came to be.

## DETERMINATE VERSUS INDETERMINATE TELEOLOGY

Two kinds of natural teleology may be distinguished: *determinate* (or *bounded* or *necessary*) and *contingent* (or *indeterminate* or *unbounded*). This distinction applies as well to purposeful objects and behaviors, but human actions are predominantly determinate, in the sense that they are consciously intended. Humans can of course walk randomly or act aimlessly and can produce objects, such as a die, that behave randomly, but for the most part these are nevertheless products of intentionality.

Determinate natural teleology exists when a specific end-state is reached in spite of environmental fluctuations. The development of an egg into a chicken and of a human zygote into a human being are examples of determinate natural teleological processes. The regulation of body temperature in a mammal is another example. In general, the homeostatic processes of organisms are instances of determinate natural teleology.

Indeterminate or unbounded teleology occurs when the end-state served is not specifically intended or predetermined, but is rather the result of a natural process selecting one among several available alternatives. For teleology to exist, the selection of one alternative over another must be deterministic and not purely stochastic. But what alternative happens to be selected may depend on environmental and/or historical circumstances, and thus the specific end-state is not generally predictable. Indeterminate teleology results from a mixture of stochastic (at least from the point of view of the teleological system) and deterministic events.

Many features of organisms are teleological in the indeterminate sense. The evolution of birds' wings requires teleological explanation: the genetic constitutions responsible for their configuration came about because wings serve for flying and flying contributes to the reproductive success of birds. But there was nothing in the constitution of the remote reptilian ancestors of birds that would necessitate the appearance of wings in their descendants. Wings came about as the consequence of a long sequence of events, at each stage of which the most advantageous alternative was selected among those that happened to be available; which alternatives were available at any one time depended at least in part on contingent events.

In spite of the role played by stochastic events in the phylogenetic history of birds, it would be mistaken to say that wings are not teleological features. As pointed out earlier, there are differences between the teleology of an organism's adaptations and the nonteleological potential uses of natural inanimate objects. A mountain may have features appropriate for skiing, but those features did not come about so as to provide ski slopes. On the other hand, the wings of birds came about precisely because they are used for flying. One explanatory reason for the existence of wings and their configuration is the end they serve—flying—which in turn contributes to the reproductive success of birds. If wings did not serve the purpose of an adaptive function they would have never come about, and would gradually disappear over the generations.

The indeterminate character of the outcome of natural selection over time is due to a variety of nondeterministic factors. The outcome of natural selection depends, firstly, on what alternative genetic variants happen to be available at any one time. This in turn depends on the stochastic processes of mutation and recombination, and also on the past history of any given population. (Which new genes may arise by mutation and which new genetic arrays may arise by recombination depend on which genes happen to be present—which depends on previous history.) The outcome of natural selection depends also on the conditions of the physical and biotic environment. Which alternatives among available genetic variants may be favored by selection depends on the particular set of environmental conditions to which a population is exposed. The historical process of evolution is contingent, but at each step there is a predominantly deterministic component provided by the natural selection of favorable variants among those that happen to be present in particular organisms at a particular time. Organisms are adapted to their environments and exhibit adaptive features owing to this deterministic component. The contingency of history and environment makes long-term evolution undetermined or unbounded. There can be little doubt that if the process of evolution on Earth were to start again from where it was 3 billion years ago, the evolved organisms would be conspicuously different from the ones that came about in the first run of the process.

## OBJECTIONS AND RESPONSES

Some evolutionists have rejected teleological explanations because they have failed to recognize the various meanings that the term "teleology" may have (eg, Pittendrigh, 1958; Mayr, 1965, 1974; Williams, 1966; Ghiselin, 1974). These biologists are correct in excluding certain forms of teleology from evolutionary explanations, but they err when they claim that teleological explanations should be excluded altogether from evolutionary theory. In fact, they themselves often use teleological explanations in their works, but fail to recognize them as such or prefer to call them by some other name, such as "teleonomic." Teleological explanations, as explained above, are appropriate in evolutionary theory, and are recognized by most evolutionary biologists and philosophers of science who have thoughtfully considered the question (Beckner, 1959; Nagel, 1961; Simpson, 1964; Dobzhansky, 1970; Ayala, 1968, 1970; Wimsatt, 1972; Hull, 1974). Which kinds of teleological explanations are appropriate and which are inappropriate with respect to various biological questions may be briefly specified.

Mayr (1965) pointed out that teleological explanations have been applied to two different sets of biological phenomena. "On the one hand is the production and perfection throughout the history of the animal and plant kingdoms of ever-new and ever-improved DNA programs of information. On the other hand is the testing of these programs and their decoding throughout the lifetime of each individual. There is a fundamental difference between end-directed behavioral activities or developmental processes of an individual or system, which are controlled by a program, and the steady improvement of the genetically coded programs. This genetic improvement is evolutionary adaptation controlled by natural selection." The "decoding" and "testing" of genetic programs of information are the issues considered by developmental biology (decoding) and functional biology (testing). The historical and causal processes by which genetic programs of information come about are the concern of evolutionary biology. Grene (1974) uses the term "instrumental" for the teleology of organs that act in a functional way, such as the hand for grasping; "developmental" for the teleology of such processes as the maturation of a limb; and "historical" for the process (natural selection) producing teleologically organized systems.

Organs and features such as the eye and the hand have determinate (and internal) natural teleology. These organs serve determinate ends (seeing or grasping), but have come about by natural processes that did not involve the conscious design of any agent. Physiological homeostatic reactions and embryological development are processes that also have determinate natural teleology. These processes lead to end-states (from egg to chicken) or maintain properties (body temperature in a mammal) that are on the whole determinate. Thus Mayr's "decoding" of DNA programs of information and Grene's "instrumental" and "developmental" teleology, when applied to organisms, are cases of determinate natural teleology (Mayr prefers to use the term "teleonomy" for this type of teleology). Human tools (such as a knife), machines (such as a car), and servomechanisms (such as a thermostat) also have determinate teleology, but of the artificial kind, since they have been consciously designed.

The process of natural selection is teleological, but only in the sense of indeterminate natural teleology. It is not consciously intended by any agent, nor is it directed toward specific or predetermined end-states. Yet the process is far from random or completely indeterminate. Among the genetic alternatives available at any one time, natural selection favors those that increase the reproductive success of their carriers in the particular environmental circumstances in which the organisms live. Reproductive success is, of course, mediated by some adaptive function, say flying, that is determined by the genetic variants favored by natural selection.

Some authors exclude teleological explanations from evolutionary biology because they believe that teleology exists only when a specific goal is purposefully sought. This is not so. Terms other than "teleology" could be used for natural (or internal) teleology, but this might in the end add more confusion than clarity. Philosophers as well as scientists use the term "teleological" in the broader sense, to include explanations that account for the existence of an object in terms of the end-state or goal that it serves.

The process of evolution by natural selection, as repeatedly stated, is not teleological in the purposeful sense. Thomas Aquinas in the 13th century and the natural theologians of the 19th century, notably William Paley, erroneously claimed that the directive organization of living beings evinces the existence of a Designer. Darwin demonstrated that the adaptations of organisms can be explained as the result of natural processes without recourse to consciously intended end-products. There is purposeful activity in the world, at least in human actions; but the existence and particular structures of organisms, including human existence, need not be explained as the result of purposeful behavior.

Lamarck (1809) erroneously thought that evolutionary change necessarily proceeded along determined paths from simpler to more complex organisms. Similarly, the evolutionary philosophies of Bergson (1907) and Teilhard de Chardin (1959), and such theories as *nomogenesis* (Berg, 1926), *aristogenesis* (Osborn, 1934), *orthogenesis*, and the like are erroneous because they all claim that evolutionary change necessarily proceeds along determined paths. These theories mistakenly take embryological development as the model of evolutionary change, regarding the teleology of evolution as determinate. Although there are teleologically determinate processes in the living world, like embryological development and physiological homeostasis, the evolutionary origin of living beings is teleological only in the indeterminate sense. Natural selection does not in any way direct evolution toward any particular organisms or any particular properties.[1]

## TELEOLOGY AND CAUSALITY

Teleological explanations are fully compatible with causal explanations (see Nagel, 1961, 1965; Ayala, 1970, 1995). It is possible, at least in principle, to give a causal account of the various physical and chemical processes in the development of an egg into a chicken, or of the physico-chemical, neural, and muscular interactions involved in the functioning of the eye. (I use the "in principle" clause to imply that any component of the process can be elucidated as a causal process if it is investigated and ascertained in sufficient detail and depth, but I know of no developmental process for which all the steps have been so investigated. The developmental processes from egg to adult are best known in the flatworm *Caenorhabditis elegans*. The development of *Drosophila* fruitflies has also become known in much detail.) It is also possible in principle to describe the causal processes by which one genetic variant eventually becomes established in a population. But these causal explanations do not make it unnecessary to advance teleological explanations where appropriate. Both teleological and causal explanations are called for in evolutionary biology.

According to Nagel, "a teleological explanation can always be transformed into a causal one." Consider a typical teleological statement in biology, "The function of gills in fishes is respiration." This statement is a telescoped argument whose content can be unraveled approximately as follows: fish respire; if fish have no gills, they cannot respire; therefore fish have gills. According to Nagel, a teleological explanation directs our attention to "the *consequences* for a given system of a constituent part or process." The equivalent nonteleological formulation focuses attention on "some of the *conditions*... under which the system persists in its characteristic organization and activities" (Nagel, 1965; see also Nagel, 1961).

"The function of gills in fishes is respiration, that is, the exchange of oxygen and carbon dioxide between the blood and the external water." A statement of this kind, according to Nagel, accounts for the presence of a certain feature $A$ (*gills*) in every member of a class of system $S$ (*fish*) which possess a certain organization $C$ (the characteristic anatomy and physiology of fish). It does so by declaring that when $S$ is placed in a certain environment $E$ (water with dissolved oxygen), it will perform a function $F$ (respiration) only if $S$ (fish) has $A$ (gills). The teleological statement, says Nagel, can be approximated as follows: when supplied with water containing dissolved oxygen, fish respire; if fish have no gills, they do not respire even if supplied with water containing dissolved oxygen; therefore fish have gills. More generally, a statement of the form "The function of $A$ in a system $S$ with organization $C$ is to enable $S$ in environment $E$ to engage in process $F$" can be formulated more explicitly: "Every system $S$ with organization $C$ and in environment $E$ engages in function $F$; if $S$ with organization $C$ and in environment $E$ does not have $A$, then $S$ cannot engage in $F$; hence, $S$ must have $A$." According to Nagel, the difference between a teleological and a nonteleological explanation is, thus, one of emphasis rather than of asserted content.

Nagel's account, however, misses an essential feature of teleological explanations, which invalidates his claim that they are equivalent to (even if less cumbersome than) causal accounts. Although a teleological explanation can be reformulated into a nonteleological one, the teleological explanation connotes something more than the equivalent nonteleological one. In the first place, a teleological explanation implies that the system under consideration is directively organized. For that reason teleological explanations are appropriate in biology and in the domain of human creations, but make no sense when used in the physical sciences to describe phenomena like the fall of a stone or the slopes of a mountain. Teleological explanations imply, while nonteleological ones do not, that there exists a distinctive or specific means-to-an-end relationship in the system under description: the eye is for seeing, the egg develops into a chicken, the knife is used for cutting.

In addition to connoting that the system under consideration is directively organized, and most importantly, teleological explanations account for the existence of specific functions in a system and more generally for the existence of the directive organization itself. A teleological explanation accounts for the presence in an organism of a certain feature, say the gills, because it contributes to the performance or maintenance of a certain function, respiration in this example. The teleological explanation also connotes, in the case of organisms, that the function came about because it contributes to the reproductive fitness of the organism. In the nonteleological translation given above, the major premise states that "fish respire." Such a formulation assumes the presence of a specified function, respiration, but it does not account for its existence. A teleological explanation implicitly (or explicitly) accounts for the presence of the function itself by connoting (or stating explicitly) that the function in question contributes to the reproductive fitness of the organism in which it exists and that *this is the reason why the function and feature came about in evolution*. The teleological explanation gives the reason why the system is directively organized. The apparent purposefulness of the ends-to-means relationship existing in organisms is a result of the process of natural selection which favors the development of any organization that increases the reproductive fitness of the organisms.

It follows that teleological explanations are not only acceptable in biology, but also indispensable to and distinctive of the discipline. It further follows that for this reason alone (and I have suggested others) biology cannot be logically "reduced" to the physical sciences (Ayala, 1968, 1977).

## TELEOLOGICAL EXPLANATIONS ARE TESTABLE HYPOTHESES

One question biologists ask about features of organisms is "What for?" that is, "What is the function or role of a particular structure or process?" The answer to this question must be formulated teleologically. A causal account of the operation of the eye is satisfactory as far as it goes, but it does not indicate all that is relevant about the eye, namely that it is useful to the organism because it serves to see. Evolutionary biologists are interested in the question of why one particular genetic alternative rather than another came to be established in a species. This also calls for teleological explanations of the type: "Eyes came into existence because they serve to see, and seeing increases reproductive success of certain organisms in particular circumstances." In fact, eyes came about in several independent evolutionary lineages: cephalopods, arthropods, and vertebrates (Dawkins, 1996, pp. 138–197).

Two questions must be addressed by a teleological account of evolutionary events. First, there is the question of whether a genetic variant contributes to reproductive success: a teleological account states that an existing genetic constitution (say, the gene coding for a normal hemoglobin beta chain) enhances reproductive fitness better than alternative constitutions. Then there is the question of how the specific genetic constitution of an organism enhances its reproductive success: a teleological explanation states that a certain genetic constitution serves a particular function (eg, the molecular composition of hemoglobin has a role in oxygen transportation).

Both questions call for specific teleological hypotheses that can be empirically tested. This point has been belabored, for example by J. Tooby and L. Cosmides (1992). It sometimes happens, however, that information is available to answer one or the other question but not both. In population genetics the fitness effects of alternative genetic constitutions can often be measured, while the mediating adaptive function responsible for the fitness differences may be difficult to identify. We know, for example, that in the fruitfly *Drosophila pseudoobscura* different versions of polymorphisms are favored by natural selection at different times of the year, but we are largely ignorant of the physiological processes involved. In a historical account of evolutionary sequences the problem is occasionally reversed: the function served by an organ or structure may be easily identified, but it may be difficult to ascertain why the development of that feature enhanced reproductive success and thus was favored by natural selection. One example is the large human brain, which makes possible culture and other important human attributes. We may advance hypotheses about the reproductive advantages of increased brain size in the evolution of man, but these hypotheses are difficult to test empirically.

Teleological explanations in evolutionary biology have great heuristic value. They are also occasionally very facile, precisely because they may be difficult to test empirically. Every effort should be made to formulate teleological explanations in a fashion that makes them readily subject to empirical testing. When appropriate empirical tests cannot be formulated, evolutionary biologists should use teleological explanations only with the greatest restraint (Williams, 1966; Gould and Lewontin, 1979).

## UTILITY AS TOUCHSTONE CRITERION

It has been argued by some authors that the distinction between systems that are goal directed and those that are not is highly vague. The classification of certain systems as teleological is alleged rather than arbitrary. A chemical buffer, an elastic solid, and a pendulum at rest are examples of physical systems that appear to be goal directed. I suggest using the criterion of utility to determine whether an entity is teleological or not. The criterion can be applied to both natural and artificial teleological systems. Utility in an organism is defined in reference to the survival and reproduction of the organism itself. A feature of a system will be teleological in the sense of natural (internal) teleology if it has utility for the system in which it exists and if such utility explains the presence of the feature in the system. Operationally, then, a structure or process of an organism is teleological if it can be shown to contribute to the reproductive efficiency of the organism itself, and if such a contribution accounts for the existence of the structure or process. Eyes, gills, and homeostatic developmental processes are features beneficial to the organisms in which they exist.

In artificial (external) teleology, utility is defined in reference to the creator of the object or system. Human-made tools and machines are teleological with external teleology if they have been designed to serve a specified purpose, which therefore explains their existence and properties. If the criterion of utility cannot be applied, a system is not teleological. Chemical buffers, elastic solids, and a pendulum at rest are not teleological systems.

The utility of features of organisms is in respect to the individual or the species in which they exist at any given time. It does not include usefulness to any other organisms. The elaborate plumage and display are teleological features of the peacock because they serve the peacock in its attempt to find a mate. The beautiful display is not teleologically directed toward pleasing our human aesthetic sense; that it pleases the human eye is accidental, because this does not contribute to the reproductive fitness of the peacock (except, of course, in the case that humans are selecting for breeding peacocks with particular plumage patterns).

The criterion of utility introduces needed objectivity in the determination of which biological mechanisms are end directed. Provincial human interests should be avoided when using teleological explanations, as Nagel has written. But he selects the wrong example when he observes that "the development of corn seeds into corn plants is sometimes said to be natural, while their transformation into the flesh of birds or men is asserted to be merely accidental" (Nagel, 1961, p. 424). The adaptations of corn seeds have developed to serve the function of corn reproduction, not to become a palatable food for birds or humans. The role of wild corn as human food is indeed accidental to the corn, and cannot be considered a biological function of corn seeds in the teleological sense. This point was clearly and repeatedly made by Darwin. For example, "Natural selection will modify the structure of the young in relation to the parent, and of the parent in relation to the young. In social animals it will adapt the structure of each individual for the benefit of the community; if each in consequence profits by the selected change. What natural selection cannot do, is to modify the structure of one species, without giving it any advantage, for the good of another species; and though statements to this effect may be found in works of natural history, I cannot find one case which will bear investigation" (Darwin, 1967, pp. 86–87).

Some features of organisms are not useful by themselves. They have arisen as concomitant or incidental consequences of other features that are adaptive or useful. In some cases, features which are not adaptive in origin may become useful at a later time. For example, the sound produced by the beating of the heart has become adaptive for modern humans because it helps the physician to diagnose the health of the patient. The origin of such features is not explained teleologically, although their preservation might be so explained in certain cases.

Features of organisms may be present because they were useful to the organisms in the past, although they are no longer adaptive. Vestigial organs, like the vermiform appendix of humans, are features of this kind. If they are neutral to reproductive fitness, these features may remain in a species indefinitely. The origin of such organs and features, although not their preservation, is accounted for in teleological terms.

## TELEOLOGY VERSUS TELEONOMY: CICERO'S TRANSLATION OF ARISTOTLE'S *AITION*

A semantic question that deserves attention concerns the terms teleology and teleonomy. It brings us to a comment about the Latin version of Aristotle's word *aition*, which Cicero (unfortunately, in my view) translated as *causa*.

Pittendrigh (1958), Simpson (1964), Mayr (1965), Williams (1966), and others have proposed using the term "teleonomic," rather than "teleological," to describe end-directed processes that do not imply that future events are active agents in their own realization, or that things or activities are conscious agents or the product of such agents. These authors argue that the term "teleology" has sometimes been used to explain the animal and plant kingdoms as the result of a preordained plan necessarily leading to the existing kinds of organisms. To avoid such connotation, the authors argue, the term teleonomy should be used to explain adaptation in nature as the result of natural selection.

Although the notion of teleology has been, and it is still being, used in the alleged sense, it is also true that other authors, like Dobzhansky (1970), Simpson (1964), and Nagel (1961) employ the term "teleology" without implying directed or intentional causation, or that the final outcome of a process is the cause of the process itself. Thus it might produce more confusion than clarity to repudiate the notion of teleology on the grounds that it connotes an intentional relationship of means to an end. The point is that what is useful is to clarify the notion of teleology by explaining the various uses of the term. One may then explicitly express in which sense the term is used in a particular context.

As I have written elsewhere, should the term "teleology" eventually be discarded from the scientific vocabulary or restricted in its meaning to pre-ordained end-directed processes, one could welcome such an event. But that is unlikely to happen. Moreover, the substitution of a term by another does not necessarily clarify the issues at stake. It would still be necessary to explicate whatever term is used instead of teleology, whether teleonomy or any other (Ayala, 1970).

Pittendrigh wrote that "It seems unfortunate that the term 'teleology' should be resurrected... The biologists' long-standing confusion would be more fully removed if all end-directed systems were described by some other term, like 'teleonomic,' in order to emphasize that the recognition and description of end-directedness does not carry a commitment to Aristotelian teleology as an efficient causal principle" (Pittendrigh, 1958, p. 394). The Aristotelian concept of teleology allegedly implies that future events are active agents in their own realization. According to other authors, Aristotelian teleology connotes that there exists an overall design in the world attributable to a Deity, or at least that nature exists only for and in relation to humans, considered as the ultimate purpose of creation (see Mayr, 1965).

Science, for Aristotle, is a knowledge of the "whys," the "reasons for" true statements. Of a thing we can ask four different kinds of questions: "What is it?", "Out of what is it made?", "By what agent?", "What for?" The four kinds of answers that can be elicited from these questions are Aristotle's four causes—formal, material, efficient, and final. Only the third type of answer is causal in the modern scientific sense. *Aition*, the Greek term that Cicero translated as "cause" (*causa* in Latin), means literally ground of explanation, ie, what can be given in answer to a question. It does not necessarily mean causality in the sense of efficient agency.

According to Aristotle, to understand an object fully we need to find out, among other things, its end, what function it serves, or what results it produces. An egg can be understood fully only if we consider it as a possible chicken. The structures and organs of animals have functions and are organized toward certain ends. Living processes proceed toward certain goals. Final causes, for Aristotle, are principles of intelligibility; they are not in any sense active agents in their own realization. For Aristotle, ends "never do anything. Ends do not act or operate, they are never efficient causes" (Randall, 1960, p. 128).

Moreover, according to Aristotle there is no intelligent Maker of the world. The ends of things are not consciously intended. Nature, humans excepted, has no purposes. The teleology of nature is objective and empirically observable.

It does not require the inference of unobservable causes (Ross, 1949; see also Randall, 1960). There is no God designer of nature. According to Aristotle, if there is a God, He cannot have purposes (Randall, 1960, p. 125).

Finally, for Aristotle, the teleology of nature is wholly "immanent." The end served by any structure or process is the good or survival of the thing in which it exists. Animals, plants, and their parts do not exist for the benefit of any other thing but themselves. Aristotle makes it clear that nutritious as acorns may be for a squirrel, they do not exist to serve as a squirrel's meal. The "natural end" of an acorn is to become an oak tree. Anything else that may happen to the acorn is accidental and may not be explained teleologically. Aristotle's insight concerning this matter surpasses Nagel's (see above).

Aristotle's main concern was the study of organisms and their processes and structures (Leroi, 2014). He observed the facts of adaptation, and explained them with considerable insight considering that he did not know about biological evolution. His error was not that he used teleological explanations in biology, but that he extended the concept of teleology to the nonliving world. In the Middle Ages Aristotle was "Christianized," particularly in the works of the great theologian Thomas Aquinas (1225–74). It was Aquinas, not Aristotle, who accounted for the teleology of organisms, and of nature in general, as the intended purpose of an Omniscient Creator.

## ERNST MAYR (1998) ON TELEOLOGY

The eminent evolutionist Ernst Mayr (1998) distinguishes "five entirely different kinds of phenomena... to which the term teleological had been applied in the past": "teleomatic processes" ("processes in inanimate nature which reach an end stage determined by the universal laws of physics"); "teleonomic processes" (a process or behavior "that owes its goal-directedness to the operation of a program... [which] is not a description of a given situation but a set of instructions"); "adapted features" (their designation as teleological is "misleading because these features are stationary systems. For me, the word teleological is not appropriate for the phenomena that do not involve movements"); "purposive behavior" (human and animal behavior "that is clearly purposive, revealing careful planning"); and "cosmic teleology" ("a belief... that changes in the world [are] teleological in nature, leading to ever greater perfection"). Mayr proposes "to restrict the use of the word teleological to cosmic teleology and to use other more specific terms instead for the other phenomena."

One need not quarrel with Mayr's distinction of five meanings of teleology. Teleology has, nevertheless, been used with other meanings as well, including in earlier pages of this chapter. A quick look at three dictionaries (*Oxford Dictionary, Webster Third New International Dictionary*, and *Merriam Webster's Collegiate Dictionary*, tenth ed.) reveals several meanings not mentioned by Mayr.

In any case, the issue at hand is not semantics, but epistemology. In this chapter the concern is to unravel patterns of explanation that are teleological in structure; ie, those patterns that (as explained above) account for the presence of a feature or behavior in a system in terms of the goal it serves or the purpose it seeks. These patterns of explanation are extensively used in biology. It is my contention that teleological explanations are appropriate only in biology among the natural sciences, and thus that they are distinctive of the discipline. Because teleological explanations cannot be translated into causal explanations without loss of explanatory content, it may be asserted as an interesting implication that biology cannot be reduced to the physical sciences.

It seems to me unlikely that the use of "teleology" will become generally limited to Mayr's fifth sense of "cosmic teleology," nor do I see a reason why it should be. I do not see why such restriction would, in practice, avoid confusion. As pointed out, the genteelism "teleonomy" was coined by Pittendrigh (1958) and has been endorsed by others (Simpson, 1964; Williams, 1966) to purge the term teleology, as P.B. Medawar and J.S. Medawar put it, from any "pretensions to providing causal explanations, and restricted exclusively to putting on record the purposes which biological structures and performances do in fact fulfill." But the substitution of a term by another does not necessarily clarify the issues at stake (Ayala, 1970). In any case "the word 'teleonomy' has not caught on, perhaps because corruption of biology by teleology is not... so grave or so imminent a danger" (Medawar and Medawar, 1983).

Mayr restricts the meaning of various terms in idiosyncratic ways that are far from common. He writes that "'ordained' is strictly theological language." I do not see why the statement, "The development of an insect from egg to adult is a precisely ordained sequence of events" should be considered a theological statement, strictly or otherwise. Dictionaries do not so restrict the meaning of "ordained," either. Mayr further writes that "purpose, as far as I can see can be attributed legitimately only to a thinking organism." Peter Medawar, who can hardly be accused of being soft-headed, has written, as quoted above: "It is folly or ignorance to deny that the purpose of nests is to protect the relatively helpless young of birds and mammals, and of the amnion to provide the embryos of land vertebrates with the aquatic environment they need in order to develop" (Medawar and Medawar, 1983).

Mayr writes that "the end point of the non-random process of elimination sooner or later is extinction [of species], but I would hesitate to call selection a teleological process because, sooner or later, it has an end point (extinction). Almost anything on earth has an end: a book does, a vacation does, an opera does, the day does, etc. Are these telic phenomena? The life of an individual has an end, is life a telic process? This is why I said it was 'dangerous' to classify telic phenomena as teleological." These statements are puzzling. The dictionary definitions of "telic" that I have seen either give it as a synonym of "teleological" and "purposive" (*Webster's Third New International Dictionary*) or do so by definition: "Expressing end or purpose" and "Directed or tending to a definite end; purposive" (*Oxford Dictionary*); "tending toward an end" (*Merriam Webster's Collegiate Dictionary*, tenth ed.). In any case, who would think of calling "selection," or anything else for that matter, a teleological process just because it has an end point in time?

Importantly, Mayr's definition of "teleonomy" (the term he proposes as a substitute for "teleology") is circular, and I have said so in the past (Ayala, 1988). His definition is: "A teleonomic process or behavior is one that owes its goal-directedness to the operation of a program"; and adds, "A program might be defined as coded or prearranged information that controls a process (or behavior) leading it toward a goal." So a process is teleonomic if governed by a program; and if we want to know what kind of programs govern teleonomic processes, Mayr says that they are those programs ("coded or prearranged information") that lead the process "toward a goal." There is not much enlightenment here. We are told, in effect, that a process is teleonomic if it seeks a goal guided by a program, namely the kind of program that guides a process toward a goal. Given this circularity, it is not surprising that the definition can be arbitrarily applied. Mayr writes: "I had first included man-made objects like loaded dice under teleonomic because they were 'programmed' to behave in a particular way, but I later excluded such objects." Why? If "loaded dice" are programmed to behave in a particular way, why should not they be considered teleonomic according to Mayr's definition? And one may wonder why only "loaded" dies might or might not be considered teleonomic. Why are dies (or cars or telephones) not considered teleonomic by Mayr, since they are programmed to behave in particular ways?

## Endnote

1. The eminent paleontologist Simon Conway Morris is one of the few distinguished biologists who disagree with this statement, if not unique. The title of his book, *Life's Solution. Inevitable Humans in a Lonely Universe* (2003), expresses Conway Morris's conviction that were evolution to occur again on our planet starting from scratch, humans (intelligent, human-like creatures) would evolve once more. "The central theme of this book depends on the realities of evolutionary convergence: the recurrent tendency of biologist organization to arrive at the same 'solution' to a particular 'need'... [this book's] main... aim is to argue that, contrary to received wisdom, the emergence of human intelligence is a near-inevitability" (Conway Morris, 2003, p. xii).

Chapter 8

# Evolution and Progress

## INTRODUCTION

The process of biological *evolution* appears as obviously progressive. The earliest organisms on Earth were no more complex than today's bacteria. Three billion years later, their descendants include orchids, ants, sharks, alligators, and eagles, which have appeared successively over several 100 million years, and our *species, Homo sapiens*, which came into existence within the last 200,000 years. The *species* that came about at later times appear to be more complex, advanced, or progressive than their primitive ancestors.

Upon reflection the issue becomes less obvious, because what do we mean when we say there has been *progress* in the evolutionary process? Some evolutionary lineages do not appear progressive at all: living bacteria are not very different from their ancestors of 2 or 3 billion years earlier. In addition, extinction can hardly be progressive; yet most evolutionary lineages have become extinct. And some organisms may be more progressive than others with respect to some features but not with respect to other features. For example, bacteria are able to synthesize all their own components and obtain the energy they need for living from inorganic compounds; human beings depend on other organisms.

The *creation* of the world as described in the *Book of Genesis* contains the explicit notion that some organisms are higher than others. The Bible's narrative of the *Creation* reflects the commonsense impression that earthworms are lower than fish and birds, and the latter are lower than humans. Similarly, the philosophers of classical Greece put forward the notion that living organisms can be classified in a hierarchy going from lower to higher forms. The idea of a "ladder of life" rising from ameba to man is present, explicitly or implicitly, in all preevolutionary biology (Lovejoy, [1936] 1960).

The theory of *evolution* adds the dimension of time and genetic continuity, or history, to the hierarchical classification of living things. The transition from bacteria to humans can now be seen as a natural, progressive *development* through time from simple to gradually more complex organisms. The expansion and diversification of life could also be judged as *progress*; some form of advance seems obvious in the transition from one or only a few kinds of living things to the several million *species* living today.

Evolution, Explanation, Ethics, and Aesthetics. http://dx.doi.org/10.1016/B978-0-12-803693-8.00008-6

**143**

The idea of *progress*, applied to human history or the world of nature, does not appear until the Renaissance. The philosophers of Ancient Greece, including Aristotle, saw the world in terms of eternal cycles. It was only in the 16th century that the first ideas about the possibility of *progress* appeared in the history of the world and of humankind. The idea of *progress*, associated with the increase in knowledge, was first entertained at that time, and became associated with the notions of change, *development*, or direction (Bury, [1932] 1955).

## EARLY IDEAS

Aristotle (384−322 BCE) is fittingly recognized as one of the great philosophers of Greek antiquity, known for his works on logic, *epistemology*, metaphysics, *ethics*, and politics. He qualifies as well as a great scientist, the first biologist in the history of the world, and dissected a variety of animals and described in detail the *development* of the chicken embryo from egg to hatching. During his three years in the Aegean Island of Lesbos (345−342), he investigated all sorts of marine and lacustrine organisms, from coral reefs and molluscs to fish, as well as land animals. In his numerous zoological works, such as *The Generation of Animals* and others, Aristotle mentions 500 animal *species*, a large number relative to the knowledge of the time (Leroi, 2014). Aristotle put forward the notion that living organisms can be classified in a hierarchy going from lower to higher forms in a sort of a natural ladder, a gradual succession of increasing complexity, from plants to animals to humans. A similar notion was elaborated around 1260 by Albert Magnus, "the first modern European to study Aristotle's zoology" (Leroi, 2014, p. 276). Albert's notion that nature proceeds by small steps became commonplace by the early 17th century and prevailed into the 18th century. It was for Linnaeus in his *Philosophia botanica* of 1751 a methodological principle: *natura non facit saltum*—nature does not make jumps. The notion that there is a ladder of nature or chain of being eventually became temporalized, according to Arthur O. Lovejoy, and "must perforce be reinterpreted so as to admit of progress in general" (Lovejoy, [1936] 1960, p. 246).

In the 17th century Francis Bacon (1561−1626) formulated a program for the advancement of knowledge, the notion that *progress* in knowledge could contribute to improving the human condition. The true purpose of knowledge, according to Bacon, was to improve human life, to provide new inventions and developments, and in the case of the natural sciences to define the supremacy of human life over the rest of nature. The true purpose of knowledge is not to pleasure the mind, or establish the superiority of some individuals over others in their reputation or power, but rather to advance the benefit, power, and domain of humanity over the rest of the universe. Bacon advances a notion of *progress* focused upon the human capacity to advance knowledge that would serve for humankind's benefit. Starting with the evidence provided by our senses, and taking advantage of technological developments, Bacon thought that knowledge could become increasingly more certain. The purpose of technology is to increase human knowledge and thus contribute to humankind's *progress*. These ideas became the core of a general doctrine of *progress* that was developed over the ensuing centuries (Ruse, 1996).

The Aristotelian and traditional notion that organisms can be classified in a hierarchy from lower to higher became further entrenched in the 17th and 18th centuries: the notion of a *scala naturae*, ranging from the inanimate world to the lower and then the higher animals and humans (Lovejoy, [1936] 1960). The chain of being is, nevertheless, perceived as static; it was created as perfect so any modification would be considered as harm or deterioration. Gradually, however, in the late 18th and early 19th centuries the notion of human *progress*—epitomized by Nicolas de Condorcet's *Sketch for a Historical Picture of the Progress of the Human Mind* (1795) in France, Adam Smith's *The Wealth of Nations* (1776) in Britain, and G.W.F. Hegel's *Philosophy of Nature* (1817) in Germany—became extended to the world of life, so that some organisms are considered more advanced or more progressive than others.

## EVOLUTIONARY PROGRESS

*Progress* as an attribute of the living world reached an early climax with the publication in 1809 of *Philosophie Zoologique* (*Philosophical Zoology*) by the great French naturalist Jean-Baptiste Lamarck (1744–1829). Lamarck proposed the first broad theory of *evolution*, with *progress* as an immanent attribute of the evolutionary process. Organisms evolve through eons of time from lower to higher forms, a process still going on and always culminating in human beings. As organisms become adapted to their environments through their habits, modifications occur. Use of an organ or structure reinforces it; disuse leads to obliteration. The characteristics acquired by use and disuse would be inherited. Organisms repeatedly evolve in a fixed sequence of progressive transitions. Today's worms will advance gradually, always eventually culminating in human beings.

The notion that *evolution* is always progressive is not part of Darwin's theory of *evolution* by *natural selection*. Darwin knew that numerous *species* became extinct without leaving descendants, and that many others have not changed for eons of time. "Some of the most ancient Silurian animals, as the Nautilus, Lingula, etc., do not differ much from living species; and it cannot on my theory be supposed, that these old species were the progenitors of all the species of the orders to which they belong, for they do not present characters in any degree intermediate between them" (Darwin, 1859, p. 306). Notable proponents of evolutionary *progress* included Ernst Haeckel, *The Evolution of Man* (1896), and Herbert Spencer, "Progress: its law and cause" (1857) and *First Principles* (1862; see also Richards, 1988).

## CHANGE, EVOLUTION, DIRECTION

The meaning of statements like "Progress has occurred in the evolutionary sequence leading from bacteria to humans" or "The evolution of organisms is progressive" is not immediately obvious. Such expressions may simply mean that evolutionary sequences have a time direction, or even more simply that they are accompanied by change. The term "progress" may be clarified by comparing it with other related terms used in biological discourse: "change," "evolution," and "direction."

The term "change" means alteration, whether in the position, the state, or the nature of a thing. *Progress* implies change, but not vice versa; not all changes are progressive. The molecules of oxygen and nitrogen in the air of a room are continuously changing positions, but such changes would not generally be regarded as progressive. The *mutation* of a *gene* from a functional allelic state to a nonfunctional one is a change, but definitely not a progressive one.

"Evolution" and "progress" can also be distinguished, although both imply that sustained change has occurred. Evolutionary change is not necessarily progressive. The *evolution* of a *species* may lead to its own extinction, a change that is not progressive, at least not for that *species*. *Progress* can also occur without evolutionary change. Assume that in a given region of the world the seeds of a certain *species* are dormant because of a prolonged drought; after a burst of rain the seeds germinate and give origin to a *population* of plants. This change might be labeled progressive for the *species*, even though no evolutionary change needs to have taken place.

The concept of "direction" implies that a series of changes has occurred that can be arranged in a linear sequence so that elements in the later part of the sequence are further from early elements of the sequence than intermediate elements are. Directional change may be uniform or not, depending on whether every later member of the sequence is further displaced than every earlier member ("uniform" change), or whether directional change occurs only on the average ("net" change). Nonuniform or net (see below) directional change occurs when the direction of change is not constant; some elements in the sequence may represent a change of direction with respect to the immediately previous elements, but later elements in the sequence are displaced further than earlier ones on the average.

"Directionality" is sometimes equated with "irreversibility" in discussions of evolution: the process of *evolution* is said to have a direction because it is irreversible. Biological *evolution* is irreversible (except perhaps in some trivial sense, as when a previously mutated *gene* mutates back to its former allelic state, or a *gene* increases in frequency for some time but decreases in later generations). Direction, however, implies more than irreversibility. Consider a new pack of cards with each suit arranged from ace to 10, then knave, queen, king, and with the suits arranged in the sequence of spades, clubs, hearts, and diamonds. If we shuffle the cards thoroughly, the *order* of the cards will change and the changes will be irreversible by shuffling. We may shuffle again and again until the cards are totally worn out, without ever restoring the original sequence. The change of order in the pack of cards is irreversible but not directional.

Directional changes occur in the inorganic as well as the organic world. The second law of thermodynamics, which applies to all processes in nature, describes sequential changes that are directional, and, indeed, uniformly directional. Within a closed system, entropy always increases; that is, a closed system passes continuously from less probable to more probable states. The concept of direction applies to what in paleontology are called "evolutionary trends." A trend occurs in a phylogenetic sequence when a feature persistently changes through time in the members of the sequence. Trends are common occurrences in all *fossil* sequences that are sufficiently long to be called "sustained" (Simpson, 1953).

The concept of direction and the concept of *progress* are distinguishable. Consider the trend in the evolutionary sequence from fish to man toward a gradual reduction over paleontological time of the number of dermal bones in the *skull* roof; or the trend toward increased molarization in the last premolar that occurred in the *phylogeny* of the *Equidae* from the early Eocene (*Hyracotherium*) to the early Oligocene (*Haplohippus*). These trends indeed represent directional change, but it is not obvious that they should be labeled progressive; to do this we would have to agree that the directional change had been for the better in some sense. Thus to consider a directional sequence progressive we need to add an evaluation, namely that the condition of the latter members of the sequence represents, according to some standard, amelioration or improvement. The directionality of the sequence can be recognized and accepted without any such evaluation being added. *Progress* implies directional change, but not vice versa.

## THE CONCEPT OF PROGRESS

Evolution, direction, and *progress* all imply a historical sequence of events that exhibit a systematic alteration of a property or state of the elements in the sequence. *Progress* occurs when there is directional change toward a better state or condition. The concept of *progress*, then, contains two elements: one descriptive, that directional change has occurred, and the other axiological (= evaluative), that the change represents betterment or improvement (Ayala, 1988). Gould (1988) argued that the idea of *progress* should be replaced by an "operational notion of directionality," so the evaluative component be avoided.

The notion of *progress* requires that a value judgment be made about what is better and what is worse, or what is higher and what is lower, according to some axiological standard. But contrary to the belief of some authors (Ginsberg, 1944; Lewontin, 1968), the axiological standard of reference need not be a moral one. Moral *progress* is possible, but not all forms of *progress* are moral. The evaluation required for *progress* may be one of better versus worse, or higher versus lower, but not necessarily one of right versus wrong. "Better" may simply mean more efficient, more abundant, or more complex, without connoting any reference to moral values or standards.

One may, then, define *progress* as "systematic change in a feature belonging to all the members of a sequence in such a way that posterior members of the sequence exhibit an improvement of that feature." More simply it can be defined as "directional change toward the better." Similarly, regress or retrogression is directional change for the worse. The two elements of the definition, namely directional change and improvement according to some standard, are jointly necessary and sufficient for the occurrence of *progress*.

Directional change, as well as *progress*, may be observed in sequences that are spatially rather than temporally ordered. Clines are examples of directional change recognized along a spatial dimension. In evolutionary discourse, however, historical sequences are of greatest interest.

To seek further clarification of the concept of *progress* and its application in evolutionary biology, it is necessary to distinguish among various kinds of *progress*. This can be accomplished according to either one of the two essential elements of the definition. I later refer to types of *progress* differentiated on the basis of axiological standards of reference. Now, I make two distinctions that relate to the descriptive element of the definition, ie, the requirement of directional change. The first distinction takes into account the *continuity* of the direction by distinguishing between "uniform" and "net" *progress*. The second refers to the *scope* of the sequence considered, and differentiates between "general" and "particular" *progress*.

*Uniform progress* takes place whenever every later member of the sequence is better than every earlier member of the sequence according to a certain feature. This may be formally stated as follows. Let $m_i$ be the members of the sequence, temporally ordered from 1 to $n$, and let $p_i$ measure the state of the feature under evaluation. There is uniform *progress* if it is the case for every $m_i$ and $m_j$ that $p_j > p_i$ for every $j > i$.

*Net progress* does not require that every member of the sequence be better than all previous members of the sequence and worse than all its successors; it requires only that later members of the sequence be better, *on the average*, than earlier members. Net *progress* allows for temporary fluctuations of value. Formally, if the members of the sequence, $m_i$, are linearly arranged over time, net *progress* occurs whenever the regression (in the sense used in mathematical sequences) of $p$ on time is significantly positive. Some authors have argued that *progress* has not occurred in *evolution* because no matter what standard is chosen, fluctuations can always be found in every evolutionary lineage. This argument is valid against the occurrence of uniform *progress*, but not against the existence of net evolutionary *progress*.

Notice also that neither uniform nor net *progress* requires that *progress* be unlimited, or that any specified goal will be surpassed if the sequence continues for a sufficiently long period of time. *Progress* requires a gradual improvement in the members of the sequence, but the rate of improvement may decrease with time. According to the definition given here, it is possible that the sequence tends asymptotically toward a definite goal, which is continuously approached but never reached.

The distinction between uniform and net *progress* is similar but not identical to the distinction between uniform and perpetual *progress* proposed by Broad (1925) and Goudge (1961). Perpetual *progress*, as defined by Broad, requires that the maximum of value increases, and the minimum does not decrease with time. In the formulation given above, Broad's perpetual *progress* requires that for every $m_i$ there is at least one $m_j$ $(j > i)$ such that $p_j > p_i$. This definition has the undesirable feature of requiring that the first element of the sequence be the worst one and the last element the best one. Neither of these two requirements is made in my definition of net *progress*. Also the term "perpetual" has connotations that are undesirable in the discussion of *progress*. The distinction between uniform and net *progress* is made implicitly, although never formally stated, by Simpson (1949), who applies terms like "universal," "invariable," "constant," and "continuous" to the kind of *progress* that I have called uniform (although he also uses these terms with other meanings).

With respect to the scope of the sequence considered, *progress* can be either general (or universal) or particular. *General* (or *universal*) *progress* is that which occurs in all historical sequences of a given domain of reality, and from the beginning of the sequences until their end. *Particular progress* is that which occurs in one or several but not all historical sequences, or that which takes place during part but not all of the duration of the sequences.

General *progress* would have occurred in *evolution* only if there were some feature or standard according to which *progress* can be predicated of the *evolution* of all life from its origin to the present. If *progress* is predicated of only one or several, but not all, lines of evolutionary descent, it is a particular kind of *progress*. *Progress* that embraces only a limited span of time from the origin of life to the present is also a particular kind of *progress*.

Some writers have denied that *evolution* is progressive on the grounds that not all evolutionary lineages exhibit advance. Some evolutionary lineages, like those leading to certain parasitic forms, are retrogressive by certain standards; and many have lineages that have become extinct without issue. These considerations are valid against a claim of general *progress*, but not against claims of particular forms of *progress*.

## EVOLUTIONARY PROGRESS

I have argued that the notion of *progress* is axiological, and therefore it cannot be a strictly scientific term: value judgments are not part and parcel of scientific discourse, which is characterized by empirically testable hypotheses and objective descriptions. Some authors have claimed, however, that there are biological criteria of *progress* that are "objective" and do not involve value judgments. I briefly review the efforts in this regard of three distinguished biologists: J.M. Thoday, M. Kimura, and J.S. Huxley.

John Thoday (1953, 1958) pointed out the obvious fact that survival is essential to life. Therefore, he argued, *progress* is the increase in *fitness* for survival, "provided only that fitness and survival be defined as generally as possible." According to Thoday, fitness must be defined in reference to groups of organisms that can have common descendants; these groups he calls "units of *evolution*." A unit of *evolution* is what *population* geneticists call a Mendelian *population*; the most inclusive Mendelian *population* is the *species*. The *fitness* of a unit of *evolution* is defined by Thoday as "the probability that such a unit of *evolution* will survive for a long period of time, such as $10^8$ years, that is to say will have descendants after the lapse of that time." According to Thoday, evolutionary changes, no matter what other results may have been produced, are progressive only if they increase the probability of leaving descendants after long periods of time. He correctly points out that this definition has the advantage of not assuming that *progress* has in fact occurred, an assumption that vitiates some other attempts to define *progress* as a purely biological concept.

The definition of *progress* given by Thoday has been criticized because it apparently leads to the paradox that *progress* is impossible; in fact, that regress is necessary since any group of organisms will be more progressive than any of their descendants. Assume that we are concerned with ascertaining whether *progress* has occurred in the evolutionary transition from a Cretaceous mammal to its descendants of 100 million ($10^8$) years later. It is clear that if the present-day mammal *species* has a probability, $P$, of having descendants $10^8$ years from now, the ancestral mammal *species* will have a probability no smaller than $P$ of leaving descendants after $2 \times 10^8$ years from the time of their existence (Ayala, 1969). The probability that the ancestral *species* will leave descendants $2 \times 10^8$ years after its existence will be greater than $P$ if it has any other living descendants besides the present-day mammal with which we are comparing it. As Thoday (1970) pointed out, such criticism is mistaken, since it confuses the probability of survival with the fact of survival. The a priori probability that a given *species* will have descendants after a given lapse of time may be smaller than the a priori probability that any of its descendants will leave progeny after the same length of time.

There is, however, a legitimate criticism of Thoday's definition of *progress*, namely that it is not operationally valid. Suppose that we want to find out whether one of today's mammal *species* is more progressive than its Cretaceous ancestor. We should have to estimate, first, the probability that today's mammal will leave descendants after a given long period of time; then we should have to estimate the same probability for the Cretaceous *species*.

Thoday enumerated a variety of components that contribute to the *fitness* of a *population* as defined by him: *adaptation*, genetic stability, genetic flexibility, phenotypic flexibility, and the stability of the environment. But it is by no means clear how these components could be quantified, nor by what sort of function they could be integrated into a single parameter. In any case, there seems to be no conceivable way in which the appropriate observations and measurements could be made for the ancestral *species*. Thoday's definition of *progress* is extremely ingenious, but lacks operational validity. If we accept his definition there seems to be no way in which we could ascertain whether *progress* has occurred in any one line of descent or in the *evolution* of life as a whole.

Another attempt to consider evolutionary *progress* as a purely biological notion was made by Motoo Kimura (1961), by defining biological *progress* as an increase in the amount of genetic information stored in an organism. This information is encoded, for the most part, in the *DNA* (deoxyribonucleic acid) of the *nucleus*. The *DNA* contains the information that in interaction with the environment directs the *development* and behavior of the individual. By making certain assumptions, Kimura estimated the rate at which genetic information accumulates in *evolution*. He calculates that in the *evolution* of "higher" organisms genetic information has accumulated from the Cambrian to the present at an average of 0.29 *bits* per generation.

This method of measuring progressive *evolution* by the accumulation of genetic information is vitiated by several flaws. First, since the average rate of accumulation of information is allegedly constant per *generation*, it follows that organisms with a shorter generation time will accumulate more information, and therefore are more progressive than organisms with a longer generation time. In the *evolution* of mammals, moles and bats would necessarily be more progressive than horses, whales, and humans. A second flaw is that Kimura is not measuring how much genetic information has been accumulated in any given organism. Rather he assumes that genetic information gradually accumulates with time, and then proceeds to estimate the rate at which genetic information could have accumulated. The assumption that more recent organisms have more genetic information and therefore they are more progressive than their ancestors is unwarranted, and completely invalidates Kimura's attempt to measure evolutionary *progress*. There is, at least at present, no precise way of measuring the amount of genetic information present in any one organism. Finally, the decision to consider the accumulation of genetic information as progressive requires a value judgment; it is not a biologically compelling notion.

Julian Huxley (1942, 1953) argued that the biologist should not attempt to define *progress* a priori, but rather should "proceed inductively to see whether he can or cannot find evidence of a process which can be legitimately called progressive." He believed that evolutionary *progress* can be defined without any reference to values. Huxley proposed first to investigate the features that mark off the "higher" from the "lower" organisms. Any evolutionary process is considered progressive in which the features that characterize higher organisms are achieved. But Huxley, like Kimura, assumed that *progress* has in fact occurred, and that certain living organisms, especially humans, are more progressive than others. Classifying organisms as "higher" or "lower" requires an evaluation. Huxley did not succeed in avoiding reference to an axiological standard. The terms he used in his various definitions of *progress*, like "improvement," "general advance," "level of efficiency," etc., are all in fact evaluative.

No attempt to define *progress* as a purely biological concept has succeeded. This is understandable in view of the analysis of the concept that I developed above. The concept of *progress* is evaluative, and hence one cannot ascertain whether *progress* has occurred without first choosing a standard against which *progress* or improvement will be assessed. Two decisions are required. First, we must choose the objective feature according to which the events of objects are to be ordered. Second, a decision must be made as to what pole of the ordered elements represents improvement. These decisions involve a subjective element, but they should not be arbitrary. Biological knowledge should guide them. There is a criterion by which the validity of a standard of reference can be judged. A standard is valid if it enables us to say illuminating things about the *evolution* of life. How much of the relevant information is available, and whether the evaluation can be made more of less precisely, should also influence the choice of values.

It is fairly apparent that there is no standard by which *uniform progress* can be said to have taken place in the *evolution* of life. Changes in direction, slackening, and reversals have occurred in all evolutionary lineages, no matter what features are considered (Simpson, 1949, 1953). The question, then, is whether *net progress* has occurred in the *evolution* of life, and in what sense.

The next question is whether there is any criterion of *progress* by which net *progress* can be said to be a general feature of *evolution*; or whether identifiable *progress* applies only to particular lineages or in particular periods. One conceivable standard of *progress* is the increase in the amount of genetic information stored in the organisms (Kimura, 1961). Net *progress* would have occurred if organisms living at a later time have, on the average, a greater content of genetic information than their ancestors. One difficulty, insuperable at least for the present, is that there is no way in which the genetic *information* present in an organism can be measured. We could choose the Shannon-Weaver solution, as Kimura (1961) did, by regarding all the *DNA* of an organism as a linear sequence of messages made up of groups of three-letter words (the codons) with a four-letter alphabet (the four *DNA* nucleotides). But the amount of information is not simply related to the amount of *DNA*, since we know that many *DNA* sequences are repetitive and even much of the nonrepetitive *DNA* may not store information in the *nucleotide* sequence. In any case, what we know about the size of the *genome* in organisms makes it unlikely that increase of genetic information could be a general feature of the *evolution* of life.

## THE EXPANSION OF LIFE

The accumulation of genetic information as a standard of *progress* can be understood in a different way. *Progress* can be measured by an increase in the *kinds* of ways in which the information is stored and an increase in the *number* of different messages encoded. Different *species* would represent different kinds of messages; individuals would be messages or units of information. Thus understood, whether an increase in the amount of information has occurred reduces to the question of whether life has diversified and expanded. This was recognized by Simpson (1949) as the standard by which what I call general *progress* has in fact occurred in the *evolution* of life. According to Simpson, we can find for *evolution* as a whole whether there has been a "tendency for life to expand, to fill in all the available spaces in the livable environments, including those created by the process of that expansion itself."

There are at least four different though related criteria by which the expansion of life can be measured: expansion in the number of kinds of organisms (that is, the number of species), expansion in the number of individuals, expansion in the total bulk of living matter, and expansion in the total rate of energy flow. Increases in the number of individuals or their bulk may be a mixed blessing, as is the case now for the human *species*, but they can be measures of biological success. By any one of these four standards of *progress*, it appears, according to Simpson (1949), that net *progress* has been a general feature of the *evolution* of life.

Reproduction provides organisms with the potential to multiply exponentially: each organism is capable of producing, on average, more than one progeny throughout its lifetime. The actual rate of increase in numbers is a net result of the balance between the rate of births and the rate of deaths of the *population*. In the absence of environmentally imposed restrictions, that balance is positive; populations have an intrinsic capacity to grow *ad infinitum*. Since the ambit in which life can exist is limited, and since the resources to which a *population* has access are even more limited, the rate of expansion rapidly decreases to zero, becomes negative, or alternates between positive and negative periods.

The tendency of life to expand encounters constraints of various sorts. Expansion is limited by the environment in at least two ways. First, the supply of resources accessible to the organisms is limited. Second, favorable conditions necessary for multiplication do not always occur. Predators, parasites, and competitors, together with the various parameters of the environment embodied by the term "weather," are the main factors interfering with the multiplication of organisms even when resources are available. Drastic and secular changes in the weather, as well as geological events, lead at times to vast decreases in the size of some populations and even the whole of life. Because of these constraints, the tendency of life to expand has not always succeeded. Nevertheless, it appears certain that life has, on average, expanded throughout most of its history.

About 1.5 million *species* now living have been described and named. Current estimates place the number of living *species* between 5 and 30 million, with most of the unidentified animal *species* being beetles and other insects. Although it is difficult to estimate the number of plant *species* that existed in the past (since well-preserved plant fossils are rare), the number of animal *species* can be roughly estimated. Approximately 150,000 animal *species* live in the seas today, probably more than the total number that existed in the Cambrian (starting about 542 million years ago), when no animal or plant *species* lived on the land. Animal life on land began in the Silurian (starting 444 million years ago—arthropods) and the Devonian (starting 416 million years ago—amphibians). The number of animal land *species* is probably at a maximum now, even if we exclude insects. Insects make up about three-quarters of all known animal *species*, and about half of all *species* if plants are included. Insects did not appear until the early Silurian, less than 444 million years ago. The number of living insect *species* has become probably greater in recent times (before the Anthropocene extinctions) than it ever was before. It seems likely that, at least on average, a gradual increase in the number of *species* has characterized the *evolution* of life (Simpson, 1949, 1953).

The number of *species* expands by a positive feedback process. The greater the number of *species*, the greater the number of environments that are created for new *species* to exploit. Once there were plants, animals could come into existence, and the animals themselves sustain large numbers of *species* of other animals that prey on them, as well as of parasites and symbionts. Thomas Huxley likened the expansion of life to the filling of a barrel. First the barrel is filled with apples until it overflows; then pebbles are added up to the brim; the space between the apples and the pebbles can be packed with sand; water is finally poured until it overflows (see Huxley and Huxley, 1947). His point is that with diverse kinds of organisms the environment can be filled in more effectively than with only one kind. But Huxley's *analogy* neglects one important aspect of life, namely that the space available for occupancy by other *species* is increased rather than decreased by some additions. A more appropriate *analogy* would have been that of a balloon or an expanding barrel.

It is difficult to estimate the number of individuals living on the Earth today with any reasonable approximation, even if we exclude microorganisms. The mean number of individuals per *species* has been estimated to be around $2 \times 10^8$, but some *species* like *Drosophila willistoni*, the tropical fruitfly with which I worked for several decades, may consist of more than $10^{16}$ individuals—and there are more than 1 million insect *species* (Ayala et al., 1972)! The number of individuals of *Euphausia superba*, the small krill eaten by whales, may be greater than $10^{20}$. It seems certain that there are more individual animals and plants living today and their bulk is greater than it was in the Cambrian, say 500 million years ago. Very likely they have become greater in recent geological times than they were at most times since the beginning of life. This is more so if we include the large number and enormous bulk of the human *population* and all the plants and animals cultivated by humans for their own use. Excluding microorganisms, it is probable that the number of living individuals has increased, on average, through the *evolution* of life. About the abundance of microorganisms, there is little that can be said with conviction. On the whole, an increase in the total bulk of living matter is even more likely than an increase in numbers because larger organisms have generally appeared later in time.

It seems likely that the rate of energy flow has increased in the living world faster than the total bulk of matter. One effect of organisms on the world is to retard the dissipation of energy. Green plants store radiant energy from the Sun that would otherwise be converted into heat. The influence of animals goes partially in the opposite direction: the living activities of animals dissipate energy, since their catabolism exceeds their anabolism (Lotka, 1945), but they store energy derived from plants that might have otherwise dissipated into heat. Animals provide a new path through which energy can flow and, moreover, their interactions with plants increase the total rate of flow through the system. An *analogy* can be used to illustrate this outcome. Suppose that a modern highway with three lanes in each direction connects two large cities. The need to accommodate an increase in the rate of travel flow can be accomplished either by addition of more lanes to the highway or by increasing the speed at which the traffic moves on the highway. In terms of the "carrying capacity" of the highway, these two approaches appear, at first sight, to work in opposite directions, but together they increase the total flow of traffic on the highway.

## STANDARDS OF PROGRESS

I argue above that the concept of *progress* involves an evaluation of better versus worse relative to some standard of reference. Many standards of reference can be chosen according to which it is possible to measure the evolutionary process of the kind I call "particular"—that is, *progress* that obtains only in certain evolutionary lineages and usually only for a limited span of time. The numerous writers on evolutionary *progress* have usually proceeded by identifying one or another attribute as the criterion of *progress* and then expanded on how *progress* has occurred in *evolution* according to the particular standard chosen. These discussions are often enlightening in that they highlight aspects of the evolutionary process that are particularly mean-ingful from a certain perspective and enhance our understanding of the pro-cess. But a common deficiency in some of these discussions is the stated or implicit conviction that *the* criterion of *progress* has been discovered, often accompanied by a lack of awareness that *progress* is a value-laden concept rather than a strictly scientific one.

I shall now mention some criteria that have been the subjects of thoughtful discussion on evolutionary *progress*. I then, by way of illustration, deal in somewhat greater detail with one specific criterion of evolutionary progress: advances in the ability of organisms to obtain and process information about the state of the environment.

Simpson (1949) examined several criteria according to which evolutionary *progress* can be recognized in particular sequences. These criteria include dominance, invasion of new environments, replacement, improvement in *adaptation*, adaptability and possibility of further *progress*, increased specialization, control over the environment, increased structural complexity, increase in general energy or level of vital processes, and increase in the range and variety of adjustments to the environment. For each of these criteria Simpson showed in which evolutionary sequences, and for how long, *progress* has taken place.

Bernard Rensch (1947) and Julian Huxley (1942, 1953) examined other lists of characteristics that can be used as standards of particular forms of *progress*. Ledyard Stebbins (1969) wrote a provocative essay proposing a law of "conservation of organization" that accounts for evolutionary *progress* as a small bias toward increased complexity of organization. Simon Conway Morris (2003) argued that, under *natural selection, progress* is a necessary attribute of the evolutionary process, culminating in intelligent humans. I have examined (Ayala, 1974, 1982, 1988, 2016) in some detail the increase in the ability of organisms to obtain and process information about the environment, as a criterion of *progress* that is particularly relevant to human *evolution*. Among the differences that mark off humans from all other animals, perhaps the most fundamental is the greatly developed human ability to perceive the environment and react flexibly to it. George Williams (1966) examined, mostly critically, several criteria of *progress*. Two brief but incisive discussions of the concept of *progress* can be found in G.J. Herrick (1956) and Theodosius Dobzhansky (1970). A philosophical study of the concept of *progress* was made by T.A. Goudge (1961). Two splendid multiauthored collections with widespread points of view are G.A. Almond, M. Chodorow, and R.H. Pearce's (Eds.) *Progress and Its Discontents* (1982) and M.H. Nitecki's (Ed.) *Evolutionary Progress* (1988). The most extensive learned exploration of the issue of evolutionary *progress* is Michael Ruse's *Mondad to Man. The Idea of Progress in Evolutionary Biology* (1996). A more concise elaboration, yet historically informative and conceptually subtle as well as profound, can be found in Chapter 4 ("Progress") of Ruse's *The Philosophy of Human Evolution* (2012, pp. 99–127).

There is no need to examine here all the standards of *progress* that have been formulated by the authors just mentioned, or to explore additional criteria. Writings about biological *progress* have involved much disputation concerning whether the notion of *progress* belongs in the realm of scientific discourse, what criterion of *progress* is "best," and whether *progress* has indeed taken place in the *evolution* of life.

These controversies can be solved once the notion of *progress* is clearly established, as I have done above. First, the concept of *progress* involves an evaluation of good versus bad, or of better versus worse. The choice of a standard by which to evaluate organisms or their features is to a certain extent subjective. However, once a standard of *progress* has been chosen, decisions concerning whether *progress* has occurred in the living world, and what organisms are more or less progressive, can be made following the usual standards and methods of scientific discourse. Second, there is no standard of *progress* that is "best" in the abstract or for all purposes. The validity of any one criterion of *progress* depends on whether the use of that standard leads to meaningful statements concerning the *evolution* of life. Which standard or standards are preferable depends on the particular context or purpose of the discussion. Third, the distinction between uniform and net *progress* makes it possible to recognize the occurrence of biological *progress* even though every member of a sequence or a group of organisms may not always be more progressive than every previous member of the sequence or every member of some other group of organisms. Fourth, the distinction between general and particular *progress* allows one to recognize *progress* that may have occurred in particular groups of organisms, or during limited periods in the *evolution* of life, but not in all of them.

Once one realizes that recognition of *progress* is only possible after a value judgment has been made as to which will be the standard against which *progress* is to be measured (and hence that there is not *a* standard of *progress*, or one that is best for all purposes), it becomes possible to seek standards of *progress* that may yield valuable insights into the study of the *evolution* of life.

## INFORMATION PROCESSING

I now, by way of illustration, discuss *progress* according to a particular standard of reference: the ability of an organism to obtain and process information about the environment. I can see two reasons that make this criterion of *progress* especially meaningful (although not, I reiterate, *the* most meaningful, because no criterion exists that is best for all purposes). First, the ability to obtain information about the environment and react accordingly is an important *adaptation*, because it allows the organism to seek out suitable environments and resources and avoid unsuitable ones. Second, because the ability to perceive the environment, and integrate, coordinate, and react flexibly to what is perceived, has attained its highest *development* in humankind. This incomparable advancement is perhaps the most fundamental characteristic that sets apart *Homo sapiens* from all other animals. Symbolic language, complex social organization, control over the environment, the ability to envisage future states and work toward them, values, and *ethics* are developments made possible by humans' greatly developed capacity to obtain and organize information about the state of the environment. This capacity has ushered in humankind's new mode of *adaptation*. Whereas other organisms become genetically adapted to their environments, humans create environments to fit their genes. It is thus that humankind has spread over the whole planet in spite of its physiological dependence on a tropical or subtropical "climate."

Increased ability to gather and process information about the environment is sometimes expressed as *evolution* toward "independence from the environment." This expression is misleading. No organism can be truly independent of the environment. The evolutionary sequence, fish to amphibian to reptile, allegedly provides an example of *evolution* toward independence from an aqueous environment. Reptiles, birds, and mammals are indeed free of the need for water as an external living medium, but their lives depend on the conditions of the land. They have not become independent of the environment, but have rather exchanged dependence on one environment for dependence on another.

The notion of "control over the environment" has also been associated with the ability to gather and use information about the state of the environment. However, true control over the environment occurs to any substantial extent only in the human *species*. All organisms interact with the environment, but they do not control it. Burrowing a hole in the ground or building a nest in a tree, like the construction of a beehive or a beaver dam, does not represent control over the environment except in a trivial sense. The ability to control the environment started with the australopithecines (or at least, with *Homo habilis* and other early *Homo* species), some of the earliest organisms that may be called human: some were able to produce devices to manipulate the environment in the form of rudimentary stone and bone tools. The ability to obtain and process information about the conditions of the environment does not provide control over the environment, but rather it enables the organisms to avoid unsuitable environments and seek suitable ones. It has developed in many organisms because it is a useful *adaptation*.

Some selective interaction with the environment occurs in all organisms. The cell membrane of a bacterium permits certain molecules but not others to enter the cell. Selective molecular exchange also occurs in the inorganic world, but this can hardly be called a form of information processing. Certain bacteria when placed on a plate with a culture medium move about in zig-zag pattern, which is almost certainly random. The most rudimentary ability to gather and process information about the environment may be found in certain single-celled eukaryotes (organisms with a true nucleus). A *Paramecium* follows a sinuous path as it swims, ingesting the bacteria that it encounters. Whenever it meets unfavorable conditions, like unsuitable acidity or salinity in the water, the *Paramecium* checks its advance, turns, and starts in a new direction. Its reaction is purely negative. The *Paramecium* apparently does not seek its food or a favorable environment, but simply avoids unsuitable conditions.

*Euglena*, also a single-cell organism, exhibits a somewhat greater ability to process information about the environment. *Euglena* has a light-sensitive spot by means of which it can orient itself toward the light. *Euglena*'s motions are directional; it not only avoids unsuitable environments but it actively seeks suitable ones. An ameba represents further *progress* in the same direction: it reacts to light by moving away from it, and also actively pursues food particles.

An increase in the ability to gather and process information about the environment is not a general characteristic of the *evolution* of life. *Progress* has occurred in certain evolutionary lines but not in others. Today's bacteria are not more progressive by this criterion than their ancestors of 2 or 3 billion years ago. In many evolutionary sequences some very limited *progress* took place in the very early stages, but without further *progress* through the rest of their history. In general, with respect to the ability to gather and process information about the environment, animals are more advanced than plants; vertebrates are more advanced than invertebrates; and mammals are more advanced than reptiles, which are more advanced than fish. The most advanced organism by this criterion is doubtless the human *species*.

The ability to obtain and process information about the environment has progressed little in the plant kingdom. Plants generally react to light and gravity. The geotropism is positive in the root, but negative in the stem. Plants also grow toward the light; some plants like the sunflower have parts that follow the course of the Sun through its daily cycle. Another tropism in plants is the tendency of roots to grow toward water. The response to gravity, to water, and to light is basically due to differential growth rates; a greater elongation of cells takes place on one side of the root or stem than on the other side. Gradients of light, gravity, or moisture are the clues that guide these tropisms. Some plants also react to tactile stimuli. Tendrils twine around what they touch; *Mimosa* and carnivorous plants like the Venus flytrap (*Dionaea*) have leaves that close upon being touched.

The ability to obtain and process information about the environment is mediated in multicellular animals by the nervous system. The simplest nervous system occurs in coelenterate hydras, corals, and jellyfishes. Each tentacle of a jellyfish reacts only if it is individually and directly stimulated. There is no coordination of the information gathered by different parts of the animal. Moreover, jellyfishes are unable to learn from experience.

A limited form of coordinated behavior occurs in the echinoderms, such as starfish and sea urchins. Whereas coelenterates possess only an undifferentiated nerve net, echinoderms possess a nerve net, a nerve ring, and radial nerve cords. When the appropriate stimulus is encountered, a starfish reacts with direct and unified actions of the whole body.

The most primitive form of a brain occurs in certain organisms like planarian flatworms, which also have numerous sensory cells and eyes without lenses. The information gathered in these sensory cells and organs is processed and coordinated by the central nervous system and the rudimentary brain; a planarian worm is capable of some variability of responses and some simple learning—that is, the same stimuli will not necessarily always produce the same response.

Planarian flatworms have progressed farther than starfish in the ability to gather and process information about the environment, and the starfish have progressed farther than sea anemones and other coelenterates. But none of these organisms has gone very far by this criterion of *progress*. The most progressive groups of organisms among the invertebrates are cephalopods and arthropods, but the vertebrates have progressed much farther than any invertebrates.

Among the ancestors of both the arthropods and the vertebrates there were organisms that, like the sponges, lacked a nervous system. These ancestors evolved through a stage with only a simple network, while later stages developed a central nervous system and eventually a rudimentary brain. With further *development* of the central nervous system and the brain, the ability to obtain and process information from the outside progressed much farther. The arthropods, which include the insects, have complex forms of behavior. Precise visual, chemical, and acoustic signals are obtained and processed by many arthropods, particularly in their search for food and their *selection* of mates.

Vertebrates are generally able to obtain and process much more complicated signals and produce a much greater variety of responses than the arthropods. The *vertebrate* brain has an enormous number of associative neurons with an extremely complex arrangement. Among the vertebrates, *progress* in the ability to deal with environmental information is correlated with increase in the size of the cerebral hemispheres and with the appearance and *development* of the "neopallium." The neopallium is involved in the association and coordination of all kinds of impulses from all receptors and brain centers. The larger brain of vertebrates, compared to that of invertebrates, permits them also to have a large number of neurons involved in information storage or memory. The neopallium appeared first in the reptiles. In the mammals it has expanded to become the cerebral cortex, which covers most of the cerebral hemispheres. The cerebral cortex in humans is particularly large, compressed over the hemispheres in a complex pattern of folds and turns. When organisms are measured by their ability to process and obtain information about the environment, humankind is, indeed, the most progressive organism on Earth.

## CONCLUDING OBSERVATIONS

One could once more reiterate that there is nothing in the evolutionary process that makes the criterion of *progress* I have just followed better or more objective than others. It may be useful because it illuminates certain features of the *evolution* of life. Other criteria may help to discern other features of *evolution*, and thus be worth examining. Particular organisms will appear more or less progressive depending on the standard that is used to evaluate *progress*. Humankind is not the most progressive *species* by many criteria. By some standards, humans (and other mammals) are among the bottom rungs of the ladder of life, for example in the ability to synthesize their own biological components from inorganic resources.

It may be properly questioned whether anything is gained by speaking of evolutionary *progress* rather than evolutionary advancement or directional change. The term "advancement" also involves an evaluation and would therefore be subject to the same pitfalls as "progress" (although it seems to elicit weaker emotional disclaimers than progress). "Directional change," however, is not an axiological concept, and thus it may be treated as other strictly scientific terms (Gould, 1988, 1997).

The notion of *progress* seems to be irrevocably ingrained among the thinking categories of modern humanity, and hence is likely to continue being used in biology, particularly in reference to the evolutionary process. Indeed, Ruse (1996) in his monumental masterpiece, *Monad to Man. The Idea of Progress in Evolutionary Biology*, argued that the concept of *progress* pervades all evolutionary writings, from Darwin to the present. I have therefore attempted to clarify the concept to demythologize it. I argue that "progressive" is an evaluative term that demands a subjective commitment to a particular standard of value. Awareness of this makes it possible to speak of *progress* in *evolution* without implying the conclusion that humans are the most progressive organisms. As I suggest, by some biologically meaningful standards of *progress* they are not.

The concept of *progress* as it has been here defined and the distinctions made may also be useful in the fields of cultural anthropology and, more generally, human history (Hoagland and Burhoe, 1962; Almond et al., 1982). It may be the case that much knowledge in these fields is largely value free, but we can predicate *progress* of human historical events only by introducing value judgments. However, once this is recognized, it becomes possible to seek criteria of *progress* that will yield valuable insights in the study of human history.

Claims that *progress* has occurred in human history need not imply that *progress* is universal, inevitable, or unlimited. Like biological *progress*, cultural *progress* may have occurred in some societies but not others, during certain periods rather than forever, and may be subject to certain limits rather than able to proceed without bounds. Where and how *progress* has taken place are matters for investigation, which, once a criterion of *progress* has been selected, can proceed according to the normal standards of scholarly inquiry.

# Chapter 9

# The Scientific Method

## INTRODUCTION

The scientific method has two stages: the first consists of formulating hypotheses, and the second consists of testing them. What differentiates science from other forms of knowledge is the second stage: subjecting hypotheses to empirical testing by ascertaining whether or not predictions derived from hypotheses are borne out in relevant observations and experiments. Hypotheses and other imaginative conjectures are the initial stage of scientific inquiry because they provide the incentive to seek the truth and a clue as to where to find it. But scientific conjectures must be subject to critical examination and empirical testing. There is a dialogue between the two episodes; observations made to test a hypothesis are the inspiration for new conjectures.

Ascertaining whether or not predictions about the world of experience deduced from the hypothesis agree with what is actually observed has been appropriately considered the "criterion of demarcation" that distinguishes science from other knowledge. But scientific hypotheses must satisfy other tests as well, eg, whether they have explanatory value and further understanding.

Science, like any human activity, is subject to error and the foibles and other failings of human beings. But rigorous attempts at empirical validation and other trials yield knowledge that stands the test of time and provides a foothold for further knowledge. Moreover, scientists have developed social mechanisms, such as peer review and publication, to evaluate their work. Because further scientific research depends on the validity of previous knowledge, it is of great consequence that they discern valid from invalid knowledge, and thus scientists are inclined to transcend ideology, nationality, friendship, monetary interest, and other prejudices when the mettle of scientific knowledge is at stake.

## BEYOND INDUCTIVISM

Francis Bacon (1561–1626) had a distinguished career as a politician, serving as Lord Chancellor of England (1618–1621). He is also recognized as a master of English prose and a philosopher. He elaborated his view of the scientific method in his best-known work, *Novum Organum* (*New Instrument*).

Evolution, Explanation, Ethics, and Aesthetics. http://dx.doi.org/10.1016/B978-0-12-803693-8.00009-8
Copyright © 2016 Elsevier Inc. All rights reserved.

He is often considered as the first important proponent of empiricism as *the* scientific method, a view with a strong British tradition, culminating perhaps with John Stuart Mill (1806–1873), who argued in his *System of Logic* (1843, 1974) and elsewhere that inductive inference is the only method to achieve valid knowledge, going as far as to claim that mathematical truths are merely very highly confirmed generalizations from experience. Empiricism prevailed in Great Britain and other English-speaking countries well into the 20th century, particularly in the form of logical positivism, with writers as distinguished and prolific as Bertrand Russell and A.J. Ayer.

Darwin bows to empiricism in the first paragraph of *Origin of Species*: "When on Board H.M.S. *Beagle*, as naturalist, I was much struck with certain facts in the distribution of the inhabitants of South America... These facts seemed to me to throw some light on the origin of species... On my return home, it occurred to me, in 1837, that something might perhaps be made out on this question by *patiently accumulating and reflecting on all sorts of facts which could possibly have any bearing on it. After 5 years' work I allowed myself to speculate on the subject, and drew up some short notes*" (Darwin, 1859, p. 1).

The facts are very different. Darwin's notebooks and private correspondence show that he came upon the hypothesis of *natural selection* in 1837, several years before he claims to have allowed himself "to speculate on the subject." Darwin rejected the inductivist claim that observations should not be guided by hypotheses. "How odd it is that anyone should not see that all observation must be for or against some view if it is to be of any service." Otherwise, he added, "a man might as well go into a gravel pit and count the pebbles and describe the colors" (Darwin, 1903, vol. 1, p. 195). In his *Auto-biography* (Darwin, 1958, p. 141) he confesses "I cannot avoid forming one [hypothesis] on every subject."

Inductivism, empiricism, and similar theories of scientific knowledge did not take much hold in continental Europe, where one finds magnificent practitioners and examples of the *hypothetico-deductive method*, such as in the 17th century Galileo's (1564–1642) laws of motion and Blaise Pascal's (1623–1662) explanation of atmospheric pressure; in the 18th century Antoine-Laurent Lavoisier's (1743–1794) theory of oxidation; and, among 19th-century biologists, Claude Bernard (1813–1878), Louis Pasteur (1822–1895), and Gregor Mendel (1823–1884).

Claude Bernard describes the two stages of the scientific method: formulation of a testable hypothesis, and testing it. Moreover, Bernard explicitly asserts that scientific theories of necessity are only partial and provisional: "A hypothesis is... the obligatory starting point of all experimental reasoning. Without it no investigation would be possible, and one would learn nothing: one could only pile up barren observations. To experiment without a preconceived idea is to wander aimlessly. Those who have condemned the use of hypotheses and preconceived ideas in the experimental method have made the mistake of confusing the contriving of the experiment with the verification of its results. When propounding a general theory in science, the one thing one can be sure of is that, in the strict sense, such theories are mistaken. They are only partial and provisional truths which are necessary... to carry the investigation forward; they represent only the current state of our understanding and are bound to be modified by the growth of science" (Bernard, 1865).

## SCIENTIFIC KNOWLEDGE

Knowledge about the world derives from many sources, which include common-sense experience, imaginative literature, artistic expression, and philosophical reflection. Scientific knowledge, however, stands apart as special. The tremendous success of science as a mode of inquiry into the nature of the universe is a matter of wonderment. The technology derived from scientific knowledge is equally wondrous: the high-rise buildings of our cities, thruways and long span-bridges, rockets that bring men to the moon, telephones that provide instant communication across continents, computers that perform complex calculations in millionths of a second, vaccines and drugs that keep bacterial parasites at bay, *gene* therapies that replace *DNA* (deoxyribonucleic acid) in defective cells. All these remarkable achievements bear witness to the validity of the scientific knowledge from which they originated. No other kind of knowledge affects human life so pervasively and drastically.

Scientific knowledge is also remarkable in the way it emerges by way of consensus and agreement among scientists, and in the way new knowledge builds upon past accomplishment rather than starting anew with each generation or each new practitioner. Certainly scientists disagree with each other on many matters; but these are issues not yet settled, and the points of disagreement generally do not bring into question previous knowledge. Modern scientists do not challenge that atoms exist, that there is a universe with a myriad stars, or that heredity is encased in *DNA*. Scientists differ from philosophers, who interminably debate the questions they seek to answer. Philosophers today focus on the same questions that were debated in antiquity or the Middle Ages or two decades ago, without ever coming to any definitive agreement. Not so with scientists, who build upon matters resolved in the past to formulate new questions and resolve them. Nor is there among scientists anything like the radically disparate and irreconcilable views held by different religions, or the ever-changing means of artistic expression.

What is it, then, that makes scientific knowledge different from all other activities by which we learn about the universe and about ourselves? We approach the matter by first identifying some distinguishing traits of scientific knowledge, then explain why science involves much more than simple inductive reasoning, and discuss the *hypothetico-deductive method* as a paradigm for understanding some distinctive features of the way in which scientists proceed to understand the world. An important consideration is the demarcation question, or how to distinguish valid from invalid scientific claims, and the social mechanisms by which scientific practice weeds out valid from invalid science. Historical examples illustrate relevant aspects of how scientific knowledge develops and demarcation works in practice.

Three characteristic traits jointly distinguish scientific knowledge from other forms of knowledge (Ayala, 1968; see Nagel, 1961). First, science seeks the systematic organization of knowledge about the world. Common sense, like science, provides knowledge about natural phenomena, and this knowledge is often correct. For example, common sense tells us that children resemble their parents and that good seeds produce good crops. Common sense, however, shows little interest in systematically establishing connections between phenomena that do not appear to be obviously related. By contrast, science is concerned with formulating general laws and theories that manifest patterns of relations between very different kinds of phenomena. Science develops by discovering new relationships, and particularly by integrating statements, laws, and theories which previously seemed to be unrelated into more comprehensive laws and theories.

Second, science strives to explain *why* observed events do in fact occur. Although knowledge acquired in the course of ordinary experience is frequently accurate, it seldom provides explanations of why phenomena occur as they do. Practical experience tells us that children resemble one parent in some traits and the other parent in other traits, and that manure increases crop yield. But it does not provide explanations for these phenomena. Science, on the other hand, seeks to formulate explanations for natural phenomena by identifying the conditions that account for their occurrence.

Seeking the systematic organization of knowledge and trying to explain why events are as observed are two characteristics that distinguish science from common-sense knowledge. But these characteristics are also shared by other forms of systematic knowledge, such as mathematics and philosophy. A third characteristic of science, and the one that distinguishes the empirical sciences from other systematic forms of knowledge, is that scientific explanations must be formulated in such a way that they can be subjected to empirical testing, a process that must include the possibility of *empirical falsification. Falsifiability* has been proposed as the *criterion of demarcation* that sets science apart from other forms of knowledge (Popper, 1959, pp. 40−42).

New ideas in science are advanced in the form of hypotheses. The tests to which scientific ideas are subjected include contrasting hypotheses with the world of experience in a manner that must leave open the possibility that anyone might reject any particular hypothesis if it leads to wrong predictions about the world of experience. The possibility of empirical *falsification* of a hypothesis is established by ascertaining whether or not precise predictions derived as logical consequences from the hypothesis agree with the state of affairs found in the empirical world. A hypothesis that cannot be subject to the possibility of rejection by observation and experiment cannot be regarded as scientific.

I return later to this matter of "empirical falsifiability" as the *criterion of demarcation* that sets apart science from other forms of knowledge. For now, I summarize my discussion of the nature of science by defining science as "knowledge about the universe in the form of explanatory principles supported by empirical observation and subject to the possibility of empirical falsification." Another definition would be: "Science is an exploration of the material universe that seeks natural, orderly relationships among observed phenomena and that is self-testing" (Simpson, 1964, p. 91). Many other definitions can be proposed, but seeking a "perfect" definition is a futile endeavor. Science is a complex enterprise that cannot be adequately captured in a compact statement. In any case, my goal here is not so much to provide an adequate definition as to identify the traits that distinguish scientific knowledge. I proceed by discussing "induction," which is sometimes said to be the method followed by scientists. I explain that *induction* is not a method by which we may establish the *validity* of scientific knowledge (although it is often a process by which we come upon new ideas, but this is an altogether different matter).

## INDUCTION IN SCIENCE

It is a common misconception that science advances by "accumulating experimental facts and drawing up a theory from them"[1] (Jacob, 1988, pp. 224–225). This misconception is encased in the much-repeated assertion that science is inductive, a notion which can be traced, as pointed out earlier, to the English statesman and essayist Francis Bacon. Bacon had an important and influential role in shaping modern science by his criticism of the prevailing metaphysical speculations of medieval scholastic philosophers. In the 19th century the most ardent and articulate proponent of inductivism was John Stuart Mill, an English philosopher and economist.

*Induction* was proposed by Bacon and Mill as a method of achieving *objectivity* while avoiding subjective preconceptions, and of obtaining *empirical* rather than abstract or metaphysical knowledge. In its extreme form this proposal holds that a scientist should observe any phenomena encountered in his/her experience, and record them without any preconceptions as to what to observe or what the truth about them might be. Truths of universal validity are expected eventually to emerge, as a result of the relentless accumulation of unprejudiced observations. The methodology proposed may be exemplified as follows. A scientist measuring and recording everything that confronts her observes a tree with leaves. A second tree, and a third, and many others are all observed to have leaves. Eventually, she formulates a universal statement, "All trees have leaves."

The inductive method fails to account for the actual process of science. First of all, no scientist works without any preconceived plan as to what kind of phenomena to observe. Scientists choose for study objects or events that, in their opinion, are likely to provide answers to questions that interest them. A scientist whose goal was to record carefully every event observed in all waking moments of his life would not contribute much to the advance of science; more likely than not, he would be considered mad by his colleagues.

Moreover, *induction* fails to arrive at universal truths. No matter how many singular statements may be accumulated, no universal statement can be logically derived from such an accumulation of observations. Even if all trees so far observed have leaves, or all swans observed are white, it remains a logical possibility that the next tree will not have leaves, or the next swan will not be white. The step from numerous singular statements to a universal one involves logical amplification. The universal statement has greater logical content—it says more—than the sum of all singular statements.

Another serious logical difficulty with the proposal that *induction* is "the" method of science is that scientific hypotheses and theories are formulated in abstract terms that do not occur at all in the description of empirical events. Mendel, the founder of genetics, observed in the progeny of hybrid plants that alternative traits segregated according to certain proportions. Repeated observations of these proportions could never have led inductively to the formulation of his hypothesis that "factors" (genes) exist in the sex cells and are rearranged in the progeny according to certain rules. The genes were not observed, and thus could not be included in statements reflecting what Mendel observed. The most interesting and fruitful scientific hypotheses are not simple generalizations. Instead, scientific hypotheses are creations of the mind— imaginative suggestions as to what might be true.

*Induction* fails in all three counts pointed out. It is not a method that ensures objectivity and avoids preconceptions, it is not a method to reach universal truths, and it is not a good description of the process by which scientists formulate hypotheses and other forms of scientific knowledge. A scientist may come upon a new idea or develop a hypothesis as a consequence of repeated observation of phenomena that might be similar or share certain traits. But how we come upon a new idea is quite a different matter from how we come to accept something as established scientific knowledge. I come back to this point later.

## THE HYPOTHETICO-DEDUCTIVE METHOD

As already stated, the validity of scientific ideas ("hypotheses") is established by deriving ("deduction") their predictions as to what should be the case in the real world, and then proceeding to ascertain whether or not the derived predictions or consequences are correct. (It is of the essence of the process, as explained later, that it should not be already known whether such consequences are the case if the observation of such consequences is to serve the purpose of validating the idea; it is also required that the consequences should be unlikely to be the case unless the hypothesis is true.) The scientific method is, accordingly, said to be *hypothetico-deductive*.

The analysis of the *hypothetico-deductive method* may be traced to William Whewell (1794–1866) and William Stanley Jevons (1835–1882) in England, and to Charles S. Peirce (1838–1914) in the United States. Its key features have been well characterized by Karl R. Popper (1959, 1963) and C.G. Hempel (1965). Scientists, of course, practiced the *hypothetico-deductive method* long before it was adequately defined by philosophers. Eminent practitioners include Blaise Pascal (1623–1662) and Isaac Newton (1624–1727) in the 17th century and, among 19th-century biologists, Claude Bernard (1813–1878) and Louis Pasteur (1822–1895) in France, Charles Darwin (1809–1882) in England, and Gregor Mendel (1822–1884) in the Czech Republic (Austria at the time of Mendel's work). These and other successful scientists practiced the *hypothetico-deductive method* even if some of them, Darwin for example, claimed to be inductivists to conform to the ideas of contemporary British philosophers.

The Nobel Laureate François Jacob, in his autobiography, describes research at the Pasteur Institute in Paris that led in the 1950s to one of the fundamental discoveries of molecular biology: "What had made possible analysis of bacteriophage multiplication, and understanding of its different stages, was above all the play of hypotheses and experiments, constructs of the imagination and inferences that could be drawn from them. Starting with a certain conception of the system, one designed an experiment to test one or another aspect of this conception. Depending on the results, one modified the conception to design another experiment. And so on and so forth. That is how research in biology worked. Contrary to what once I thought, scientific progress did not consist simply in observing, in accumulating experimental facts and drawing up a theory from them. It began with the invention of a possible world, or a fragment thereof, which was then compared by experimentation with the real world. And it was this constant dialogue between imagination and experiment that allowed one to form an increasingly fine-grained conception of what is called reality" (Jacob, 1988, pp. 224–225).[1]

Science is a complex enterprise that essentially consists of two interdependent episodes, one imaginative or creative, and the other critical. To have an idea, advance a hypothesis, or suggest what might be true is a creative exercise. But scientific conjectures or hypotheses must also be subject to critical examination and empirical testing. Scientific thinking may be characterized as a process of invention or discovery followed by validation or confirmation. One process concerns the formulation of new ideas ("acquisition of knowledge"), and the other concerns their validation ("justification of knowledge").

Scientists, like other people, come upon new ideas, *acquire* knowledge, in all sorts of ways: from conversation with other people, reading books and newspapers, inductive generalizations, and even from dreams and mistaken observations. Newton is said to have been inspired by a falling apple. The German chemist August Kekulé (1829–1896) had been unsuccessfully attempting to devise a model for the molecular structure of benzene. One evening he was dozing in front of the fire. The flames appeared to Kekulé as snake-like arrays of atoms. Suddenly one snake seemed to bite its own tail and then whirled mockingly in front of him. The circular appearance of the image inspired in him the model of benzene as a hexagonal ring. The model to explain the evolutionary diversification of *species* came to Darwin while riding in his coach and observing the countryside. "I can remember the very spot in the road… when to my joy the solution came to me… The solution, as I believe, is that the modified offspring… tend to become adapted to many and highly diversified places in the economy of nature" (Darwin, 1958).

Hypotheses and other imaginative conjectures are the initial stage of scientific inquiry. It is the imaginative conjecture of what might be true that provides the incentive to seek the truth and a clue as to where we might find it (Medawar, 1967).[2] Hypotheses guide observation and experiment because they suggest what to observe. The empirical work of scientists is guided by hypotheses, whether explicitly formulated or simply in the form of vague conjectures or hunches about what the truth might be. But imaginative conjecture and empirical observation are mutually interdependent episodes. Observations made to test a hypothesis are often the inspiring source of new conjectures or hypotheses. As described by Jacob, the results of an experiment often inspire the modification of a hypothesis and the *design* of new experiments to test it (Jacob, 1988).

The starting point of scientific inquiry is the conception of an idea—a process that is, however, not a subject of investigation for logic or *epistemology*. The complex conscious and unconscious events underlying the creative mind are properly the interest of empirical psychology. The creative process is not unique to scientists. Philosophers, novelists, poets, and painters are also creative; they too advance models of experience and they also generalize by *induction*. What distinguishes science from other forms of knowledge is the process by which this knowledge is justified or validated.

## THE CRITERION OF DEMARCATION

Testing a hypothesis (or theory) involves at least four different activities. First, the hypothesis must be examined for internal consistency. A hypothesis that is self-contradictory or not logically well formed in some other way should be rejected. Second, the logical structure of the hypothesis must be examined to ascertain whether it has explanatory value, ie, whether it makes the observed phenomena intelligible in some sense, and provides an understanding of why the phenomena do in fact occur as observed. A hypothesis that is purely tautological should be rejected because it has no explanatory value. A scientific hypothesis identifies the conditions, processes, or mechanisms that account for the phenomena it purports to explain. Thus hypotheses establish general relationships between certain conditions and their consequences or between certain causes and their effects. For example, the motions of the planets around the Sun are explained as a consequence of gravity, and respiration as an effect of red blood cells that carry oxygen from the lungs to various parts of the body.

Third, the hypothesis must be examined for its consistency with hypotheses and theories commonly accepted in the particular field of science, or to see whether it represents any advance with respect to well-established alternative hypotheses. Lack of consistency with other theories is not always ground for rejection of a hypothesis, although it will often be. Some of the greatest scientific advances occur precisely when a widely held and well-supported hypothesis is replaced by a new one that accounts for the same phenomena that were explained by the preexisting hypothesis, as well as other phenomena it could not account for. One example is the replacement of Newtonian mechanics by the theory of relativity, which rejects the conservation of matter and the simultaneity of events that occur at a distance—two fundamental tenets of Newton's theory (Schwartz, 1992).

Examples of this kind are pervasive in rapidly advancing disciplines, such as molecular biology at present. The so-called "central dogma" holds that molecular information flows only in one direction, from *DNA* to *RNA* (ribonucleic acid) to *protein*. The *DNA* contains the genetic information that determines what the organism is, but that information has to be expressed in enzymes (a particular *class* of proteins) that guide all chemical processes in cells. The information contained in the *DNA* molecules is conveyed to proteins by means of intermediate molecules, called messenger *RNA*. David Baltimore and Howard Temin were awarded the Nobel Prize for discovering that information could flow in the opposite direction, from *RNA* to *DNA*, by means of the *enzyme* reverse transcriptase. They showed that some viruses, as they infect cells, are able to copy their *RNA* into *DNA*, which then becomes integrated into the *DNA* of the infected cell, where it is used as if it were the cell's own *DNA* (Temin and Mizutani, 1970; Baltimore, 1970).

As another example, until very recently it was universally thought that only the proteins known as enzymes could mediate (technically "catalyze") the chemical reactions in cells. However, as pointed out in Chapter 3, Thomas Cech and Sidney Altman received the Nobel Prize in 1989 for showing that certain *RNA* molecules act as enzymes and catalyze their own reactions (Cech, 1985). One more example concerns the so-called "colinearity" between *DNA* and *protein*. It was generally thought that the sequence of nucleotides in the *DNA* of a *gene* is expressed consecutively in the sequence of amino acids in the *protein*. This conception was shaken by the discovery that genes come in pieces, separated by intervening *DNA* segments that do not carry code for amino acids; Richard Roberts and Philip Sharp received the 1993 Nobel Prize for this discovery (Crick, 1979; Chambon, 1981).

These revolutionary hypotheses were published after their authors had subjected them to severe empirical tests. Theories that are inconsistent with well-accepted hypotheses in the relevant discipline are likely to be ignored if they are not supported by convincing empirical evidence. The microhistory of science is littered with farfetched or *ad hoc* hypotheses, often proposed by individuals with no previous or posterior scientific achievements. Theories of this sort usually fade away because they are ignored by most of the scientific community, although on occasion they engage their interest because the theory may have received attention from the media or even from political or religious bodies. The flop over "cold fusion" is an example of an unlikely and poorly tested hypothesis that received some attention from the scientific community because its proponents were well-established scientists.[3]

The fourth and most distinctive test consists of putting on trial an empirically scientific hypothesis by ascertaining whether or not predictions about the world of experience derived as logical consequences from the hypothesis agree with what is actually observed. This is the critical element that distinguishes the empirical sciences from other forms of knowledge: the requirement that scientific hypotheses be empirically falsifiable. Scientific hypotheses cannot be consistent with all possible states of affairs in the empirical world. A hypothesis is scientific only if it is consistent with some but not with other possible states of affairs not yet observed in the world, so it may be subject to the possibility of *falsification* by observation. The predictions derived from a scientific hypothesis must be sufficiently precise that they limit the range of possible observations with which they are compatible. If the results of an empirical test agree with the predictions derived from a hypothesis, the hypothesis is said to be provisionally corroborated; otherwise it is falsified.

The requirement that a scientific hypothesis be falsifiable has been appropriately called the *criterion of demarcation* of the empirical sciences, because it sets apart these sciences from other forms of knowledge. A hypothesis that is not subject to the possibility of empirical *falsification* does not belong in the realm of science (Popper, 1959, 1963).

## VERIFIABILITY AND FALSIFIABILITY

The requirement that scientific hypotheses be falsifiable rather than simply verifiable seems surprising at first. It might seem that the goal of science is to establish the "truth" of hypotheses rather than attempt to falsify them, but it is not so. There is an asymmetry between the *falsifiability* and the *verifiability* of universal statements that derives from the logical structure of such statements. A universal statement can be shown to be false if it is found inconsistent with even one singular statement, ie, a statement about a particular event. But as I pointed out in the discussion of *induction*, a universal statement can never be proven true by virtue of the truth of particular statements, no matter how numerous these may be.

Consider a particular hypothesis from which a certain consequence is logically derived. Consider the argument: if the hypothesis is true, then the specific consequence must also be true; it is the case that the consequence is true; therefore the hypothesis is true. This is an erroneous kind of inference called by logicians the "fallacy of affirming the consequent." The error of this kind of inference may be illustrated with the following trivial example: if apples are made of iron, they should fall to the ground when they are cut off a tree; apples fall when they are cut off; therefore apples are made of iron. The conclusion is invalid even if both premises are true. The reason is that there may be some other explanation or hypothesis from which the same consequences or predictions are derived. The observed phenomena are true because they are consequences from this different hypothesis, rather than from the one used in the deduction.

The proper form of logical inference for conditional statements is what logicians call the modus tollens (manner of taking away). It may be represented by the following argument. If a particular hypothesis is true, then a certain consequence must also be true; but evidence shows that the consequence is not true; therefore the hypothesis is false. By way of simple example, consider the following argument. If apples are made of iron, they will sink in water; they do not sink; therefore they are not made of iron. The modus tollens is a logically conclusive form of inference. If both premises are true, the conclusion falsifying the hypothesis necessarily follows.

It follows from this reasoning that it is possible to show the falsity of a universal statement concerning the empirical world; but it is never possible to demonstrate conclusively its truth. This asymmetry between verification and *falsification* is recognized in the statistical methodology of testing hypotheses. The hypothesis subject to test, the *null hypothesis*, may be rejected if the observations are inconsistent with it. If the observations are consistent with the predictions derived from the hypothesis, the proper conclusion is that the test has failed to falsify the null hypothesis, not that its truth has been established.

The requirement that scientific hypotheses be falsifiable has a parallel in statistical inference, namely in the demand that the power of the test be greater than zero. Statisticians recognize two kinds of errors: a Type I error, the probability of rejecting the null hypothesis when it is true, usually represented as $\alpha$; and a Type II error, the probability of not rejecting the hypothesis when it is false, symbolized as $\beta$. Scientists pay considerable attention to Type I errors and thus choose $\alpha$ levels sufficiently low, but pay less attention to Type II errors. Yet the power of the test depends on the probability, $1-\beta$, of rejecting the null hypothesis when it is wrong. Thus small levels for both $\alpha$ and $\beta$ are desirable. Although for any given test the magnitudes of $\alpha$ and $\beta$ are inversely related, the value of $\beta$ may be reduced by increasing the sample size or the number of replications in a test.

## EMPIRICAL CONTENT OR "TRUTHFULNESS"

Tests of a scientific hypothesis must have a positive probability of resulting in the rejection of the hypothesis if this is false. A scientific hypothesis divides all particular statements of fact into two subclasses. First, we have the *class* of all statements with which it is inconsistent; this is the *class* of the "potential falsifiers" of the hypothesis. Second, there is the *class* of all statements that the hypothesis does not contradict, the *class* of "permitted" statements. A hypothesis is scientific only if the *class* of its potential falsifiers is not empty, because the hypothesis makes empirically meaningful assertions only about its potential falsifiers—it asserts that they are false. "Not for nothing do we call the laws of nature 'laws': the more they prohibit the more they say" (Popper, 1959, pp. 40–42; see also pp. 91–92, 119–121).

The empirical or information content of a hypothesis (the "truthfulness" conveyed by a scientific statement) is measured by the *class* of its potential falsifiers. The larger this *class*, the greater the information content of the hypothesis. As already stated, a hypothesis asserts that its potential falsifiers are false; if any of these is true, the hypothesis is proven false. A hypothesis or theory consistent with all possible states of affairs in the natural world (eg, "birds have wings because God made them so; fish do not for the same reason") lacks empirical content and hence is not scientific.

## CONTINGENCY AND CERTAINTY

Scientific hypotheses can only be accepted contingently, since their truth can never be conclusively established. This does not mean we have the same degree of confidence in all hypotheses that have not yet been falsified. A hypothesis that has passed many empirical tests may be said to be "proven" or "corroborated." The degree of *corroboration* is not simply a matter of the number of tests, but rather their severity. Severe tests are precisely those that are very likely to have outcomes incompatible with the hypothesis if the hypothesis is false. The more precise the predictions being tested, the more severe the test. A so-called critical or crucial test is an experiment for which competing hypotheses predict alternative, mutually exclusive outcomes. A critical test will thus corroborate one hypothesis and falsify the others.

One example is the experiment by Matthew Meselson and Franklin Stahl (1958) testing the double-helix model of *DNA* proposed by James Watson and Francis Crick (1953) that marks the beginning of molecular biology, one of the great scientific revolutions of all times. The double-helix model predicts that the replication of *DNA* is "semiconservative," that is, each daughter *DNA* molecule will consist of one parental strand (the conserved strand) and a newly synthesized strand. Two other possible models of *DNA* replication are the *conserved* model, according to which the parental *DNA* molecule is fully conserved and the daughter molecule consists wholly of newly synthesized *DNA*; and the *dispersive* model, according to which both daughter *DNA* molecules are newly synthesized and the parental molecule becomes degraded into its component fragments (nucleotides), which are then used, together with additional nucleotides, in the synthesis of the daughter *DNA* molecules.

Meselson and Stahl produced bacteria with heavy nitrogen (the isotope $^{15}$N) in their *DNA*, then transferred these bacteria to a medium containing light ($^{14}$N) nitrogen. They also had a method to determine precisely the density of the *DNA* in the bacteria. The double-helix model predicted that after one generation of replication all the *DNA* will be intermediate in density (because one strand of each molecule would have heavy nitrogen and the other strand light nitrogen). This was also predicted by the dispersive model (because each molecule would have about equal number of heavy and light *nucleotide* components); but not by the conserved model (which predicted that half the *DNA* molecules would be heavy and half light). The double-helix model predicted that after a second round of replication, half the *DNA* molecules would be intermediate in density and half would be light. The other two models made different predictions for the second-generation molecules. In particular, the dispersive model predicted that all *DNA* molecules would be identical to one another, with density one-quarter of the way between the light and the heavy molecules. (The predictions of the three models were also different for the third and later rounds of *DNA* replication.) Meselson and Stahl carried out this critical experiment, corroborated the double-helix model, and rejected the other two.

The larger the variety of severe tests withstood by a hypothesis, the greater its degree of *corroboration*. Hypotheses or theories may thus become established beyond reasonable doubt. The double-helix model of *DNA*, for example, was also corroborated by an experiment performed by J. Herbert Taylor and his colleagues using autoradiographically labeled *DNA* from plant roots (Taylor et al., 1957, p. 122), and by direct microscopic observation of replicating chromosomes (the cell bodies containing the DNA) (Cairns, 1963, p. 43). Since the 1960s the observations and experiments corroborating the double-helix model (and falsifying alternative models) of *DNA* are so numerous and consistent as to defy any challenge.

## "FACT" AND "THEORY" IN SCIENTIFIC USE

Scientific hypotheses or models that have become established beyond reasonable doubt are sometimes referred to by scientists as "facts." For example, the molecular composition of matter, the *DNA* double helix, and the *evolution* of organisms are said to be facts. The theoretical possibility that these and other hypotheses or explanations might be wrong remains as an abstraction, but they have been confirmed in so many ways, and so much knowledge has been built upon such hypotheses, that it would be totally unreasonable to think they will be proved wrong at some future time. We simply do not expect that the Sun will stop rising or that snow will melt into something other than water.

Scientists, however, sometimes refer to a well-established hypothesis or explanation by calling it a "theory" or a "model." Scientists, for example, speak of the "molecular theory of matter" or of the "theory of evolution." These expressions do not challenge that the knowledge in question is well corroborated. Rather, in scientific usage, the term "theory" often implies a body of knowledge, a set of interrelated principles and explanations and the facts that support them. Scientific usage differs in this, as in many other cases, from common usage. In common language, a "theory" is often an imperfect fact, an explanation for which there is little or no evidence—as in "I have my own theory as to why so many people become addicted to smoking tobacco."

## ERROR AND FRAUD IN SCIENCE

The procedure by which scientific hypotheses are empirically tested and rejected (the modus tollens) is a logically conclusive method—if a necessary consequence of a premise is false, then the premise must also be false. Nevertheless, the process of *falsification* is subject to human error. It is possible, for example, that an observation or experiment contradicting a hypothesis may have been erroneously performed or wrongly interpreted. Thus it is often required, particularly in the case of important or well-corroborated hypotheses, that the falsifying observation be repeatable or that other falsifying tests be performed.

The modus tollens may also lead to an erroneous conclusion if the prediction tested is not a necessary logical consequence of the hypothesis. The connection between a hypothesis and specific predictions derived from it is often not a simple matter. The logical validity of an inference may depend not only on the hypothesis being tested, but also on other hypotheses, whether explicitly stated or not, as well as on assumptions concerning the particular conditions under which the deduced inferences obtain (the so-called "boundary conditions"). If a particular prediction is falsified, it follows that the hypothesis tested as well as other hypotheses necessarily implied and the boundary conditions cannot all jointly be correct. The possibility exists that one of the subsidiary hypotheses or some assumed condition may be false. Thus a proper test of a hypothesis assumes (and in some cases tests) the validity of all other hypotheses and conditions involved in the *design* and performance of the experiment or observation by which the hypothesis is tested.

Erroneous conclusions in science are frequently a consequence of erroneous assumptions in the *design* or performance of experiments. The erroneous assumptions may be erroneous hypotheses assumed to be correct, or mistakes in the materials or conditions used. One reason why scientists invest so much time and effort in the peer-review process (discussed later) is that they want to weed out erroneous hypotheses as well as erroneous procedures.

An experiment (as it might be performed in a laboratory investigating, say, issues on *population* genetics) may take several months and require the investment of tens of thousands of dollars in materials, labor, and equipment. For this reason, scientists must specify in full detail the materials, conditions, and procedures used in their experiments. In the standard format of a scientific paper there is a detailed section, often entitled "Materials and methods," that usually follows the introduction setting up the problem but precedes the presentation of results. (Some scientific journals have in recent years relegated the materials and methods to the last part of an article, or to the "Supporting information," which in some cases is available only online.) Because a scientist's work depends on the validity of the work of others, the scientific profession is self-policing. Certainly abuses occur, but usually scientists are the ones who discover the violations of scientific mores. Their stakes are high.

Failure to test adequately is usually the most flagrant reason accounting for erroneous scientific conclusions. But whenever these conclusions are of theoretical or practical import, other scientists will perform additional tests and uncover the error. Improper or inadequate testing is sometimes accompanied by other violations of the canons of science. As we shall see later, Robert Koch, the discoverer of the tuberculosis bacillum, took advantage of his considerable prestige to avoid submitting his claim of having found a cure to proper peer review. The proponents of cold fusion made the same error of inadequate testing, but also sought extensive publicity and financial backing by communicating their claims to the media, instead of submitting them to peer review and publication in scientific journals (Taubes, 1993).

Errors in science are not always due to mistaken assumptions, nor are they often fraudulent. There are four stages in what is a continuous progression from unavoidable error to fraud (National Academy of Sciences, 1989; see also Black et al., 1994).[4] First, there are "quirks of nature," events that may happen because of unknown laws of nature, or that occur even though quite improbable. This situation may be illustrated with an example, which is only a caricature. Assume that a scientist is asked to find out whether heads and tails are equally probable for a particular coin. The scientist throws the coin 20 times, obtains heads every time, and concludes that the coin is biased. Yet this outcome is compatible with a fair coin: the probability of all 20 throws yielding heads is only one in a million, but it may indeed happen. The example is a caricature, because an experiment so simple should be repeated many more times before reaching any conclusion. The possibility that quirks of nature may occur is one reason why experiments are replicated by scientists.

Errors may also be due to "honest" mistakes. A scientist may have mistakenly used the wrong material, made the wrong measurement, or assumed the wrong conditions. These errors are usually discovered by repetition. But a scientist does not have unlimited resources or time, so even the most conscientious scientist can make a mistake. Errors of this kind are corrected when other scientists reproduce the experiments or test the same hypothesis in some other way.

A third source of error is negligence. A scientist may reach the wrong conclusion because of haste, inattention, or sloppiness. These and similar faults are violations of the standards expected in science and are condemned by scientists, even though the erroneous results are not intentional.

Finally, there is outright fraud, when a scientist conceals, modifies, or fabricates results. This is an even more grievous violation of scientific standards than carelessness, and is accordingly penalized when discovered. Sloppiness and fraud can both do countless harm to the scientific enterprise. However, the conclusions based on them are unlikely to persist if they are significant, because other scientists will seek to corroborate or falsify any results of interest. Sometimes the errors will be discovered, at great personal cost, by other scientists who had assumed their validity in performing their own experiments.

## THE SCIENTIFIC METHOD IN PRACTICE

The model of scientific practice I have sketched can be exemplified *ad infinitum* in the history of science. Generally known examples are Galileo's and Newton's experiments demonstrating the laws of motion, Blaise Pascal's measurements of atmospheric pressure, William Harvey's demonstration of the circulation of the blood, Antoine Lavoisier's rejection of the *phlogiston* theory and demonstration of the existence of oxygen, Louis Pasteur's experiments on fermentation and putrefaction, showing that they are caused by living organisms, and many others (Goldstein and Goldstein, 1978; Schwartz, 1992). I mentioned earlier the experiment of Meselson and Stahl demonstrating that *DNA* replicates as predicted by the double-helix model. The two episodes that characterize scientific knowledge can be seen in every case: the formulation of a daring hypothesis is associated with experiments cleverly designed to falsify the hypothesis if it were not true.

I now describe in somewhat more detail another example: Mendel's discovery of the laws of heredity and his formulation of a theory that remains the core of the science of genetics. Mendel's example is telling because it shows the dialogue between hypothesis and experiment. Initial experiments designed to test simple hypotheses (eg, whether both the maternal and paternal traits are passed on to progeny) lead to the formulation of new hypotheses (the first and second laws of heredity), which are further tested and stimulate a general theory of heredity, which is then subject to critical experiments. It is notable that all of this is accomplished in a single scientific paper, the author of which was an obscure schoolteacher.

## A HISTORICAL PARADIGM: MENDEL'S DISCOVERY OF THE LAWS OF HEREDITY

Gregor Mendel (1822–1884) was an Augustinian monk living in the Austrian city of Brünn (now Brno, Czech Republic). He studied under distinguished scientists at the University of Vienna and became a high school science teacher. Mendel succeeded where better-known contemporary scientists and distinguished predecessors had failed: he discovered the laws of inheritance and formulated the theory upon which all of modern genetics is built.

Mendel performed experiments with pea plants and reported his discoveries in a paper published in 1866, "Experiments in Plant Hybridization," remarkable for his lucid awareness of the requirements of the scientific method.[5] Mendel formulated hypotheses; examined their consistency with previous results; then submitted the hypotheses to severe critical tests and suggested additional tests that might be performed.

Mendel's genius is evident in his recognition of the conditions required to formulate and test a theory of inheritance: different traits in a plant (such as flower color or seed shape) should be considered individually; alternative states of the traits should differ in clear-cut ways (such as white and purple flower color); and ancestry of the plants should be precisely known by using only true-breeding lines in the experiments. (In modern jargon, these are "boundary conditions" that must obtain to ascertain the patterns by which parental traits are inherited by their offspring.) Mendel's hypotheses were formulated in probabilistic terms; accordingly, he obtained large samples and subjected them to statistical analysis.

Mendel studied the transmission of seven different traits in the garden pea, *Pisum sativum*, including the color of the seed (yellow versus green), the configuration of the seed (round versus wrinkled), and the height of the plant (tall versus dwarf). The results of Mendel's experiments are too well known to need detailed presentation here, but it is worth analyzing the various stages of his methodology. His first series of experiments was with plants that differ in a single trait. The regularities observed led to certain generalizations having the form of law-like statements: only one of the two traits (the *dominant* trait) appears in the first-generation progenies; after self-fertilization, three-fourths of the second-generation progenies exhibit the *dominant* trait, and one-fourth exhibit the other (*recessive*) trait; the second-generation plants exhibiting the *recessive* trait breed true in the following generations, but the plants exhibiting the *dominant* trait are of two kinds, one-third breed true, the other two-thirds are hybrids. Mendel tested these generalizations by repeating his experiments for each of the seven characters. These generalizations were summarized in a law, later called the principle of segregation: hybrid plants produce seeds that are one-half hybrid, one-fourth pure breeding for the *dominant* trait, and one-fourth pure breeding for the *recessive* trait.

Mendel tested the hypothesis of segregation by deriving and verifying additional predictions. For example, he predicted that after $n$ generations of self-fertilization the ratio of true-breeding to hybrid plants in the progeny of a hybrid should be $2^n-1$ to 1. He explicitly stated that this prediction would obtain only if the condition obtained that all plants have "equal average fertility... in all generations" (which is an interesting insight on the consequences of *natural selection*, a notion that was elucidated by his 13-years-older contemporary, Charles Darwin).

The study of the offspring of crosses between plants differing in two traits (eg, round and yellow seeds in one parent, wrinkled and green seeds in the other parent) led Mendel to formulate a second law, later called the principle of independent assortment: "The principle applies that in the offspring of the hybrids in which several essentially different characters are combined... the relation of each pair of different characters in hybrid union is independent of the other differences in the two original parental stocks." He corroborated this principle by examining progenies of plants differing in three and four traits. He correctly predicted and corroborated experimentally that in the progenies of plants hybrid for $n$ characters there will be $3^n$ different classes of plants.

The formulation and experimental testing of the two principles stated (also known as the first and second laws of inheritance) take up only approximately the first half of Mendel's paper. Midway through the paper Mendel advanced what he properly called a "hypothesis" or theory to account for his previous results and the two laws. The second half of the paper is dedicated to the derivation of predictions from the theory and testing them.

Mendel's theory of inheritance contains five elements: (1) for each character in any plant, whether hybrid or not, there is a pair of hereditary factors ("genes"); (2) these two factors are inherited one from each parent; (3) the two factors of each pair segregate during the formation of the sex cells, so that each *sex cell* receives only one factor; (4) each *sex cell* receives one or the other factor of a pair with a probability of one-half; (5) alternative factors for different characters associate at random in the formation of the sex cells.

Mendel's well-deserved eminence as one of the greatest scientists of all times rests particularly on the formulation of this theory of heredity. He was also quite aware of the logical status of his proposal, namely that it was a hypothesis that required experimental *corroboration*. The theory was eventually shown to be correct in its entirety, bar meaningful exceptions particularly with respect to (5). The slightest reflection would suffice to see, in any case, that Mendel's theory was a daring creative leap, not an *induction* by generalizing observed facts. Mendel's "factors" were not observed, nor was their behavior in the formation of the sex cells and their mutual associations. I willingly take this opportunity to set forward that the common image of Mendel as a good monk who through sheer repetition of careful experiments stumbled upon some important discovery could not be farther from the truth. Not only was Mendel extremely imaginative in the careful *design* of his experiments and in developing a theory to account for their result, but he had a remarkable insight into the nature of scientific inquiry. Halfway through his classic paper and just after formulating the elements of his theory that I have enumerated, he writes that "this hypothesis would fully suffice to account for the development of the hybrids in the separate generations" (ie, for the 3:1 and 9:3:3:1 ratios he described in detail). Then he goes on as follows: "In order to bring these assumptions to an experimental proof the following experiments were designed." The experiments he designs are two series of "back-crosses" that gave previously unobserved and unexpected (except on the basis of his theory's predictions) ratios and confirmed segregation and independent assortment separately in the egg cells and the pollen cells. Mendel's awareness of the requirements of scientific proof would do credit to the best scientists of our day. He knew that the agreement of a theory with the results it seeks to explain carries little weight if the theory has been created precisely to fit the results. Independent tests are necessary, and will be significant to the extent that the predictions of the theory are (quantitatively) precise and unexpected on any grounds other than the correctness of the theory.

## DESTRUCTION OF KNOWLEDGE BY IDEOLOGY: LYSENKO AND GENETICS IN THE SOVIET UNION

An egregious example of scientific malpractice and outright fraud is the case of Lysenkoism, which violated virtually every canon of scientific practice and stands as dramatic counterexample to Mendel's achievements (Medvedev, 1969). In February 1935 the agronomist Trofim Denisovich Lysenko—an opportunist charlatan with pretensions of being a great revolutionary scientist—addressed the Second Soviet Congress of Collective Farms on the shameful status of Soviet agriculture. Lysenko castigated Soviet geneticists, accusing them of being enemies of the people who were destroying Soviet agriculture by relying on abstract theories imported from the capitalistic West. Stalin, presiding over the event, expressed his approval: "Bravo, comrade Lysenko, bravo!"

Stalin's public approval consummated Lysenko's meteoric rise to fame and power. For three long decades, until the fall of Khrushchev in October 1964, Lysenko and his partisans presided over Soviet agriculture, imposed their ideas on biology, and completed the elimination of Soviet genetics (and of numerous Soviet geneticists, who were sentenced to death, sent to concentration camps, or at best removed from their research and teaching positions).[6] The Soviet Union, a country with enormous agricultural potential, would as a consequence become for many years, extending into the last third of the 20th century, agriculturally insufficient and backward in biology (contrary to its successes in other disciplines, like physics and mathematics).

Lysenko denounced genetics as a capitalistic science perpetuating the notion that there are qualitative differences—claimed to be rooted in the genes—in plants, animals, or people. Such immutable differences do not exist, according to Lysenko; rather, differences between individuals are due to environmental effects and can be radically modified by exposing organisms to appropriate environmental challenges. Therefore the production of new crops, or their *adaptation* to new habitats, need not be the long process of *selection* of suitable genotypes claimed by the capitalists, but can be simply and rapidly accomplished by exposing seeds or young plants to suitable conditions. At the height of his power, under Stalin's protecting approval, Lysenko's absurd utterances included the claim that in the appropriate environment wheat plants produce rye seeds.

Lysenko promised rapid increases in crop yields and the transformation of barren or poor lands into agricultural windfalls. He introduced practices such as the "vernalization" method of seed *adaptation* to harsh climates and the grassland system of crop rotation, which proved to be gigantic agricultural catastrophes. He suppressed genetics research and eliminated the teaching of genetics from universities and agricultural institutes.

How could absurd claims of such enormous magnitude and economic consequence persist for decades? Social, political, and other factors came into play, of course. The relevance to my present purposes is that Lysenko completely abrogated the traditional practices of science. He avoided properly designed tests that could falsify his theories and, instead, supported his claims with crude experiments that could be interpreted at will. Contrary evidence was denied or denounced on the grounds that nothing could possibly be right that contradicted the superior ideology of Marxism–Leninism. The large-scale failure of Lysenko's agricultural practices was attributed to subversion by farmers and enemies of the people. Any evidence, any practice, any theory was measured by its congruence with Marxist ideology; all, and only those, actions and results were acceptable that served the cause of the Soviet state.

The extent to which political considerations rather than scientific practice dominated the Lysenko affair is apparent in the stenographic record of a session of the Lenin Academy of Agricultural Sciences of the USSR (July 31–August 7, 1948). On this occasion Lysenko routed the remnants of genetics (and the geneticists) in the Soviet Union. In the opening address, Lysenko stated: "The party, the Government and J.V. Stalin personally, have taken an unflagging interest in the further *development* of the Michurin teaching. There is no more honorable task for us as Soviet biologists than creatively to develop Michurin's teachings." Ivan Vladimirovich Michurin (1855–1935) was the Russian horticulturist whose ideas concerning the inheritance of acquired characteristics Lysenko was consecrating.

The transcript of the closing meeting of the Academy's session includes Lysenko's concluding remarks:

> *Comrades, before I pass to my concluding remarks, I consider it my duty to make the following statement. The question is asked in one of the notes handed to me, 'What is the attitude of the Central Committee of the Party to my report?'*

*I answer: The Central Committee of the Party examined my report and approved it. (Stormy applause. Ovation. All rise.)*

*Long live the Michurin teaching, which shows how to transform living nature for the benefit of the Soviet people! (Applause.)*

*Long live the Party of Lenin and Stalin which discovered Michurin for the world (Applause) and created all the conditions for the progress of advanced materialist biology in our country. (Applause.)*

*Glory to the great friend of science, our leader and teacher, Comrade Stalin! (All rise, prolonged applause.)[7]*

## THE CURIOUS CASE OF DARWIN, OR THE DISCREPANCY BETWEEN WHAT SCIENTISTS SAY AND WHAT THEY DO

Few scientists in the 19th century or at any earlier time equal Mendel's clear delineation of the scientific method he was pursuing. In the English-speaking countries scientists advanced hypotheses and tested them in their work, but often claimed in their writings, particularly in Great Britain, to be following the orthodoxy of *inductionism*, proclaimed by some philosophers as the method of good science. Charles Robert Darwin (1809–1882) is a remarkable example of this discrepancy.

In his *Autobiography* Darwin says that he proceeded "on true Baconian principles and without any theory collected facts on a wholesale scale" (Darwin, 1958, p. 119). The opening paragraph of *Origin of Species*, partially quoted earlier, conveys the same impression:

*When on board H.M.S. Beagle, as naturalist, I was much struck with certain facts in the distribution of the inhabitants of South America, and in the geological relations of the present to the past inhabitants of that continent. These facts seemed to me to throw some light on the origin of species—that mystery of mysteries, as it has been called by one of our greatest philosophers. On my return home, it occurred to me, in 1837, that something might perhaps be made out on this question by patiently accumulating and reflecting on all sorts of facts which could possibly have any bearing on it. After five years' work I allowed myself to speculate on the subject, and drew up some short notes; these I enlarged in 1844 into a sketch of the conclusions, which then seemed to me probable: from that period to the present day I have steadily pursued the same object.*

Darwin also claims in other writings to have followed the inductivist canons. The facts are very different from these claims, however. Darwin's notebooks and private correspondence show that he entertained the hypothesis of the evolutionary transmutation of *species* shortly after returning from the voyage of the *Beagle*, and that the hypothesis of *natural selection* occurred to him in late 1837 or early 1838—several years before he claims to have allowed

himself for the first time "to speculate on the subject." Between the return of
the *Beagle* on October 2, 1836, and publication of *Origin of Species* in 1859
(and, indeed, until the end of his life), Darwin relentlessly pursued empirical
evidence to corroborate the evolutionary origin of organisms and test his
theory of *natural selection*.

Why this disparity between what Darwin was doing and what he
claimed? There are at least two reasons. First, in the temper of the times,
"hypothesis" was a term often reserved for metaphysical speculations
without empirical substance. This is the reason why Newton, the greatest-
ever theorist among scientists, also claimed hypotheses *non fingo* ("I fabri-
cate no hypotheses"). Darwin expressed distaste and even contempt for
empirically untestable hypotheses. He wrote of Herbert Spencer: "His
deductive manner of treating any subject is wholly opposed to my frame of
mind. His conclusions never convince me... His fundamental generalizations
(which have been compared in importance by some persons with Newton's
Laws!) which I daresay may be very valuable under a philosophical point of
view, are of such a nature that they do not seem to me to be of any strictly
scientific use. They partake more of the nature of definitions than of laws of
nature. They do not aid me in predicting what will happen in any particular
case" (Darwin, 1958, p. 109).

There is another reason, a tactical one, why Darwin claimed to proceed
according to inductive canons: he did not want to be accused of subjective bias
in the evaluation of empirical evidence. Darwin's true colors are shown in a
letter to a young scientist written in 1963: "I would suggest to you the
advantage, at present, of being very sparing in introducing theory in your
papers (I formerly erred much in Geology in that way); *let theory guide your
observations*, but till your reputation is well established, be sparing of publishing theory. It makes persons doubt your observations" (Darwin, 1903, vol.
2, p. 323; Hull, 1973). Nowadays scientists, young or not, also often report
their work so as to make their hypotheses appear as afterthoughts, conclusions
derived from the evidence at hand, rather than as preconceptions tested by
empirical observations.

Darwin rejected the inductivist claim that observations should not be
guided by hypotheses. The statement quoted earlier, "A man might as well go
into a gravel-pit and count the pebbles and describe the colors," is followed by
this telling remark: "How odd it is that anyone should not see that all observation must be for or against some view if it is to be of any service!" (Darwin,
1903, p. 195). He acknowledged the heuristic role of hypotheses, which guide
empirical research by telling us what is worth observing, what evidence to
seek. He confesses: "I cannot avoid forming one [hypothesis] on every subject" (Darwin, 1958, p. 141).

Darwin was an excellent practitioner of the *hypothetico-deductive method*
of science, as modern students of Darwin have abundantly shown.[8] He
advanced hypotheses in multiple fields, including geology, plant morphology

and physiology, psychology, and *evolution*, and subjected his hypotheses to empirical tests. "The line of argument often pursued throughout my theory is to establish a point as a probability by *induction* and to apply it as a hypothesis to other parts and see whether it will solve them."[9] Popper (1959) not only made it clear that *falsifiability* is the *criterion of demarcation* of the empirical sciences from other forms of knowledge, but also that *falsification* of seemingly true hypotheses contributes to the advance of science. Darwin recognized the same: "False facts are highly injurious to the *progress* of science, for they often endure long; but false views, if supported by some evidence, do little harm, for every one takes a salutary pleasure in proving their falseness; and when this is done, one path toward error is closed and the road to truth is often as the same time opened" (Darwin, 1871, 1889).

Some philosophers of science have claimed that evolutionary biology is a historical science that does not need to satisfy the requirements of the *hypothetico-deductive method*. The *evolution* of organisms, it is argued, is a historical process that depends on unique and unpredictable events, and thus is not subject to the formulation of testable hypotheses and theories. Such claims emanate from a monumental misunderstanding. There are two kinds of questions in the study of biological *evolution*.[10] One concerns history: the study of *phylogeny*, the unraveling and description of the actual course of *evolution* on Earth that has led to the present state of the biological world. The scientific disciplines contributing to the study of *phylogeny* include *systematics*, paleontology, *biogeography*, comparative anatomy, comparative embryology, and comparative molecular biology. The second kind of question concerns the elucidation of the mechanisms or processes that bring about evolutionary change. These questions deal with causal, rather than historical, relationships. *Population* genetics, *population* ecology, paleobiology, and many other branches of biology are the relevant disciplines.

There can be little doubt that the causal study of *evolution* proceeds by the formulation and empirical testing of hypotheses, according to the same hypothetico-deductive methodology characteristic of the physicochemical sciences and other empirical disciplines concerned with causal processes. But even the study of evolutionary history is based on the formulation of empirically testable hypotheses. Consider a simple example. For many years specialists proposed that the evolutionary lineage leading to humans separated from the lineage leading to the great apes (chimpanzee, gorilla, orangutan) before the lineages of the great apes separated from each other. Some recent authors have suggested instead that humans, chimpanzees, and gorillas are more closely related to each other than the chimpanzee and the gorilla are to the orangutan and other Asian apes. A wealth of empirical predictions can be derived logically from these competing hypotheses. One prediction concerns the degree of similarity between enzymes and other proteins from different *species*. It is known that the rate of amino acid substitutions is approximately constant when averaged over many proteins and long periods of time. If the older hypothesis is correct, the average amount of *protein* differentiation should be greater between humans and the African apes than among these and orangutans. On the other hand, if the newer hypothesis is correct, human, gorilla, and chimpanzee should have greater *protein* similarity than any of the three has with orangutans. These alternative predictions provide a critical empirical test of the hypotheses. The available data favors the second hypothesis: human, chimpanzee, and gorilla appear to be phylogenetically more closely related to each other than any one of them is related to orangutans. Indeed, investigations of *DNA* and proteins have shown that humans and chimpanzees are phylogenetically more closely related to each other than either one is to the gorillas or any other primates.

Certain biological disciplines relevant to the study of *evolution* are largely descriptive and classificatory. Description and classification are necessary activities in all branches of science, but play a greater role in certain biological disciplines, such as *systematics* and *biogeography*, than in other disciplines, such as *population* genetics. Nevertheless, even *systematics* and *biogeography* use the *hypothetico-deductive method* and formulate empirically testable hypotheses.

## THEORY REPLACEMENT: PHLOGISTON AND LAVOISIER

Science is progressive. Theories that are accepted at one time may later be rejected because they are found to be wrong. More often, however, particularly in well-developed scientific disciplines, a theory that accounts for much that is known at a time is only rejected when it becomes replaced by a different theory that accounts for the same phenomena as well as others that the former theory left unexplained. Two examples illustrate these two situations: the *phlogiston* theory, replaced by Lavoisier's discovery of oxygen and his theory of combustion, and Newton's theory of motion, which was replaced by the theory of relativity.

Johann Becher in 1669 proposed that matter consisted of three kinds of earth: the vitrifiable, the mercurial, and the combustible. A substance such as wood consisted of combustible earth plus ashes. When the wood was burned, combustible earth was liberated. The hypothesized combustible earth was named "phlogiston" half a century later by Georg Stahl, who claimed that the corrosion of metals was also a form of combustion, and that *phlogiston* was lost in the process. The *phlogiston* theory was accepted by Joseph Priestly and other eminent 18th-century scientists.

The *phlogiston* theory was demolished by Antoine Lavoisier (1743–1794) in a series of experiments published in 1787.[11] This publication was followed in 1789 by his *Traité élémentaire de chimie* (*Elements of Chemistry*), which may very well be considered the foundational treatise of modern chemistry. Lavoisier rejected the *phlogiston* theory on the grounds that it led to erroneous predictions. He first noticed that the ash of wood (or other burned organic substances) weighed less than these substances did before burning, whereas sulfur and phosphorus weighed more, although *phlogiston* had been liberated in both cases according to the theory. Lavoisier then tested the *phlogiston* theory by systematically weighing all matter involved in the combustion or calcination of a variety of organic substances as well as metals. These experiments manifested the presence of two substances in air: one (which he named oxygen) was absorbed by burning, and the other was the "nonvital" air (nitrogen) that remained behind. He proposed that combustion was not the result of liberation of the hypothetical *phlogiston*, but the combination of the burning substance with oxygen. He tested this theory with carefully designed experiments in which all substances involved were weighed before and after burning, but also by extending the theory to other processes involving oxidation, such as rusting, and to a variety of natural phenomena. Thus Lavoisier explained water as the product of the combination of oxygen and hydrogen. He applied this methodology of testing theories by predicting events and precisely measuring their outcome to the resolution of numerous matters of public interest. In a well-known instance, he collaborated with Benjamin Franklin in debunking Franz Anton Mesmer's claim that he was able to cure by means of "animal magnetism."

The *phlogiston* story illustrates an important dimension of the scientific process: the reluctance of scientists to reject an accepted theory until another becomes formulated that accounts for the phenomena explained by the preexisting theory. Joseph Priestley and other contemporary scientists continued for a time to accept the *phlogiston* theory even in the face of falsifying experiments. The *phlogiston* theory became generally rejected only toward the end of the 18th century, after Lavoisier had developed and corroborated his own theory of combustion.

## THEORY REPLACEMENT: NEWTONIAN MECHANICS AND EINSTEIN

Scientific advance occurs not only, as with *phlogiston*, through the replacement of an erroneous theory by a correct one, but also by the replacement of a largely correct theory by a more precise or more inclusive theory. The examples are numerous. One famous instance is the replacement of Newtonian mechanics by Einstein's theory of relativity. As is often the case in the *progress* of scientific knowledge, the predictions made by the earlier theory are largely correct, which is why the theory, in this case Newtonian mechanics, had passed numerous tests and become generally accepted. But the newer scientific theory is able to account for phenomena left unexplained by the previous theory. In some instances this happens because the new theory has much greater generality and is able to account for phenomena explained by different theories or even different disciplines. One example is statistical mechanics, which became able to account for many conclusions of thermodynamics once it was discovered that the temperature of a gas reflects the mean kinetic energy of its molecules.

In the case of Einstein *vis-à-vis* Newton, it is particularly interesting that fundamental assumptions of the Newtonian theory, eg, that mass is constant and space and time are absolute realities, are rejected by the theory of relativity. Yet with respect to bodies with intermediate mass and intermediate velocities (that is, the bodies and motions encountered in the course of ordinary experience), Newton's and Einstein's theories make practically identical predictions (Schwartz, 1992).

Isaac Newton (1642—1727) is one of the greatest scientists of all times. He formulated the laws of motion and the law of gravity, developed a theory of light, invented the calculus, and much more. Newton's myriad discoveries include solutions of the so-called "two-body problem," ie, the shape and size of the planetary orbits; the mass of the Moon (1—80th that of the Earth), calculated from the heights of the tides; the tilt of the Earth's axis (23.5°) that accounts for the seasons; the size of the Earth's bulge at the equator; and showing that the periods of the orbits of the planets should be proportional to the square of their distance from the Sun, rather than to three halves as predicted by Descartes's theory.

Albert Einstein (1879—1955) is another scientific giant who, like Newton, made discoveries of monumental importance. In 1905 he formulated the special theory of relativity, which sets that the mass of a body is not constant, as assumed by Newton's theory and common-sense experience, but increases with the speed of the body and becomes nearly infinite as a body approaches the speed of light. (The equation is $m = m_o \sqrt{1 - v^2 / c^2}$, where $m_o$ is the mass at rest, $v$ is the speed of the body, and $c$ is the speed of light in vacuum.) Einstein's general theory of relativity (1916) sets that mass is not constant, but rather can be converted into energy, as famously expressed by the equation $E = mc^2$; that, also contrary to common-sense experience and the Newtonian theory, space and time are not absolute; that the same two events may be simultaneous for one observer, but not so for a different observer; that the speed of light is the maximum possible velocity in the universe; that the "timing" of a moving object decreases as its velocity increases (and thus that if a space traveler were to leave his twin brother on Earth while he traveled at great speed for a year, he would discover upon return that he was younger than his twin brother); and so on. The special theory of relativity is now well confirmed and the general theory has been shown consistent with some critical experiments designed to test it. Concerning phenomena of ordinary experience, the results predicted by relativity and Newtonian mechanics are virtually identical, although the two theories greatly deviate in their predictions of phenomena occurring at velocities near the speed of light.

As in the case of relativity theory, scientific knowledge often advances by the substitution and supplementation of one theory by another more complete, more precise, or more inclusive. Thus the modern theory of genetics, for example, has identified conditions that are exceptions to Mendel's second law, has defined the chemical composition of genes, has subsumed much that had been earlier formulated by the cell theory, and has integrated Darwin's theory of *natural selection* in the subdiscipline known as *population* genetics.

## HURRIED SCIENCE: ROBERT KOCH'S FAILED TUBERCULOSIS VACCINE

Robert Koch (1843—1910) by his middle 30s was already considered a distinguished scientist. While a practicing physician and working in a modest laboratory that he had built in his own home (in the small northern German town of Wollstein), he developed methods to culture and photograph bacteria. These methods led to the discovery of the life cycle of anthrax (which made it possible to explain the recurrence of the disease in long-unused pastures). He later acquired a scientific post in Berlin, where he started investigating tuberculosis, the major cause of mortality among young adults in 19th-century Europe. On March 24, 1882, Koch announced that he had discovered the cause of tuberculosis, the tubercle bacillus, a discovery that brought him further fame and later the Nobel Prize.[12]

Koch isolated and cultured the tubercle bacillus, and set about finding a cure for tuberculosis. He soon announced that he had discovered a substance that could protect against tuberculosis and even cure it. This announcement was received as a bombshell by the medical world. British journals like *The Lancet* and the *British Medical Journal* published complete translations of the article, and the *Review of Reviews* dedicated nearly a complete issue to the subject. Arthur Conan Doyle, who was still practicing medicine although already a well-known writer (creator of the literary detective Sherlock Holmes), arrived in Berlin shortly after the announcement and soon published an article on Koch and his discovery.

Two matters were troublesome with Koch's announcement. One was that he refused at first to reveal the nature of the curative substance, although he did so a year later under the pressure of public criticism. The second matter may have been related to the first: Koch's experimental testing of the vaccine was virtually lacking, and the vaccine was eventually to prove ineffective as either prevention or cure. Koch had anticipated, on the basis of limited evidence, that an injection of dead bacilli into a person who was later infected with living ones would result in a local reaction that might protect the person. In any case the local reaction would serve for diagnostic purposes. Perhaps because of his early successes (which included the discovery of the agent of cholera and its mode of transmission), Koch had become persuaded that his hypothesis for diagnosis and cure would prove to be correct. Thus he proceeded to announce it as a curative method without adequate testing. The *British Medical Journal*, which had earlier celebrated the original announcement, published a devastating article condemning Koch for having attempted to keep secret the composition of the substance and having recommended it as a remedy without adequate testing (Brock, 1999).

## EXPLANATORY CONTEXT OF DISCOVERY, OR WHY EMPIRICAL TESTING IS NOT ENOUGH: AVERY'S DNA AND WEGENER'S CONTINENTAL DRIFT

Empirical testing may be necessary, but is not sufficient for the scientific community to accept a new hypothesis. A hypothesis that has withstood even the most severe attempts to falsify it will not be accepted unless it has explanatory value; ie, unless it can be understood within the contemporary scientific context and it makes the problem at hand intelligible. There are several notable cases of scientific discoveries that were not accepted at the time because they were "premature," and were not contextually intelligible. Mendel's discovery of the laws of inheritance may fit this situation. Two more recent examples are the discovery by Oswald Avery (1877−1955) and his colleagues that *DNA* is the hereditary substance (rather than *protein*, as was generally believed at the time), and the theory of continental drift proposed by Alfred Wegener (1880−1930).

Avery was a distinguished scientist at a leading research institution, the Rockefeller Institute for Medical Research in New York. In 1944 he published a paper with his colleagues C.M. MacLeod and M. McCarty showing that the "transforming factor" responsible for the hereditary specificity of *Pneumococcus* bacteria (agents of severe pneumonia) was *DNA* and *protein* was not involved at all (Avery et al., 1944). Avery had performed a careful series of diverse and very specific tests that definitely identified *DNA* as the transforming factor and excluded other molecular *species*. There was no challenge to the experimental results, but the scientific community refused for several years to accept that *DNA* is the substance of heredity. This reluctance derived precisely from what was known about *DNA*, "knowledge" that made it impossible for *DNA* to encode hereditary information. It turned out eventually that "what was known" about *DNA* was wrong, or at least one seemingly inconsequential fact was. *DNA* became accepted as the hereditary substance only after the erroneous "detail" was corrected.

*Nucleic acid* was discovered in 1869 by Johann Friedrich Miescher, a 25-year-old Swiss. By the 1920s two kinds of *nucleic acid* (RNA and DNA) had become known, and their composition was soon thereafter elucidated. *DNA* was shown to be made up of four relatively simple components (nucleotides) similar to each other in all respects except their *nitrogen base*, which could be one of four: adenine, guanine, cytosine, and thymine (usually represented by A, G, C, and T). Much of the relevant knowledge came from Phoebus Aaron Levene, an organic chemist of towering reputation also working at the Rockefeller Institute. Levene had proposed that *DNA* was made up of long repetitions of the four nucleotides following one another in an invariant fashion. This was called "the tetranucleotide hypothesis," and was accepted without challenge—largely because accurate measurement of the proportions of the four nucleotides was not possible with the analytical methods of chemistry available at the time, but also because it was incorporated in the model for the composition of *DNA* elucidated by the highly reputed Levene.

The tetranucleotide hypothesis entailed that *DNA* could not be the carrier of hereditary information. An endless repetition of the same four components in the same order could not encode information of any kind, for the same reason that a repetition of the same four letters of the English alphabet cannot convey semantic information, no matter how long the sequence. *Protein*, in contrast, was known to be made up of some 20 different kinds of amino acids, which varied in proportion from one to another *protein*. *Protein* therefore could be an informative molecule, whereas *DNA* was a "stupid" molecule. Since both *protein* and *DNA* were present in the *nucleus* of the cell, it was generally assumed that *protein* would prove to be the carrier of hereditary information. In any case, *DNA* could not be, the experiments of Avery notwithstanding, because it could not convey information. Later on, after the chemist Erwin Chargaff at Columbia University showed that the proportions of the four bases, A, T, C, and G, vary from one to another organism and the tetranucleotide hypothesis was rejected, *DNA* became promptly accepted as the hereditary chemical.[13] The race to determine its structure was on, a feat that was accomplished in 1953 by James Watson and Francis Crick (1953).

A somewhat different state of affairs, but grounded on the same need for explanatory value, is the case of Alfred Wegener, a respected meteorologist and geologist who first proposed in 1912 and developed in 1915 the hypothesis of continental drift. He noted the complementary shape of the coastlines on both sides of the Atlantic, and reviewed geological and paleontological evidence scattered in the literature that led him to conclude that during the late Paleozoic (225—350 million years ago) all continents were assembled into a single supercontinent, which he named "Pangea."

Wegener tested his hypothesis that the continents had drifted by searching the literature for relevant geological, biogeographical, and paleoclimato-logical evidence. The evidence was striking, showing for example that strata and folds on opposite sides of the Atlantic fitted precisely with each other and extended beyond the coastlines in complementary patterns. Wegener, however, was unable to produce a convincing explanation of how the continents could move. His hypothesis was rejected with disbelief and the evidence relegated to a curiosity. It was only three decades later that continental drift became accepted, after the theory of plate tectonics provided a plausible mechanism for continental displacement (Bullen, 1976; Romano and Cifelli, 2015).[14]

## SOCIAL MECHANISMS: PEER REVIEW AND PUBLICATION

The process of testing a scientific hypothesis may corroborate it or falsify it. *Corroboration* may later be overturned. *Falsification* is a logically conclusive method: if a necessary consequence of a premise is false, then the premise must also be false. But both *falsification* and *corroboration* are subject to human error. For example, the modus tollens may lead to erroneous conclusions if the prediction tested does not in fact logically follow from the hypothesis. Moreover, an observation or experiment contradicting a hypothesis may have been performed or interpreted erroneously. Thus scientists require that their experiments be made public with sufficient detail so that they can be repeated.

The actual replication of experiments is, nevertheless, selective. It is usually reserved for experiments of unusual significance or those that conflict with established knowledge. Confronted with a new result that impacts their own work, scientists usually do not proceed to check it by repeating it, but rather will build upon the result, modify their own hypotheses, and *design* their own experiments accordingly. If something goes wrong with their work, they may turn back to the original result and check it by repeating the test; but time, resources, and prestige will have been lost along the way (National Academy of Sciences, 1989).

To minimize such problems, review mechanisms have become an integral part of science. The scientific community simultaneously seeks to encourage innovative thinking and ensure that new ideas are subjected to rigorous review. On the one hand, science is a creative process in which advances occur only if researchers are encouraged to develop and test innovative ideas. On the other hand, because science is a cumulative subject in which each scientist must build on the work of others, the scientific community has great stakes in weeding out false ideas. Accordingly, creativity is tempered by the need for rigorous review of new results.

Peer review represents an effort both to police scientific claims and to ensure their widest possible dissemination. The pressure on scientists to publish derives not only from narrow concerns for recognition and career advancement, but also from the desire of all scientists to learn of new developments that may guide their own work. Because submitting a paper for peer review is the best way to disseminate and establish priority for a new discovery or idea, the process serves to get new information out fast as well as to control its quality. The comments of peer reviewers contribute to the advancement of science by helping proponents of new hypotheses to improve their research and interpretations.

Peer-review scrutiny of science takes place in a variety of contexts. Informal review can occur when scientists discuss their work with one another at the laboratory bench, during conversations and seminars, and at scientific meetings. Formal peer review is generally an integral part of the scientific publication process and the process by which funds are granted for the conduct of research. Any claim that would significantly add to or change the body of scientific knowledge must be regarded skeptically if it has not been subject to some form of peer scrutiny, preferably by submission to a reputable journal. Publication in a peer-reviewed journal does not by itself guarantee the validity of the published results, nor is there reason for outright rejection of every work that has not been published in a reputable journal. But one should treat with utmost suspicion a proposition that has not been subjected to peer review.

Peer review delays somewhat the publication of results, but the delay and the large investment of time that reviewers and journal editors dedicate to the process are justified by the need to weed out erroneous results. The process of peer review is subject to human error and prejudices, but it is the most accessible and often most dependable element of the process of invention, validation, and refinement by which scientific knowledge advances.

Peer review does not thwart new ideas. Journal editors and the "scientific establishment" are not hostile to new discoveries. Science thrives on discovery, and scientific journals compete to publish new breakthroughs. Indeed, the most prestigious prizes are awarded to those scientists who make the most daring and dramatic discoveries, even when these contradict revered theories. I referred earlier to the revolutionary character of Einstein's relativity theory and the explosive advances of molecular biology, some of which were triggered by a sequence of unanticipated discoveries and many rewarded by the Nobel Prize and other awards, that contradicted previous assumptions.

Mistakes, errors, failures, and prejudices infect science as well as other human activities. But the large and ever-expanding body of scientific knowledge and its useful applications attest to the success of the scientific enterprise. The distinctive methodology of science accounts for much of this success, but the institutional mechanisms that have been developed contribute to that success.

## Endnotes

1. Jacob writes that he had started his scientific research under the naive misconception that science proceeds by *induction*, but soon realized that what was going on in a laboratory was quite different.

2. *The Art of the Soluble* (Methuen, London, 1967). This small book provides a very eloquent, yet profound, discussion of the scientific method as a dialogue between the two essential episodes of science: conjectures and refutations.

3. The hapless protagonists of the cold-fusion fiasco are Martin Fleishmann and B. Stanley Pons. The tale is well told by Gary Taubes, *Bad Science: The Short Life and Weird Times of Cold Fusion* (1993).

4. National Academy of Sciences, *On Being a Scientist* (1989) by the Committee on the Conduct of Science. According to this document, "Instances of scientific fraud have received a great deal of public attention in recent years, which may have exaggerated perceptions of its apparent frequency. Over the past few decades, several dozen cases of fraud have come to light in

science. These cases represent a tiny fraction of the total output of the large and expanding research community. Of course, instances of scientific fraud may go undetected, or detected cases of fraud may be handled privately within research institutions. But there is a good reason for believing the incidence of fraud in science to be quite low. Because science is a cumulative enterprise, in which investigators test and build on the work of their predecessors, fraudulent observations and hypotheses tend eventually to be uncovered. Science could not be the successful institution it is if fraud were common. The social mechanisms of science, and in particular the skeptical review and verification of published work, act to minimize the occurrence of fraud" (pp. 14−15).

5. Mendel's paper has been reprinted in English translation in numerous publications. The one herein used is from E.W. Sinnot, L.C. Dunn, and T. Dobzhansky, *Principles of Genetics* (McGraw Hill, New York, 1958), Appendix, pp. 419−443. A short biography of Mendel, as well as an annotated edition of his classic paper, can be found in Alain F. Coreos and Floyd V. Monaghan, *Gregor Mendel's Experiments on Plant Hybrids. A Guided Study* (Rutgers University Press, New Brunswick, NJ, 1993). A splendid biography of Mendel and statement of his work is Vítěslav Orel's *Gregor Mendel. The First Geneticist* (Oxford University Press, New York, 1996).

6. The distinguished Russian evolutionist Nikolai N. Vorontsov has written of Lysenko's impact on Soviet biology: "I remember 1948 very well, that fall, in all universities, in all institutions 3000 biologists lost their jobs and all possibility of research—three *thousand*. See Current State of Evolutionary Theory in the USSR," in L. Warren and H. Koprowski (Eds.), *New Perspectives in Evolution* (John Wiley, New York, 1991), p. 68.

7. The excerpts are cited in L. Warren and H. Koprowski (Eds.), *New Perspectives in Evolution* (John Wiley, New York, 1991), footnote 27, p. 76.

8. See, eg, Gavin De Beer, *Charles Darwin, A Scientific Biography* (Doubleday, Garden City, New York, 1964); M.T. Ghiselin, *The Triumph of the Darwinian Method* (University of California Press, Berkeley, 1969); D. Hull, *Darwin and His Critics* (Harvard University Press, Cambridge, MA, 1973); E. Mayr, "Introduction," in Charles Darwin, *On the Origin of Species* (Harvard University Press, Cambridge, MA, 1964).

9. Charles Darwin, 1960. *Darwin's Notebooks on Transmutation of Species*, G. De Beer (Ed.), Bulletin British Museum 2, 23−200.

10. See Theodosius Dobzhansky, *Genetics and the Origin of Species*, third ed. (Columbia University Press, New York, 1951), pp. 11−12.

11. See Henry Guerlac, "Lavoisier, Antoine Laurent," in *Dictionary of Scientific Biography*, vol. VIII (Scribner's Sons, New York, 1973), pp. 66−99.

12. A brief, readable, and well-documented biography is C.E. Dolman, "Koch, Heinrich Hermann Robert," in *Dictionary of Scientific Biography*, vol. VII (Scribner's Sons, New York, 1973), pp. 420−435.

13. Chargaff showed that the frequencies of the four nucleotides making up *DNA* were not identical. Rather, he established what became known as "Chargaff's rules," namely that the amount of A is equal to the amount of T, and the amount of C is equal to the amount of G; but the amount of A + T was often not equal to the amount of G + C. E. Chargaff, "Structure and function of nucleic acids as cell constituents," *Fed. Proc.* 10, 1951, 654−659.

14. See K.E. Bullen, "Wegener, Alfred Lothar," in *Dictionary of Scientific Biography*, vol. XIV (Scribner's Sons, New York, 1976), pp. 214−217.

Chapter 10

# Reduction and Emergence

## INTRODUCTION

One outstanding characteristic of life is complexity of organization. There is a hierarchy of complexity that runs from atoms and molecules through cells, tissues, individual organisms, populations, communities, and ecosystems to the whole of life on Earth. Different biological disciplines concentrate on the study of one or several levels of hierarchical organization. The analytical method so successful in the scientific study of the physical world has also proved fruitful in biology. Phenomena at one level of complexity are illuminated by the study of the component elements and underlying processes. The most impressive achievements of biological research during the last few decades are those of molecular biology, which works at the level of the molecular components of organisms, particularly deoxyribonucleic acid (*DNA*) and proteins.

Because of the success of the analytical method, and in particular the spectacular accomplishments of molecular biology, some molecular biologists and philosophers have gone so far as to contend that the only biological research that is ultimately significant, and indeed truly "scientific," is that pursued at the molecular level. In contrast, "conservative" biologists have claimed that molecular biology may be good physics or good chemistry, but has contributed little to the understanding of the most important biological problems. Most biologists and philosophers of science reject these two contentions as extreme and, indeed, unreasonable. Yet there are genuine questions concerning what has been called the "problem of reduction" in biology: the extent to which biological phenomena can be explained in terms of the underlying components of organisms, and ultimately in terms of atoms and molecules and the laws of physics and chemistry. In 1989, the announcement of the Human Genome Project by the United States was received with much acclaim by many molecular geneticists and other biologists, as well as by many members of the public who became aware of the project and had some understanding of what its objectives were and the possible implications. The enthusiasm reached levels of paroxysm, even among eminent scientists who should have been more thoughtful and had better understanding of the enormous possibilities that the Human Genome Project would open up, but also

Evolution, Explanation, Ethics, and Aesthetics. http://dx.doi.org/10.1016/B978-0-12-803693-8.00010-4

that knowing the *DNA* sequence by itself would not bring about much understanding of what human beings are, or contribute to much medical or any other kind of meaningful *progress* (Ayala, 1999, 2007; see Chapter 5).

## REDUCTIONISM

The distinguished evolutionist Stephen Jay Gould has written that the study of *evolution* embodies "a concept of hierarchy—a world constructed not as a smooth and seamless continuum, permitting simple extrapolation from the lowest level to the highest, but as a series of ascending levels, each bound to the one below it in some ways and independent in others ... 'emergent' features not implicit in the operation of processes at lower levels, may control events at higher levels" (Gould, 1980, p. 121).

The world, and not only the world of life, is hierarchically structured, and so are many human contraptions. A steam engine may consist only of iron and other materials, but it is something more than iron and the other components. Similarly a computer is more than just a pile of semiconductors, wires, plastic, and other materials. Organisms are made up of atoms and molecules, but they are highly complex systems and systems of systems of these atoms and molecules. Cellular, physiological, developmental, and other living processes are highly special and highly improbable patterns of physical and chemical processes. The issue arises whether complex entities can be explained in terms of the properties of their constituent components. This is the issue of *reductionism*, which arises in the world of physical, nonliving nature, but acquires extended magnitude in the world of life.

Organisms are complex, self-organizing entities made up of hierarchically constituted components of increasing complexity: organs, tissues, cells, organelles, and ultimately molecules and atoms. A question arises concerning the relationship between the whole and its component parts. The issue at stake has been called "the question of reduction" or "the problem of reductionism." Few if any questions in the philosophy of science have received more attention and been more actively debated than the question of reduction (Ayala, 1968, 2008a; Ayala and Arp, 2010; Edelman, 1974; Gould, 1982a,b; Lange, 2007; Nagel, 1961; Ruse, 1996, 2008a,b; Searle, 1998; Stebbins and Ayala, 1981, 1985). The debates, however, often involve several different issues, not always properly distinguished. Issues about the relationship between organisms and their physical components, or between biology and the physical sciences, arise in three domains: ontological, methodological, and epistemological.

## THREE ISSUES: ONTOLOGICAL, METHODOLOGICAL, AND EPISTEMOLOGICAL REDUCTION

Reductionistic questions arise, first, in what may be called the ontological, the structural, or the constitutive domain. The issue here is whether or not physicochemical entities and processes underlie all living phenomena. Are organisms constituted of the same components as those making up inorganic matter? Or is it the case that organisms consist of other entities besides molecules and atoms? There are other related questions. Are organisms nothing more than aggregations of atoms and molecules? Do organisms exhibit properties other than those of their constituent atoms and molecules?

Secondly, there are reductionist questions that might be called methodological, procedural, or strategic. These concern the strategy of research and the acquisition of knowledge, the approaches to be followed in the investigation of living beings. The general question is whether a particular biological problem should always be investigated by studying the underlying (ultimately physical) processes, or whether it should also be studied at higher levels of organization, such as the cell, the organism, the *population*, and the community.

The third type of reductionistic questions concerns issues that may be called epistemological, theoretical, or explanatory. The fundamental issue here is whether or not the theories and laws of biology can be derived from the laws and theories of physics and chemistry. *Epistemological reductionism* is concerned with whether biology may be dispensed with as a separate science because it represents simply a special case of physics and chemistry.

Distinction of the various kinds of questions being asked in debates about *reductionism* is the first step toward solving the issues. Much argumentation and confusion have resulted from failure to identify the question being argued in particular instances. It is not untypical, for example, to see a reductionist concerned primarily with the ontological question accusing a self-proclaimed antireductionist of having vitalistic ideas, while the latter may be an epistemological antireductionist but also in fact an ontological reductionist.

## ONTOLOGICAL REDUCTION

In the ontological or constitutive domain, the reductionist—antireductionist controversy in its extreme form resolves into the mechanism versus vitalism issue. The mechanist position is that organisms are ultimately made up of the same atoms that make up inorganic matter, and of nothing else. Vitalists argue that organisms are made up not only of material components (atoms, molecules, and aggregations of these) but also of some nonmaterial entity, variously called by different authors entelechy, vital force, *e'lan vital*, radial energy, and the like. Aristotle (384—322 BCE), the great Greek philosopher who was also the best biologist of his time, is sometimes said to have been the first systematic proponent of vitalism. The modern mechanism—vitalism controversy dates from the 17th century, when René Descartes (1596—1650) proposed that animals are nothing more than complex machines. Early in the 20th century, vitalism was defended by philosophers, such as Henri Bergson (1859—1941) and some biologists, notably Hans Driesch (1867—1941). At present, this extreme form of vitalism has no distinguished proponents among biologists, and few among philosophers.

Vitalism has been excluded from science primarily because it does not meet the requirements of a scientific hypothesis: it is not subject to the possibility of empirical *falsification* and therefore leads to no fruitful observations or experiments. Moreover, all available evidence indicates that organisms and life processes can be explained without recourse to any substantive nonmaterial entity.

*Ontological reductionism* claims that organisms are exhaustively composed of nonliving parts; no substance or other residue remains after all atoms making up an organism are taken into account. *Ontological reductionism* also implies that the laws of physics and chemistry fully apply to all biological processes at the level of atoms and molecules.

Ontological reductionists do not necessarily claim, however, that organisms are nothing but atoms and molecules. The inference that because something consists only of something else it is nothing but this "something else" is an erroneous inference, called by philosophers the "nothing but" fallacy. Organisms consist exhaustively of atoms and molecules, but it does not follow that they are nothing but heaps of atoms and molecules: they are highly complex patterns, and patterns of patterns, of these atoms and molecules. Living processes are highly complex, highly special, and highly improbable patterns of physical and chemical processes.

A much debated reductionist question that belongs in the ontological domain is whether organisms exhibit "emergent" *properties*, or whether their properties are simply those of their physical components. For example, are the functional properties of the kidney simply the properties of the chemical constituents of that organ? It must be pointed out, first, that the question of emergent properties is not exclusive to biology, but arises for all complex systems with respect to their parts. The general formulation of this question is whether the properties of a certain kind of object are simply the properties of other kinds of objects, namely their component parts, organized in certain ways. The issue of *emergence*, as pointed out later, is largely spurious and can be solved as a matter of definition.

## METHODOLOGICAL REDUCTION

One outstanding characteristic of living beings is a complexity of organization recognized in their common name, "organisms." As noted, there is a hierarchy of levels of complexity that runs from atoms to molecules, macromolecules, organelles, cells, tissues, multicellular organisms, populations, and communities. Some biological disciplines focus on one or a few of these levels of organizational complexity. Cytology is the study of cells, histology is the study of tissues, and ecology studies populations and communities. Yet, the biological disciplines are distinguished more by the kinds of questions asked and the kinds of answers sought than by the levels of organization investigated.

*Methodological reductionism* is the claim that the best strategy of research is to study living phenomena at increasingly lower levels of complexity, and ultimately at the level of atoms and molecules. For example, genetics should seek to understand heredity ultimately in terms of the behavior and structure of *DNA*, *RNA*, enzymes, and other macromolecules, rather than in terms of whole organisms, which is the level at which the Mendelian laws of inheritance are formulated. *Methodological reductionism* has its counterpart in what may be called methodological compositionism (Simpson, 1964). This is the claim that to understand organisms we must explain their organization—how organisms and groups of organisms come to be organized, and what functions the organization serves. According to methodological compositionism, organisms and groups thereof must be studied as wholes and in the multiplicity of their interactions, not only in their component parts.

*Methodological reductionism* in its extreme form would be the claim that biological research should be conducted only at the level of the physiochemical component parts and processes. Research at other levels is, allegedly, not worth pursuing or at best of only provisional value, since biological phenomena must ultimately be understood at the molecular and atomic levels. Methodological compositionism in its extreme form makes the opposite claim, namely that the only biological research worth pursuing is that at the level of whole organisms, populations, and communities. Research at lower levels of organization may, allegedly, be good physics or good chemistry but it has no biological significance.

It is unlikely that any scientist would thoughtfully sponsor the extreme forms of either compositionism or *reductionism* advanced in the previous paragraph. Extreme *methodological reductionism* would imply the unreasonable claim that genetic investigations should not have been undertaken until the discovery of *DNA* as the hereditary material, or that a moratorium should be declared in ecology until we can investigate the physiochemical processes underlying ecological interactions. Similarly, extreme methodological compositionism would imply that understanding the structure of *DNA* or the enzymatic processes involved in replication is of no significance to the study of heredity, or that the investigation of physicochemical reactions in the transmission of nerve impulses is of no interest to the understanding of animal behavior.

A moderate position of *methodological reductionism* points to the success of the analytical method in science, and the obvious fact that the understanding of living processes at any level of organization is much advanced by knowledge of the underlying processes. Moderate methodological reductionists claim that the best strategy of research is to investigate any given biological phenomenon at increasingly lower levels of organization as this becomes possible, and ultimately at the level of atoms and molecules.

The positive claims of moderate methodological reductionists are legitimate. The analytical method is of great heuristic value; much is often learned about a phenomenon through the investigation of its component elements and processes. In biology, the most impressive achievements of the last few decades are those of molecular biology. But there is little justification for any exclusionist claim that research should always proceed by investigation of lower levels of integration. The only criterion of validity of a research strategy is its success. Compositionist as well as reductionist approaches, synthetic as well as analytic methods of investigation, are justified if they further our understanding of a phenomenon, if they increase knowledge. Reductionist and compositionist approaches to the study of a biological problem are complementary; often the best strategy of research is an alternation between analysis and synthesis.

Investigation of a biological phenomenon in terms of its significance at higher levels of complexity often contributes to the understanding of the phenomenon itself; compositionist investigations are also heuristic. For example, it is doubtful that the structure and functions of *DNA* would have been known as readily as they were if there had been no previous knowledge of Mendelian genetics. The problem of the specificity of the immune response of antibodies proved refractory to a satisfactory solution as long as antibodies, antigens, and their structure alone were taken into consideration. The *natural selection* theory of antibody formation emerged only when antibodies were considered in their organismic milieu. Although the idea of clonal *selection* was first logically inadequate and quite vague, it had an enormous heuristic value in helping to understand how the specificity of antibodies comes about (Edelman, 1974).

*Methodological reductionism* and compositionism are sometimes based on convictions about the possibility of epistemological reduction. A methodological reductionist might claim that research should be pursued at increasingly lower levels of organization because she is convinced that ultimately all biological phenomena will be explained by the laws and theories of the physical sciences, and thus the only knowledge of lasting value is that acquired at the level of atoms and molecules. A methodological compositionist might claim that *epistemological reductionism* is impossible, and therefore full understanding of a biological phenomenon requires knowledge of its significance for higher levels of integration. *Methodological reductionism* and *epistemological reductionism* are nevertheless separate issues. *Methodological reductionism* is concerned with the strategies of research and the acquisition of knowledge. *Epistemological reductionism* deals with the organization of knowledge and the logical connections between theories. An epistemological reductionist, for example, might accept compositionist approaches to research because of their heuristic value. Similarly, epistemological antireductionists often claim that biological research should be pursued at all levels of integration of the living systems, including atomic and molecular levels.

## EPISTEMOLOGICAL REDUCTION

When philosophers of science speak of *reductionism* they generally refer to neither ontological nor methodological issues, but to epistemological reduction. In biology, the question of epistemological reduction (also known as theoretical or explanatory reduction) is whether the laws and theories of biology can be shown to be special cases of the laws and theories of the physical sciences.

Science seeks to discover patterns of relations among vast kinds of phenomena in such a way that a small number of principles explain a large number of propositions concerning those phenomena. Science advances by developing gradually more comprehensive theories, ie, by showing that theories and laws that had hitherto appeared as unrelated can in fact be integrated in a single theory of greater generality. For example, the theory of heredity proposed by Mendel can explain in many kinds of organisms diverse observations, such as the proportions in which traits are transmitted from parents to offspring, why progeny exhibit some traits inherited from one parent and some from the other, and why the offspring may exhibit traits not present in their parents. The discovery that the behavior of chromosomes during meiosis is connected with the Mendelian principles made possible the explanation of many additional observations concerning heredity; for example, why certain traits are inherited independently from each other, while other traits are transmitted together more often than not. Further discoveries made possible the *development* of a unified theory of inheritance of great generality which explains many diverse observations, including the distinctness of individuals, the adaptive nature of organisms and their traits, and the discreteness of *species*.

The connection among theories has sometimes been established by showing that the tenets of a theory or branch of science can be explained by the tenets of another theory or branch of science of greater generality. The more specific theory (or branch of science), called the secondary theory, is then said to have been reduced to the more general or primary theory. Epistemological reduction of one branch of science to another takes place when the theories or experimental laws of a branch of science are shown to be special cases of the theories and laws formulated in some other branch of science. The integration of diverse scientific theories and laws into more comprehensive ones simplifies science and extends the explanatory power of scientific principles, and thus conforms to the goals of science. Epistemological reductions are of great value to science because, as special cases of the integration of theories, they greatly contribute to the advance of scientific knowledge.

The reduction of a theory or even a whole branch of science to another has been repeatedly accomplished in the history of science (Nagel, 1961; Popper, 1974; Ayala, 1968, 1987a,b, 2008a,b). One of the most impressive examples is the reduction of thermodynamics to statistical mechanics, made possible by the discovery that the temperature of a gas reflects the mean kinetic energy of its molecules. Several branches of physics and astronomy have been to a large extent unified by their reduction to a few theories of great generality, such as quantum mechanics and relativity. A large sector of chemistry was reduced to physics after it was shown that the valence of an element bears a simple relation to the number of electrons in the outer orbit of the atom. Parts of genetics have been to some extent reduced to chemistry after discovery of the structure and mode of replication and action of the hereditary material, *DNA*.

The impressive successes of these reductions, and in particular the spectacular achievements of molecular biology, have led some authors to claim that the ideal of science is to reduce all natural sciences, including biology, to a comprehensive physical theory that would provide a common set of principles of maximum generality capable of explaining all observations about natural phenomena. Some authors have gone so far as to claim that the only biological research worth pursuing is that contributing to the explanation of biological phenomena in physicochemical terms.

Epistemological reductions are of great value to science because, as special cases of the integration of theories, they greatly contribute to the advance of scientific knowledge. In biology, the reduction of Mendelian genetics to molecular genetics has been far from completely successful. Yet there can be little doubt that much has been learned from what has been accomplished to date. Nagel (1965) and Popper (1974; see also Ayala, 1987a) elaborated the conditions that must be met to accomplish epistemological reduction of one theory to another and analyzed claims of successful reduction. Popper showed that no major case of epistemological reduction (including such model cases as the reduction of thermodynamics to statistical mechanics) "has ever been *completely* successful: there is almost always an unresolved residue left by even the most successful attempts at reduction" (Popper, 1974, p. 260). The same can be said of the reduction of genetics to molecular biology. Although much has been learned, and continues to be learned at an increased pace, by the study of heredity by investigating *DNA* and related molecular processes, much remains to be learned about biological heredity and how it explains the makeup and behavior of organisms (Ayala, 1968, 1987a, 2008a,b).

## EMERGENCE

One way in which some authors express the irreducibility of the properties of a complex entity to its components is to point out, as Gould (1980, p. 121) does, that organisms, like other complex entities, exhibit "emergent" features. But here again, as in my discussion of *reductionism*, it may be helpful to distinguish between the ontological and the epistemological domains. Consider the following question: are the properties of common salt, sodium chloride, simply the properties of sodium and chlorine when they are associated according to the formula NaCl? If among the properties of sodium and chlorine we include their association into table salt and the properties of the latter, the answer is "yes." In general, if among the properties of an entity we include the properties that the entity has when associated with other entities, it follows that the properties of complex systems, including organisms, are also the properties of their component parts. This is simply a definitional maneuver that contributes little to understanding the relationships between complex systems and their parts.

In common practice, we do not include among the properties of an entity all the properties of the systems resulting from its association with any other entities. There is a good reason for that. No matter how exhaustively an entity or component is studied in isolation, there is usually no way to ascertain all the properties that it may have in association with any other entities. Among the properties of hydrogen we do not usually include the properties of water, ethyl alcohol, proteins, and humans. Nor do we include among the properties of iron those of the steam engine.

The question of emergent properties may be formulated in a somewhat different manner, namely epistemologically. Can the properties of complex systems be *inferred* from knowledge of the properties that their component parts have in isolation? For example, can the properties of benzene be predicted from knowledge about oxygen, hydrogen, and carbon? Or, at a higher level of complexity, can the behavior of a cheetah chasing a deer be predicted from knowledge about the atoms and molecules making up these animals? Formulated in this manner, the issue of emergent properties is an epistemological question, not an ontological one; we are now asking whether the laws and theories accounting for the behavior of complex systems can be derived as logical consequences from the known laws and theories that explain the behavior of their component parts. This is the informative way of formulating questions about the relationship between complex systems and their component parts.

The issue of *emergence* cannot be productively settled by discussion of the *nature* of things or their properties, but it is informative by reference to our *knowledge* of those entities. Assume that by studying the components in isolation we can infer the properties they will have when combined with other component parts in certain ways. In such a case, it would seem reasonable to include the "emergent" properties of the whole among the properties of the component parts. (Notice that this solution to the problem implies that a feature which may seem emergent at a certain time might not appear as emergent any longer at a more advanced state of knowledge; and vice versa, a feature that may not seem to be emergent at a certain time may become emergent due to new knowledge.) Often, no matter how exhaustively an entity (or component part) is studied in isolation, there is no way to ascertain the properties it will have (as a component) in association with other entities (or component parts). We cannot infer the properties of ethyl alcohol, proteins, or human beings from the study of hydrogen, and thus it makes no good sense to list their properties among those of hydrogen. The important point, however, is that in the ontological sense the issue of emergent properties is spurious and it needs to be reformulated in terms of propositions expressing our knowledge. It is legitimate to ask whether the *statements* concerning the properties of organisms, and of their components at increasing levels of hierarchical organization, can be logically deduced from statements concerning the properties of their components at lower levels of organization.

I further illustrate the distinction between the ontological and epistemological domains made here, as well as earlier when discussing reduction, with a culinary *analogy*. Consider a favorite Spanish cold soup, gazpacho, made of tomatoes, peppers, cucumbers, celery, carrots, garlic, and other pureed vegetables mixed with oil, vinegar, a dash of lemon, and so forth. No sensible person would argue that gazpacho is made of anything else other than these ingredients (read: ontological reduction or identity at the level of components), or that the flavors of gazpacho come from anything other than its components. A different question is whether we can predict gazpacho's magic flavors from what we know about the flavors of its ingredients. I do not think so. But be that as it may, my point here is to distinguish the different issues at stake when speaking of reduction or *emergence*. Are the flavors of gazpacho the same as the flavors of its components? The answer would be "yes" if among the components' flavors we include the flavors they yield when suitably combined with the other components in gazpacho soup. But if we cannot predict gazpacho's flavors from what we know by tasting each component separately, the appropriate answer would be "no." A chef may discuss in detail the components used to prepare gazpacho soup, and talk about the distinctive flavors of the soup, but is not likely to assert that the distinctive flavors of gazpacho can be fully attributed to the ingredients in isolation. After all, those ingredients are used to prepare other delicacies. To say that table salt is *nothing else* than sodium and chlorine (or similarly for organisms or gazpacho that they are nothing else than their components) is to present, as stated earlier, the nothing-but fallacy.

## CONDITIONS FOR EPISTEMOLOGICAL REDUCTION

Nagel (1961) formulated the two conditions that are necessary and jointly sufficient to effect the reduction of one theory or branch of science to another: the condition of *derivability* and the condition of *connectability*. Epistemological reduction takes place when the experimental laws and theories of a branch of science are shown to be special cases of the experimental laws or theories of some other branch of science. The condition of derivability simply states that to reduce a branch of science to another it is necessary to show that the laws and theories of the secondary science can be derived as logical consequences from the laws and theories of the primary science.

No terms can appear in the conclusion of a demonstrative argument that do not appear in the premises. The reduction of one theory to another takes the form of a deductive argument in which one of the premises is the primary theory and the conclusion is the secondary theory. For the deduction to be logically valid, there must be another premise that establishes the connection between the terms of the primary theory and those of the secondary theory. This is the condition of connectability. Generally the experimental laws and theories of a branch of science contain distinctive terms that do not appear in other branches of science. To accomplish an epistemological reduction it is necessary that suitable connections be established between the terms of the secondary science and those used in the primary science. This may be accomplished by redefining the terms of the secondary science using terms of the primary science. The (partially successful) reduction of thermodynamics to statistical mechanics required the definition of terms, such as "temperature" by means of terms, such as "kinetic energy." The reduction of theories or experimental laws of genetics to physicochemistry requires that terms, such as "gene" and "chromosome" be defined by means of terms, such as "hydrogen bond," "nucleotide," "DNA," "histone protein," and the like.

Whenever the conditions of connectability and derivability are satisfied, the epistemological reduction of one theory to another becomes logically feasible. If all the experimental laws and theories of a branch of science can be reduced to experimental laws and theories of another branch of science, the former science will have been completely reduced to the latter.

It bears repetition that epistemological reduction is not a question whether the *properties* of a certain kind of objects, such as organisms, result from the properties of other kinds of objects, such as their component parts. This problem is an issue of *ontological reductionism*, which can be solved by a convention as to what is to be included among the properties of component parts. The reduction of one science to another is rather a matter of deriving a set of *propositions* from another such set. Scientific laws and theories consist of propositions about the natural world. The question of epistemological reduction can only be settled by concrete investigation of the logical consequences of propositions, and not by discussions about the nature of things or their properties. It is of course a legitimate epistemological question to ask whether *statements* concerning the properties of organisms can be deduced logically from statements concerning the properties of their physical components.

It follows from the previous comments that questions of epistemological reduction can only be properly answered by concrete reference to the actual state of *development* of the disciplines involved. Certain parts of chemistry were reduced to physics after the theory of atomic structure was advanced half a century ago. That reduction could not have been accomplished before such a *development*. If the reduction of one science to another is not possible at the present stage of *development* of two given disciplines, claims that such a reduction will be possible in the future carry little weight since these claims depend on the hypothetical *development* of as-yet nonexistent theories.

## FROM BIOLOGY TO PHYSICS?

There are some extreme positions about the question of *epistemological reductionism* that can be quickly discounted. Some substantive vitalists have claimed that biology is irreducible in principle to the physical sciences because living phenomena are the manifestation of nonmaterial principles, such as vital forces, entelechies, and so on. Epistemological antireductionism is then predicated on ontological antireductionism. Vitalism is not an empirical hypothesis because it does not lend itself to the possibility of empirical *falsification*. Moreover, the origin, structure, and functions of organisms can be explained without recourse to nonmaterial components or principles.

At the other end of the spectrum is the claim at the epistemological reduction of biology to the physical sciences is possible, and indeed the most important task of biologists at present. The impressive successes of molecular biology over recent decades have moved some people to claim that the only worthy and truly scientific biological investigations are those leading to the explanation of biological phenomena in terms of the underlying physico-chemical components and processes. Nevertheless, the epistemological reduction of biology to the physical sciences is not possible at present. In the current stage of scientific *development*, a great many biological terms, such as organ, *species*, consciousness, mating propensity, *fitness*, competition, predator, and many others cannot adequately be defined in physicochemical terms. Nor is there any *class* of statements and hypotheses in physics and chemistry from which every biological law could logically be derived. Neither the condition of connectability nor the condition of derivability—the two necessary conditions for epistemological reduction—is satisfied.[1]

A moderate reductionist position is probably not uncommon among biologists. It is claimed that although the reduction of biology to physicochemistry cannot be effected at present, it is possible in principle and is indeed a goal to be actively sought. The factual reduction of biology to the physical sciences is made contingent upon further *progress* in the biological or physical sciences, or both. This moderate form of *epistemological reductionism* is often based on convictions about *ontological reductionism*. It is generally accepted by biologists that living beings are exhaustively made up of physical components. It does not follow, however, that organisms *are nothing but* physical systems. *Ontological reductionism* does not entail *epistemological reductionism*. From the fact that organisms are exhaustively composed of atoms and molecules, it does not logically follow that the behavior of organisms can be exhaustively explained by the laws advanced to explain the behavior of atoms and molecules.

The claim that the reduction of biology to physicochemistry will eventually be possible is contingent upon unspecified, and at present unspecifiable, scientific advances. It is, therefore, a position that cannot be convincingly argued. Moreover, there are reasons to believe that complete reduction of biology to physics will never be possible. Popper argued that no major case of epistemological reduction "has ever been *completely* successful" (Popper, 1974, p. 260; see also Hull, 1974). It does not follow, however, that biologists should not attempt to reduce their theories to those of the physical sciences whenever such an undertaking seems likely to be successful. On the contrary, epistemological reductions are very successful forms of scientific explanation. A great deal is learned from epistemological reductions even when they are incomplete or not fully successful, because much understanding is gained by the partial success and valuable problems arise from the partial failure (Popper, 1974). In biology, the reduction of Mendelian genetics to molecular genetics has been far from completely successful (Hull, 1974). Yet there can be little doubt that much has been learned from what has been accomplished up to the present.

Some authors have claimed that the reduction of biology to physics and chemistry is impossible in principle because biological disciplines employ patterns of explanation that do not occur in the physical sciences (Simpson, 1964; Ayala, 1968, 2008a,b). Historical explanations are sometimes mentioned as distinctive of the biological and social sciences. Historical explanations play a role in evolutionary theory, although this theory is primarily concerned with causal explanations of evolutionary processes. Historical patterns of explanation, however, occur in some physical sciences, such as astronomy and geology. On the other hand, as I have proposed elsewhere (Ayala, 1970), teleological patterns of explanation are appropriate in biology but appear to be neither necessary nor appropriate in the explanation of natural physical phenomena. There seems no possible way in which teleological explanations, which are so fundamental for the understanding of evolutionary *adaptation*, could be reduced to evolutionary explanations in the physical sciences, where teleological explanations are not appropriate (Chapter 7).

## CONSCIOUSNESS AND THE BRAIN

A notable case where the issue of *emergence* versus *reductionism* arises is the brain–mind puzzle discussed in Chapter 5. The brain is the most complex human organ. It consists of 30 billion neurons; nerve cells each connected to many others through synapses between their two kinds of extensions, called the axon and the multiple dendrites. (In addition to the neurons, the human brain includes glial cells, about equally or even more numerous than the neurons.) As pointed out in Chapter 5, there are not enough bits of information in the *DNA* sequence of the human *genome* to specify the trillions of possible synapses and pathways among the 30 billion neurons in the human brain.

Concerning the mind—brain relationship, the question of reduction may be formulated as how the physical phenomena happening in the brain become the mind, the mental experiences that we call consciousness; how the electrochemical signals transmitted across the synapses, between exons and the dendrites of multiple neurons, become ideas, feelings, desires, and all sorts of sensations and perceptions, called "qualia" (singular quale) by philosophers; how out of the multiplicity of these experiences emerges the mind, a reality with unitary properties; and how, through the synapses between neurons, the mind creates language, art, music, science, mathematics, philosophy, and all other mental activities distinctive of human beings.

One possible answer is to deny that ontological reduction obtains in this case; that is, the phenomena occurring inside the brain do not account for the phenomena of consciousness, the qualia, which make up the human mind. Rather, consciousness or the mind would emerge from a spiritual, nonmaterial entity. Religious believers might assert that the soul accounts for the mind, thus providing the answer to the brain—mind puzzle. Most scientists and philosophers would, however, assert that consciousness is an outcome of the neural processes happening in the brain, so at the ontological level there is no entity accounting for consciousness other than the brain itself and neurophysiological processes.

It is obvious that epistemologically consciousness cannot, at least at present, be reduced to brain processes. The philosopher John Searle states both *ontological reductionism* and epistemological independence: "First, consciousness and indeed all mental phenomena are *caused by lower level neurological processes in the brain*; and, second, consciousness and other mental phenomena are *higher level features of the brain*" (Searle, 1998, p. 379). Answering the question of how neurobiological processes in the brain cause consciousness "would be the most important scientific discovery of the present era" (Searle, 1998, p. 379).

Methodologically, with respect to the conduct of research, the mind—brain problem requires reduction as well as composition. Neurobiology has developed over the past few decades into one of the most exciting biological disciplines. As pointed out in Chapter 5, a great deal has been learned about how light, sound, temperature, resistance, and chemical impressions received in our sense organs trigger the release of chemical transmitters and electric potential differences that carry the signals through the nerves to the brain and elsewhere in the body. Much has also been learned about how neural channels for information transmission become reinforced by use or may be replaced after damage; about which neurons or groups of neurons are committed to processing information derived from a particular organ or environmental location; and many other matters. Surely the increasing commitment of financial and human resources, including the *creation* and expansion of academic departments focusing on neurobiology in many research universities, will extend the unprecedented rate of discovery witnessed in recent decades.

Philosophy, psychology, and other disciplines will persist in investigating the qualia, the elements that contribute to consciousness, as well as consciousness itself and the products of the mind, such as language, art, literature, science, and philosophy. Whether epistemological reduction of consciousness and the mind to the underlying neurobiological processes will be ultimately accomplished at some future time remains at best an optimistic desideratum. "[The] new biology postulates that consciousness and mental processes are biological processes that, in due time, will be explained in terms of signaling molecular pathways used by nerve circuits that interact among them" (Kandel, 2003).

## Endnote

1. This view was challenged by J. S. Wicken (1987) in an ambitious book that purports to advance a new evolutionary theory encompassing prebiotic as well as organismal evolution, grounded as it is on thermodynamic principles. It is Wicken's thesis "(a) that the emergence and evolution of life are phenomena causally connected with the Second Law; and (b) that thermodynamics allows for the understanding of organic nature from organisms to ecosystems" (p. 5). The claims are sweeping: "By maintaining the part-whole relationality of ecosystem dynamics, thermodynamics addresses these problems by *extending* Darwinian tenets through prebiotic evolution and into life's first murmurings" (p. 120). "[T]his book aims at joining a century of Darwinian thought with a century of thermodynamic thought ... [I]t attempts to provide *unifying* principles for evolution—from the prebiotic generation of molecular complexity through the self-organization of living systems through their phylogenetic diversification" (p. 9). For Wicken, the claim that biology is an autonomous discipline is a "most unproductive attitude" (p. 55; see also p. 120) and indeed the nemesis of biology, particularly evolutionary studies. "Evolutionary theory has ... grown up as an 'autonomous' science, unconnected with either origins or physical dynamics. Sometimes, this sad truth is voiced in triumphant tones, as in Mayr's ... proclamation that 'the past twenty-five years has also seen the final emancipation of biology from the physical sciences' ... Evolution is a science of *connection*, not separation" (p. 55). If you wonder what in the world may be meant by the last sentence just quoted, you may as well be told that Wicken's book abounds in such vague language as well as in logical fallacies, inconsistencies, and equivocal terminology. Moreover, poetic language and metaphor are hopelessly woven into the fabric of the argument. Consider, by way of relevant example, the issue of *reductionism*. Wicken states: "I am therefore an ontological antireductionist and a methodological reductionist" (p. 11). This statement would seem surprising in view of the sentences quoted previously. Hence one might think that this is a matter of semantics: he is simply using the terms "ontological" and "methodological" in some idiosyncratic way. The statement just quoted is immediately followed by this: "That living systems operate, and came into being, according to materialistic principles does not imply that all those principles can be discovered at any given level of complexity" (p. 11). The initial clause of this sentence would seem a statement of *ontological reductionism*; the main clause an affirmation of methodological antireductionism, since it refers to the "discovery" of principles. In the next paragraph he concludes: "Thus [the] rejection of ontological reductionism. Organization provides boundary conditions on the operation of physicochemical processes that *cannot be deduced* from the principles governing those processes" (p. 11). This last sentence would seem to refer to

*epistemological reductionism.* Elsewhere on the same page, he writes: "Life can't, however, be reduced to thermodynamics, or to any combination of physical sciences" (p. 11). We are in a hopeless muddle. Does he now mean that "biology" cannot be reduced to the physical sciences? Or that "life" consists of more than just physical components and processes? The next sentence introduces additional confusion: "What makes the biotic realm unique is that the beings that inhabit it have 'sensitivity.' They carry subjective maps to which sensations can be referred, and their adaptive maneuverings are predicated on utilizing these maps in interpreting environments" (p. 11). What is the meaning of organisms "utilizing [subjective] maps in interpreting environments"? And if living beings are characterized by "sensitivity" and "sensations," are not plants living? Wicken sees the second law of thermodynamics as the unifying notion that explains how organisms came about and their *evolution.* "Thermodynamic principles operate at different levels of the organic hierarchy ... [L]ife is both operationally and functionally related to the rest of nature, and must be understood in terms of this relationality" (p. 9). Organisms are best defined as "autocatalytic organizations" (p. 17). Indeed, "A living system is an example of a 'dissipative structure'—a system that maintains a high degree of internal order by dissipating entropy to its surroundings" (p. 31). I have no quarrel with these notions, but they do not say much about organisms and their assemblages in populations, communities, and ecosystems. Thermodynamics has a role to play in the explanation of the function and *evolution* of organisms, but it is not enough. Thermodynamic principles may be in some biological disciplines necessary, but they are not sufficient.

Chapter 11

# Microevolution and Macroevolution: A New Evolutionary Synthesis?

## INTRODUCTION

In his widely read book *Chance and Necessity*, the Nobel Prize laureate Jacques Monod wrote that "the elementary mechanisms of evolution have been not only understood in principle but identified with precision... the problem has been resolved and evolution now lies well to this side of the frontier of knowledge" (Monod, 1972, p. 139).

Monod's unbridled optimism is unwarranted. The causes of *evolution* and the patterning of the processes that bring it about are far from completely understood (for example, Cela-Conde and Ayala, 2007, 2016; Dobzhansky et al., 1977; Mayr, 1982; Pigliucci and Müller, 2010; Reznick, 2010; Valentine, 2004). But neither is there any justification for the condemnation of the modern theory of *evolution* voiced particularly by a few paleontologists. Stephen J. Gould has written that "The modern synthesis, as an exclusive proposition, has broken down on both of its fundamental claims: extrapolationism (gradual allelic substitution as a mode for all evolutionary change) and nearly exclusive reliance on selection leading to adaptation" (Gould, 1980, p. 119); and, further, "the synthetic theory... is effectively dead, despite its persistence as textbook orthodoxy" (p. 120).

Gould's critique of the modern theory of *evolution* is grounded on a distorted version of the modern synthesis and has been adequately refuted (see, eg, Levinton and Simon, 1980; Stebbins and Ayala, 1981; Charlesworth et al., 1982). After the publication of these rebuttals, and possibly because of them, Gould had second thoughts. He explained that "Nothing about microevolutionary population genetics, or any other aspect of microevolutionary theory, is wrong or inadequate at its level... But it is not everything" (Gould, 1982a, p. 104). The criticisms, he qualified, propose "much less than a revolution... The modern synthesis is incomplete, not incorrect" (Gould, 1982b, p. 382). That microevolutionary theory "is not everything" and "the modern synthesis is incomplete" are tame propositions with which one can only agree.

Evolution, Explanation, Ethics, and Aesthetics. http://dx.doi.org/10.1016/B978-0-12-803693-8.00011-6

## RECTANGULAR EVOLUTION VERSUS GRADUALISM

Gould (1980), Eldredge (1971), Stanley (1979, 1982), Vrba (1980), and other paleontologists criticized neo-Darwinism to set the stage for a positive proposition, namely that *macroevolution*—the *evolution* of *species*, genera, and higher taxa—is an autonomous field of study, independent of microevolutionary theory (and the intellectual turf of paleontologists). This claim for autonomy was expressed as a "decoupling" of *macroevolution* from *microevolution* (eg, Stanley, 1979, pp. x, 187, 193) or a rejection of the notion that microevolutionary mechanisms can be extrapolated to explain macroevolutionary processes (eg, Gould, 1980, p. 383).

The paleontologists who argue for the autonomy of *macroevolution* base their claim on the notion that large-scale *evolution* is "punctuated" rather than "gradual." The model of punctuated equilibrium proposes that morphological *evolution* happens in bursts, with most phenotypic change occurring during *speciation* events, so new *species* are morphologically quite distinct from their ancestors but do not thereafter change substantially in *phenotype* over a lifetime that may encompass many millions of years. The punctuational model is contrasted with the gradualistic model, which sees morphological change as a more or less gradual process, not always associated with *speciation* events.

According to the punctuated model, *speciation* events are geologically "instantaneous"; and *species* may persist for many millions of years without change. According to the gradualist model, morphological *evolution* occurs over the lifetime of a *species*, with rapidly divergent *speciation* occurring occasionally. *Gradualism* does not imply that phenotypic change occurs continuously at a more or less constant rate throughout the life of a lineage, or that some acceleration does not take place during *speciation*, but rather that phenotypic change may occur at any time throughout the lifetime of a *species* and often it spreads throughout the *species*' persistence, even though not necessarily or usually at a very homogeneous rate through time.

Lest there is any doubt that punctualists predicate the autonomy of *macroevolution* on the alleged punctuational nature of large-scale *evolution*, I offer two quotations. "If rapidly divergent speciation interposes discontinuities between rather stable entities (lineages), and if there is a strong random element in the origin of these discontinuities (in speciation), then phyletic trends are essentially *decoupled* from phyletic trends within lineages. Macroevolution is decoupled from microevolution" (Stanley, 1979, p. 187). "Punctuated equilibrium is crucial to the independence of macroevolution—for it embodies the claim that species are legitimate individuals, and therefore capable of displaying irreducible properties" (Gould, 1982a, p. 94).

Whether phenotypic change in *macroevolution* occurs in bursts or is more or less gradual, is a question to be decided empirically. Examples of rapid phenotypic *evolution* followed by long periods of morphological *stasis* are known in the *fossil* record. However, there are instances as well in which phenotypic *evolution* appears to occur gradually within a lineage. The question is the relative frequency of one or the other mode, and paleontologists disagree in their interpretation of the *fossil* record: some favor punctualism (Eldredge, 1971; Eldredge and Gould, 1972; Hallam, 1978; Raup, 1978; Stanley, 1979; Gould, 1980; Vrba, 1980), while others favor *phyletic gradualism* (Kellogg, 1975; Gingerich, 1976; Levinton and Simon, 1980; Cronin et al., 1981; Reznick, 2010).

Whatever the paleontological record may show about the frequency of smooth or jerky evolutionary patterns, there is one fundamental reservation that must be raised over the theory of punctuated equilibrium. This evolutionary model argues not only that most morphological change occurs in rapid bursts followed by long periods of phenotypic stability, but also that the bursts of change occur during the origin of new *species*. Stanley (1979, 1982), Gould (1982a,b), and other punctualists have made it clear that what is distinctive in the theory of punctuated equilibrium is this association between phenotypic change and *speciation*. One quotation should suffice: "Punctuated equilibrium is a specific claim about speciation and its deployment in geological time; *it should not be used as a synonym for any theory of rapid evolutionary change at any scale...* Punctuated equilibrium holds that accumulated speciation is the root of most major evolutionary change, and that what we have called anagenesis is usually no more than repeated cladogenesis (branching) filtered through the net of differential success at the *species* level" (Gould, 1982a, pp. 84−85).

*Species* are groups of interbreeding natural populations that are reproductively isolated from any other such groups (Mayr, 1963; Dobzhansky et al., 1977). *Speciation* involves, by definition, the *development* of reproductive isolation between populations previously sharing a common *gene* pool. But it is no way apparent how the *fossil* record could provide evidence of the *development* of reproductive isolation. Paleontologists recognize *species* by their different morphologies as preserved in the *fossil* record. New *species* that are morphologically indistinguishable from their ancestors (or from contemporary *species*) go totally unrecognized. *Sibling species* are common in many groups of insects, rodents, and other well-studied organisms (Mayr, 1963; Dobzhansky, 1970; Nevo and Shaw, 1972; Dobzhansky et al., 1977; White, 1978; Benado et al., 1979). Moreover, morphological discontinuities in a time series of fossils are usually interpreted by paleontologists as *speciation* events, even though they may represent *phyletic evolution* within an established lineage, without splitting of lineages.

Thus when paleontologists use evidence of rapid phenotypic change to favor the punctuational model, they are committing a definitional fallacy.

*Speciation* as seen by these paleontologists always involves substantial morphological change because they identify new *species* by the eventuation of substantial morphological change. If *species* emerge that are not morphologically distinguishable, they could not be recognized, based on the paleontological record, as different *species*. In addition, Stanley (1979, p. 144) argued that "rapid change is concentrated in small populations and... such populations are likely to be associated with speciation and unlikely to be formed by constriction of an entire lineage." But the two points he makes are arguable. First, rapid (in the geological scale) change may occur in populations that are not small. Second, bottlenecks in *population* size are not necessarily rare (again, in the geological scale) within a given lineage.

Punctualists speak of evolutionary change "concentrated in geologically instantaneous events of branching speciation" (Gould, 1982b, p. 383). But events that appear instantaneous in the geological timescale may involve thousands, even millions of generations. Gould (1982a, p. 84), for example, has made operational the fuzzy expression "geologically instantaneous" by suggesting that "it be defined as 1 percent or less of later existence in stasis. This permits up to 100,000 years for the origin of a species with a subsequent life span of 10 million years." But 100,000 years encompasses 1 million generations of an insect, such as *Drosophila*, and tens or hundreds of thousands of generations of fish, birds, or mammals. *Speciation* events or morphological changes deployed during thousands of generations may occur by the slow processes of allelic substitution that are familiar to the *population* biologist. Hence, the problem faced by microevolutionary theory is not how to account for rapid paleontological change, because there is ample time for it, but why lineages persist for millions of years without apparent morphological change. Although other explanations have been proposed, it seems that stabilizing *selection* may be the process most often responsible for the morphological *stasis* of lineages (Stebbins and Ayala, 1981; Charlesworth et al., 1982). Whether microevolutionary theory is sufficient to explain punctuated as well as gradual *evolution* is, however, a different question.

## THE THEORY OF EVOLUTION

The current theory of biological *evolution* (the "*synthetic theory*" or "modern synthesis") may be traced to Theodosius Dobzhansky's *Genetics and the Origin of Species*, published in 1937: a synthesis of genetic knowledge and Darwin's theory of *evolution* by *natural selection*. The excitement provoked by Dobzhansky's book was soon reflected in many important contributions which incorporated into the modern synthesis relevant fields of biological knowledge. Notable landmarks are Ernst Mayr's *Systematics and the Origin of Species* (1942), Julian S. Huxley's *Evolution: The Modern Synthesis* (1942), George Gaylord Simpson's *Tempo and Mode in Evolution* (1944), and G. Ledyard Stebbins's *Variation and Evolution in Plants* (1950). It seemed to many scientists that the theory of *evolution* was essentially complete and all that was left was to fill in the details.

There can be little doubt that some components of *synthetic theory* are well established. In a nutshell, the theory proposes that *mutation* and sexual recombination furnish the raw materials for change; that *natural selection* fashions from these materials genotypes and *gene* pools; and that, in sexually reproducing organisms, the arrays of adaptive genotypes are protected from disintegration by reproductive isolating mechanisms (*speciation*). But Monod's (1970, p. 139) fervor is unwarranted. Indeed, "the causes of evolution, and the patterning of the processes that bring it about, are far from completely understood. We cannot predict the future course of evolution except in a few well-studied situations, and even then only short-range predictions are possible. Nor can we, again with a few isolated exceptions, explain why past evolutionary events had to happen as they did. A predictive theory of evolution is a goal for the future. Hardly any competent biologist doubts that natural selection is an important directing and controlling agency in evolution. Yet, one current issue hotly debated is whether a majority or only a small minority of evolutionary changes are induced by selection" (Dobzhansky et al., 1977, pp. 129–130).

Perhaps no better evidence can be produced of the unfinished status of the theory of *evolution* than pointing out some of the remarkable discoveries and theoretical developments of recent years. Molecular biology has been one major source of *progress*. One needs only mention the early acquired ability to obtain quantitative measures of genetic variation in populations and of genetic differentiation during *speciation* and *phyletic evolution*; the discontinuous nature of the coding sequences of eukaryotic genes with its implications concerning the evolutionary origin of the genes themselves; the dynamism of deoxyribonucleic acid (*DNA*) increases or decreases in amount and changes in position of certain sequences, etc. But notable advances have occurred as well, and continue to take place, in evolutionary ecology, theoretical and experimental *population* genetics, and other branches of evolutionary knowledge.

## GOULD'S THEORY

As I noted, Gould's 1980 criticisms of the *synthetic theory* of *evolution* were acerbic and extreme: "The modern synthesis... has broken down on both of its fundamental claims" (1980, p. 119) and "the synthetic theory... is effectively dead" (1980, p. 120). Those criticisms were largely later withdrawn and replaced by tame statements: "Nothing about... microevolutionary theory is wrong" (1982a, p. 104) and "The modern synthesis is incomplete, not incorrect" (1982b, p. 382).

*The Structure of Evolutionary Theory*, published in early 2002 a few months before Gould's death on May 20, 2002, is a monumental work, weighing in at 1343 pages of text (plus 44 pages of references), where he develops what he sees as an original and comprehensive new theory of *evolution*. "I wrote *The Structure of Evolutionary Theory*... to devise a... sufficiently novel theoretical structure that then yielded a sufficient number of original insights" (Gould, 2002, p. 48). "This book attempts to expand and alter the premises of Darwinism, in order to build an enlarged and distinctive evolutionary theory... within the tradition, and under the logic of Darwinian argument" (Gould, 2002, p. 1339). Gould's monumental masterpiece deserves extended commentary.

## Darwinian Logic

The "essence" or core of evolutionary theory is embodied by three "fundamental principles of Darwinian logic" that may be represented as the three main branches emerging from a tree (or coral) trunk. The trunk is Darwin's theory of *natural selection*. The three branches or "fundamental principles of Darwinian logic" are, according to Gould: agency (*natural selection* acting on individual organisms), efficacy (producing new *species*), and scope (accumulation of changes that through geological time yield diversity and morphological complexity). According to Gould, then, *natural selection* acting on individual organisms produces new *species* (agency), which are adapted to their environments (efficacy), causing changes that accumulate over time (scope), yielding through geological eras the panoply of taxonomic diversity and increased morphological complexity manifest in the living world and evinced by the *fossil* record. But Gould's efforts to contribute something important to each of these three fundamental components of Darwinian theory are far from successful.

Gould announces that he has something important to contribute to each branch of Darwinian logic. These contributions are buttressed by "new techniques and conceptualizations [which during the last third of the 20th century] opened up important sources of data that challenged orthodox formulations for all three branches of essential Darwinian logic" (Gould, 2002, pp. 25–26). For the first branch, the relevant *development* is the theory of punctuated equilibrium, Gould's most distinctive contribution to evolutionary theory (originally with Niles Eldrege), "which allowed us... to rethink macroevolution as the differential success of species rather than the extended anagenesis [ie, gradual change, as anagenesis is interpreted by Gould] of organismal adaptation." "On the second branch of full efficacy for natural selection... [are] the stunning discoveries of extensive deep homologies across phyla separated by more than 500 million years." Gould refers here to *Homeobox (Hox) gene* clusters which largely control the body plans of diverse kinds of animals and are similarly organized and recognizably *homologous* in organisms as different as beetles and monkeys.

"On the third branch of extrapolation," Gould asserts that the discovery "of a truly catastrophic trigger for at least one great mass extinction (the K–T

[Cretaceous—Tertiary] event of 65 million years ago), fractured the uniformitarian consensus" (Gould, 2002, p. 26). The dinosaurs and 90% of all animal *species* then existing became extinct at the K/T event, the transition from the Cretaceous to the *Tertiary* geological periods. Uniformitarianism interprets geological history as caused by more or less steady forces of nature. Uniformitarianism was forcefully argued by James Hutton (1726—1797) and particularly by Charles Lyell (1797—1875), who had considerable influence on his younger contemporary, Charles Darwin. The French comparative anatomist and paleontologist Georges Cuvier (1769—1832) argued instead that only the occurrence of catastrophic events could account for the sharp discontinuities observed in the geological record. It is now commonly accepted that the impact of a kilometer-wide meteor on the Yucatan peninsula caused, at least in part, the mass extinctions associated with the K/T event. However, extensive efforts have failed to discover similar extraterrestrial agencies, or major "sudden" events, as causes of other mass extinctions that occurred in the geological history of the Earth, some of which, such as at the transition between the Permian and Triassic periods, were even more extensive than the K/T event.

Gould says that his book "cycles through the three central themes of Darwinian logic [agency, efficacy, and scope] at three scales—by brief mention of a framework in this chapter [Chapter 1], by full exegesis of Darwin's presentation in Chapter 2, and by lengthy analysis of the major differences and effects in historical (Part 1) and modern critiques (Part 2) of these three themes in the rest of the volume" (Gould, 2002, p. 13). Chapter 1, which introduces the three themes and provides a summary of the remaining 11 chapters, runs to 89 large pages.

In Part I, the three themes are revisited as they are differently embraced by previous authors—Darwin first and foremost, but also the likes of Lamarck, Haeckel, Cuvier, Richard Owen, Hugo de Vries, Weisman, and the three great authors of the "modern synthesis," Theodosius Dobzhansky, G. G. Simpson, and Ernst Mayr. This historical Part I embraces Chapters 2—7, with a total of 500 pages.

Part II is Gould's own "Revised and Expanded Evolutionary Theory," Chapters 8—12, weighing in at 750 pages. Not surprisingly for a book of this length written over 20 years, which "cycles through three central themes," numerous and extensive repetitions occur, not only of substantive and relevant issues but also in the form of tedious, long commentaries about Gould's personal biography, love of baseball and of Gilbert and Sullivan, his friendships and dislikes. None of his previous critics escapes Gould's acrid scrutiny and damming censure, nor is any previous collaborator or supporter ignored or left verbally unrewarded.[1] The book and Gould's message would have greatly benefited from severe editing and drastic reduction in size, to less than half its current 1343 pages.[2]

## Theoretical Propositions

The chief theoretical propositions in *The Structure of Evolutionary Theory* will not be new for those familiar with Gould's previous writings, although the joint use of the labels for the three principles encompassing the "essence" of Darwinian logic ("agency," "efficacy," and "scope") is new. The arguments are developed at greater length than ever before.

Toward the end of the book Gould summarizes his objectives: "The most adequate one-sentence description of my intent in writing this volume flows best as a refutation to the claim of paradox just above: This book attempts to expand and alter the premises of Darwinism, in order to build an enlarged and distinctive evolutionary theory that, while remaining within the tradition, and under the logic, of Darwinian argument, can also explain a wide range of macroevolutionary phenomena lying outside the explanatory power of extrapolated modes and mechanisms of microevolution, and that would therefore be assigned to contingent explanation if these microevolutionary principles necessarily build the complete corpus of general theory in principle" (Gould, 2002, p. 1339).

Is Gould claiming an expansion with some modification, of the "modern synthesis" of evolutionary theory or is his claim more ambitious, namely the advance of a new theory, even if within the Darwinian tradition? Gould's statements are inconsistent, if not contradictory. At times he seems to be making the latter, more revolutionary claim. Thus: "I also hold... that substantial changes, introduced during the last half of the 20th century, have built a structure so expanded beyond the original Darwinian core, and so enlarged by new principles of macroevolutionary explanation, that the full exposition, while remaining within the domain of Darwinian logic, must be construed *as basically different from the canonical theory of natural selection, rather than simply extended*" (Gould, 2002, p. 3). In other places, as in his extended metaphor grounded on Milan's cathedral, Gould seems to claim something more modest: important theoretical additions to the existing theory. Incremental additions to preexisting theory are something to be expected as part and parcel of the growth of any scientific theory holding currency. Is Gould proposing a radical theoretical replacement, as in the replacement of Newtonian mechanics by general relativity, which denies previous fundamental premises, such as the constancy of mass in the universe or the radical and unbridgeable distinction between mass and energy? Gould's extended metaphors suggest that his ambitions are more modest, and his achievements certainly are.

The construction of Milan's Duomo began in the late 14th century in late flamboyant Gothic style, but much of the main western façade and entranceway was added later, in the baroque style of the 16th century. The additions were important, but incremental rather than substantive. The metaphor implies growth rather than replacement of evolutionary theories.[3]

The core of *The Structure* and Gould's main claim to theoretical innovation is the theory of punctuated equilibrium (PE). This theory pervades the whole book and is the subject of Chapter 9, "Punctuated Equilibrium and the Validation of Macroevolutionary Theory," which runs for 280 pages. A six-page summary is included in Chapter 1.

The theory of PE was first advanced in 1971 by Niles Eldredge and received its moniker in 1972 in a paper coauthored by Eldredge and Gould.[4] It has been the subject of some argumentation among scientists and much misrepresentation in the media and by fundamentalist creationists. The PE theory proposes that the frequently observed scarcity or absence in the *fossil* record of specimens that are intermediate in morphology between successive *fossil* forms (each with sustained presence in the *fossil* record) is not always or even generally due to the incompleteness of the record. According to PE theory, the record should be taken at face value. The abrupt appearance of new *fossil species* reflects their *development* in bursts of *evolution*, after which *species* remain unchanged in their morphology for the *species'* duration, which may extend for millions of years. The theory proposes that the prevailing view of morphological *evolution* as predominantly gradual must be replaced with a model of *speciation* with two distinct sequential components, a burst of change during the origination of a *species*, followed by a long period of *stasis* for the remaining duration of the *species*. Gould acknowledges that gradual and punctuational change are both represented in the *fossil* record, but he affirms that the punctuational mode appears at much higher frequency.

## The Autonomy of Macroevolution

The PE theory provides, according to Gould, the foundation on which he builds the claim that *macroevolution* (ie, *evolution* on a large scale with respect to time and morphological diversification) is an autonomous subject of evolutionary investigation, given that the punctuational pattern is not predictable on the basis of the small and gradual genetic changes investigated by *population* geneticists and other students of microevolutionary processes, such as they occur in living organisms.[5]

I return to this claim of macroevolutionary autonomy later, but two conceptual clarifications are first needed.

In sexually reproducing organisms, *species* are groups of interbreeding natural populations that are reproductively isolated from any other such groups.[6] In sexually reproducing organisms, *speciation* involves, by definition, the *development* of reproductive isolation between populations previously sharing a common *gene* pool. However, it is no way apparent how the *fossil* record could provide evidence of the *development* of reproductive isolation. Paleontologists recognize *species* by their different morphologies as preserved in the *fossil* record. New *species* that are morphologically indistinguishable from their ancestors (or from contemporary *species*) go totally unrecognized.

As pointed out earlier, sibling *species*, that is, *species* that are morphologically indistinguishable from one another, are common in many groups of insects, marine bivalves, rodents, and other well-studied organisms.[7] Thus, when Gould uses evidence of rapid phenotypic change in favor of the punctuational model of *speciation*, he commits the fallacy of definitional circularity. *Speciation* as seen by the paleontologist always involves morphological change because paleontologists identify new *species* by the eventuation of substantial morphological change.

The second conceptual clarification concerns the "sudden" appearance of new *species* in the *fossil* record, which indeed does not require unusual genetic mechanisms nor imply abrupt change of any sort when examined at the scale of the duration of the organisms' life cycle. The succession of *fossil* forms is associated with the succession of stratigraphic geological deposits, which accumulate for millions of years, separated by discontinuous transitions. The discontinuities reflect periods during which sediments failed to accumulate that typically last 50,000—100,000 years or longer. However, a time span of 100,000 years encompasses 1 million generations of insects, such as *Drosophila* or snails, such as *Cerion* (Gould's subject of empirical research), and thousands of generations for many *species* of fish, reptiles, birds, and mammals. *Speciation* events and morphological changes deployed during thousands of generations may occur by the slow processes of *gene* substitution that are familiar to the *population* geneticist. *Speciation* typically involves a few thousand generations, although it can occur considerably faster. The well-documented evolutionary diversification of *Drosophila species* and land snails in Hawaii, the largest and most recent island of the archipelago, shows that scores of sequential *speciation* events and extensive morphological diversification can occur in much less than 1 million years, by the gradual processes of *gene* substitution. There are more than 500 *Drosophila species* in Hawaii and they exhibit much morphological (as well as ecological and behavioral) diversification. Whether patterns of morphological *evolution* are rapid or slow is determined by environmental opportunities and pressures, and both patterns can be accomplished by gradual accumulation of *gene* substitutions.

## Tempo and Mode in Evolution

In spite of the relative scarcity of *fossil* remains that were available to Darwin in the mid-19th century, he took notice in *The Origin of Species* and elsewhere of the so-called "living fossils," which give no evidence of morphological change over millions of years. In one of the great books that originated the modern theory of *evolution, Tempo and Mode in Evolution*, published in 1944, the paleontologist George Gaylord Simpson wrote: "Their [evolutionary patterns'] seemingly infinite variety is so bewildering that generalization appears impossible at first, yet through them all there run three major styles, the basic modes of *evolution*. Thus, despite their complexity and peculiarity in

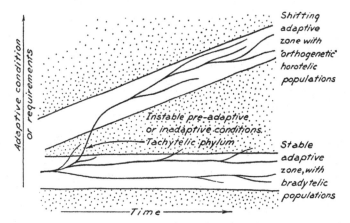

**FIGURE 11.1**  Three rates of evolution. Diagrammatic representation of the adaptive grip of a bradytelic group, a tachytelic line arising from it, and the subsequent deployment and further evolution of this line as a horotelic group. *Simpson, G.G., 1944. Tempo and Mode in Evolution. Columbia University Press, New York, p. 193.*

each case, almost all evolutionary events can be considered either as exemplifying one or another of these three modes or, more often, as susceptible to analysis as compounds of two or of all three."[8] Simpson depicted in his Figure 28 (here reproduced as Fig. 11.1) two of these "major styles," which he named "tachytelic," for fast morphological change, and "bradytelic" for absence or reduced morphological *evolution*. Lineages with average rates of *evolution* are called "horotelic." Simpson proposed that rapid change would typically be associated with the invasion of a new "adaptive zone," which might happen because of rapid environmental change in a locality inhabited by the organisms, or because the organisms had colonized a new and different environment uninhabited by organisms related to the invader. As Simpson indicated (as shown by the pointing arrow and the label in the center of Fig. 11.1, "Instable preadaptive or inadaptive conditions. Tachytelic phylum"), morphological change would in these cases occur over a short time interval, as a *population* of organisms shifted to a new adaptive zone. This very rapid, or tachytelic, pattern of change is often followed in the *fossil* record by a much slower or absent rate of change, which then persists over long stretches of paleontological time. This alternation of bursts of rapid change and long periods of morphological stability, described with supporting evidence by Simpson, is what Gould and Eldredge would later call PE. Gould's distinctive contribution is not the discovery of the alternating patterns of morphological change, but the claim that this style of *evolution* prevails in the *fossil* record, a claim disputed by many paleontologists. The controversy among paleontologists is not whether the punctuated mode of *evolution* exists, but whether it is more common in the record than other more or less gradual modes, as well as those exhibiting irregular or oscillating change.

Creationists have argued that *punctuated evolution* manifests the intervention of God in the evolutionary process. The sudden appearance of new
*species* would indicate divine acts of special *creation*. In *The Structure*, as he
had done many times before, Gould negates this implication and verbally
castigates its proponents. Gould reiterates, as he has stated elsewhere, that
the geological "instants" during which "sudden" change occurs typically
encompass 50,000–100,000 years, and these bursts of change result from the
well-known processes studied by evolutionary geneticists, genetic *mutation*
and *natural selection*, yielding adaptive evolutionary change. The creationist
claim is based on an additional and truly monumental misunderstanding. The
bursts of morphological change noticed by Gould and others do not involve
new body plans, the *emergence* of radically different kinds of organisms, or the
appearance of new limbs or organs, such as wings or lungs. Rather the traits
manifesting PE are traits, such as the shell flatness of oysters, irregular patterns
of coiling in ammonites, or the configuration of the head bones in lungfish
(see, for example, the next subsection and Fig. 11.2).

### *Eocoelia*: Punctuated or Gradual Evolution?

The observation just made deserves to be elaborated for the benefit of those
unfamiliar with the *fossil* record, whether scientists, philosophers, theologians,
or anyone interested in the evolutionary process. I can see no better way of
simply illustrating the *fossil* patterns of evolutionary change than by Fig. 11.2.[9]
The trait examined is rib strength in a group of brachiopods. These are marine
animals with shells, abundantly represented in ancient *fossil* beds: "rib
strength" is the ratio (ranging from 0% to 60%, as shown at the bottom of the
figure) of the height to the width of the shell ribs. The figure spans from 415 to
405 million years ago (see dates on the left). There are 13 samples (obtained
from three stratigraphic sites) at the times indicated by the dots on the right.
The observations are summarized in the middle of the figure for each of the 13
samples. For each sample, three numbers are graphically given (I refer in
parentheses to the values in the bottom sample, by way of example): range of
variation of individuals in the sample, represented by the horizontal line
(observed values ranging from about 22% to 54%); the mean or average value
for all individuals in the sample, represented by the vertical line (about 45%):
and the confidence interval of the mean, represented by the box (from about
39% to 52%). The confidence interval is a statistical statement that, on the
basis of the sampled individuals, the true mean value of the *population*
sampled has a probability of 95% of lying somewhere within the confidence
interval. Thus for the bottom sample we are 95% "confident" that the mean lies
somewhere between 39% and 52%.

I follow the logic of the paleontologist seeking to identify how many
*species* can be defined among the 13 samples. The mean of the five bottom
samples oscillates between 41% (middle sample) and 49% (second sample

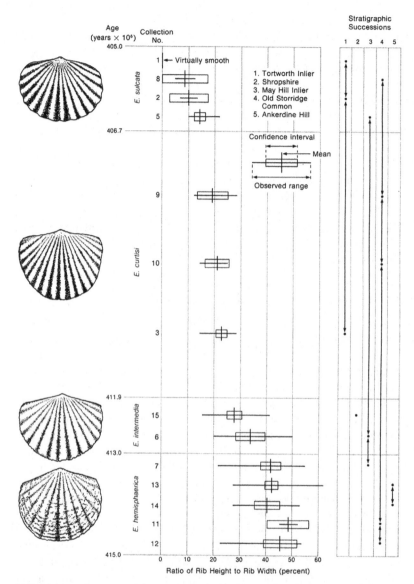

**FIGURE 11.2**  Evolution of rib strength in the brachiopod *Eocoelia* between 415 (bottom) and 405 (top) million years ago (Dobzhansky et al., 1977, p. 329).

from below), but the five confidence intervals considerably overlap. That is to say, these five bottom samples are not statistically different from one another, and thus they are identified as members of one *species, Eocoelia hemisphaerica* (see label sideways on the left). The *species* so defined persists without (statistically evinced) change for about 2 million years, between 415 and 413 million years ago. The sixth sample from the bottom has a mean displaced to the left and, although statistically no different from three of the samples below (their confidence interval boxes overlap), it overlaps with the seventh sample just on top of it, which in turn does not overlap with any of the bottom five samples. Thus, these two samples are considered a new *species*, named *Eocoelia intermedia*. The transition from *E. hemisphaerica* to *E. intermedia* has occurred over about 200,000 years (between samples 5 and 6), which is a time span relatively short when compared to the 2 million years' duration attributed to *E. hemisphaerica*. Samples 8, 9, and 10 are not statistically different from one another (ie, their confidence intervals overlap), but they are statistically different from all samples below and, accordingly, are placed in a new *species*, *Eocoelia curtisi*, which is attributed a duration somewhat greater than 5 million years (between 411.9 and 406.7 million years ago). The confidence intervals of the three top samples overlap with one another and, although they also overlap with the two samples of *E. curtisi* just below them, they do not overlap with the bottom sample of *E. curtisi*. The top three samples are therefore placed in a distinct *species*, *Eocoelia sulcata*, which is attributed a duration of nearly 2 million years (from 406.7 to 405.0 million years ago).

The logic used is scientifically sound. It follows accepted conventions and practices in the field of paleontology. But it should be painfully obvious that the claim that morphological change is associated with the origination of new *species* evinces circularity, since a new *species* is described whenever there is, and only if there is, a change in the mean value of the trait under consideration. Similarly, the claim of *stasis*, namely absence of change for the duration of each *species*, is a necessary consequence of an operational convention. In the sequence represented in Fig. 11.2, there is no known observation or experiment that could establish whether or not individuals assigned to one *species* could have intercrossed with individuals assigned to a different *species*. Nor is there any known procedure to determine that individuals assigned to the same *species*, whether they lived at the same or different times, could have interbred with one another. The ability or capacity to interbreed and produce fertile progeny is the criterion used to define *species* among living organisms with sexual reproduction.

## Gene Evolution Versus Species Selection

According to Gould, *phyletic evolution* proceeds at two levels. First, there is change within a *population* that is continuous through time. This, as pointed out earlier, consists largely of allelic substitutions prompted by *natural selection*, *mutation*, *genetic drift*, and the other processes familiar to the *population* geneticist operating at the level of the individual organism. This is *evolution* within established lineages, which, according to Gould, rarely if ever yields any substantial morphological change.[10] Second, there is the process of origination and extinction of *species*. According to PE, most morphological change is associated with the origin of new *species*. The theory claims, therefore, that evolutionary trends result from the patterns of origination and extinction of *species*, rather than from *evolution* within established lineages. Hence, the relevant unit of macroevolutionary study is the *species* rather than the individual organism. It follows from this PE argument that the study of microevolutionary processes provides little, if any, information about macroevolutionary patterns, the tempo and mode of large-scale *evolution*. Thus according to Gould and PE theory, *macroevolution* is autonomous relative to *microevolution*, much in the same way as biology is autonomous relative to physics.

Thus Gould's most innovative (or revolutionary, if one agrees with the view inconsistently expressed by Gould that his *The Structure* amounts to a theoretical revolution or full replacement, even if emerged from Darwinian logic) claim is the theory of PE and its chief implication of *species selection*, rather than *selection* between individuals, as the driving process (*agency*) of evolutionary change. By reference to Fig. 11.2, the PE claim would be that whichever evolutionary shift we perceive in the *evolution* of *Eocoelia* from 415 to 405 million years ago, it is not the outcome of *natural selection* acting among individuals (although Gould concedes that *natural selection* has consistently occurred throughout the full time spanned). Rather, following Gould, we would conclude that the changes observed are fully accounted by *species selection*—that is, the survival of some *species* and the extinction of others. But this conclusion cannot be warranted.

The claim of *species selection* as an important evolutionary process was repeated over 30 years by Gould and other proponents of PE. But where is the evidence that *species selection* occurs? This certainly cannot be convincingly inferred from the observations displayed in Fig. 11.2, which are more parsimoniously interpreted as outcomes of individual *selection*. Nor can it be inferred (and for the same reason) from any typical descriptions of *fossil* morphological *evolution*. Instances of *species selection* have been proposed over the past three decades, but in no case known to me—or known to Gould, he now admits—have they survived critical scrutiny. Gould argues that this is because scientists have not looked hard enough. So, the case needs to be made by hypothetical examples, of which he provides one.

The decisive characteristic in his example is the different degree of variability within each hypothetical *species*. *Natural selection*, Gould says, acts among *species* because the *species* that survives is the one *species* with the greater variability, while others become extinct. The character—variability within each *species*—"*does not exist* at the organismal level, and each species develops only one state of the (emergent) character because the character belongs to the species as a whole. Therefore, selection for this character can only occur among species" (Gould, 2002, p. 665). Gould's "one hypothetical example that I have often used to illustrate this issue and to argue for species selection" proceeds as follows (Gould, 2002, pp. 665–666):

*Suppose that a wondrously optimal fish, a marvel of hydrodynamic perfection, lives in a pond. This species has been honed by millennia of conventional Darwinian selection, based on fierce competition, to this optimal organismic state. The gills work in an exemplary fashion, but do not vary among individual organisms for any option other than breathing in well-aerated, flowing water. Another species of fish—the middling species—ekes out a marginal existence in the same pond. The gills don't work as well, but their structure varies greatly among organisms. In particular, a few members of the species can breathe in quite stagnant and muddy waters.*

*Organismic selection favors the optimal fish, a proud creature who has lorded it over all brethren, especially the middling fish, for ages untold. But now the pond dries up, and only a few shallow, muddy pools remain. The optimal fish becomes extinct. The middling species persists because a few of its members can survive in the muddy residua. (Next decade, the deep, well aerated waters may return, but the optimal fish no longer exists to reestablish its domination.)*

Gould continues: "Can we explain the persistence of the middling species, and the death of the optimal form, only by organismic selection? I don't think so. The middling species survives, in large part, as a result of the greater variability that allowed some members to hunker down in the muddy pools." I fully disagree with Gould. Because of changed environmental conditions, *natural selection* has favored the few individuals of the "middling species" capable of breathing in muddy waters. Other individuals of this *species* and those of the other *species* have not survived because they lack this capacity. The middling *species* has survived not because it had more variability but because some of its individuals are capable of surviving in muddy waters. The trait under *selection* is not degree of variability within *species*, but the breathing properties of individual fish.

Notice also that Gould's account is paradoxical. If the trait under *selection* were variability within *species*, *selection* would have reduced (rather than maintained or enhanced) the trait. At the end of the process, the middling *species* has less variability than before *selection*: only those few members able to breathe in muddy waters have survived. The other variability originally present in the *species* has now disappeared.

If this is the best "evidence" of *species selection* that Gould can marshal in his imagination in support of *species selection*, one may wonder about the emperor's new clothes. The monumental theoretical edifice built in *The Structure* crumbles on such flimsy foundations. The exuberance of verbal acrobatics comes to naught. Thirty years after it was first postulated as a pivot of PE theory, it remains to be demonstrated that *species selection* occurs at all. Moreover, the concept of *species selection* has not shown much, if any, heuristic value as a hypothesis guiding decisive observations or experiments, or as a theoretical construct adding to our understanding of the evolutionary process.

## The Spandrels of San Marco

Gould's second branch of Darwinian logic, *efficacy*, splits into two main subbranches that go in disparate directions. One is the argument that many features of organisms do not arise as the direct target of *natural selection*, but as "exaptations"—that is, as incidental consequences of the *evolution* of adapted features, just as the spandrels of the Basilica of San Marco in Venice were not created to depict the four evangelists (although they were used for this purpose), but came about because they are necessary architectural features to build a circular dome over a square base defined by columns at the four corners. The point that not all features of organisms come about by *natural selection* is well taken, but familiar and abundantly elaborated by evolutionists. Nobody would claim that the beating of the human heart came about as an *adaptation* because of its usefulness for ascertaining the state of health of an individual (by applying a stethoscope to the patient's chest and listening to the regularity of the beating). The spandrels argument has been salutary in recent evolutionary dialogue, because it has served as an antidote to the facile predisposition of some ethologists and evolutionary psychologists to attribute imagined functions to every trait, anatomical or behavioral, of an organism.

The other subbranch of *efficacy* points out to the historically determined constraints that frame the range within which an organism can evolve. A trivial but valid example is that the anatomy and physiology of elephants, determined by their evolutionary history, make it impossible for elephants to evolve wings that will enable them to fly. Gould treasures the story of the *Hox* genes and other amazing discoveries made by developmental geneticists in the previous two decades, advances that prompted the appearance of the subdiscipline known as evolutionary developmental biology, or "*evo-devo*" for short. There is little, if anything, that paleontology has contributed to these conceptual advances and empirical discoveries. Nevertheless, the new *evo-devo* knowledge fits well with Gould's emphasis on the significance of historically evolved morphological and functional constraints. One might perversely point out that Gould's emphasis here is all on the genes and organismal *selection*, and not at all on *species selection* or *species* interactions.

Nevertheless, Gould seeks to gain some advantage from these developments, and in so doing he overplays his hand. He attacks the great evolutionary geneticist Theodosius Dobzhansky for his theory of "adaptive peaks," because this grants too much ground to *natural selection* and the ecological landscape. Dobzhansky uses the example of cats and dogs, which exist as discrete types, with nothing in between, because cats and dogs exploit distinct life styles and ecological niches. Gould explains that the differentiation between cats and dogs is not due to the existence of distinct ecological niches, but rather it is histor-ically determined through the separate *evolution* of and inheritance from dog-like and cat-like ancestors. But *evolution* provides abundant evidence of ecological niches that impact the separate convergent *evolution* of distinctive adaptations. The cactus *family*, which evolved in the Americas, and the euphorbia *family*, which evolved in the Old World, encompass *species* with similar features which evolved as adaptations to the dry conditions of deserts and other arid climates. The two families had divergent ancestors, different in both lineages and different from the living *species* which are now more similar in morphology and functionality than their ancestral *species* were. Cat-like, dog-like, and other paired types of *species* have separately evolved among marsupials (in Australia and South America) and placental mammals (in the Old World and North America). *Species* pairs with similar morphologies and lifestyles (placental mammals first) include wolf—Tasmanian wolf; ocelot—*Dasyurus* cats; flying squirrel—flying phalanger; ground hog—wombat; *Myrmecophaga* anteater—*Myrmecobius* anteater; *Talpa* moles—*Notoryctes* moles; and the mice *Mus* and *Dasycerceus*.

## Hierarchy and Epistemological Emergence

The third principle of Darwinian logic to which Gould claims to have significantly contributed is *scope*, evolutionary change and diversification at the largest scales. This subject, central to Gould's professional interests as a paleontologist, is surprisingly treated much more briefly (Chapter 12, 48 pages) than the other two "agencies." Gould expresses exasperation with his own prolixity and even exhaustion: "And yet, as an epilog to this epilog and, honest to God, a true end to this interminable book, I risk a final state-ment" (Gould, 2002, p. 1340). Three-and-a-half pages later the book comes to an end.

Much of this final chapter defends the disciplinary autonomy of paleon-tology and macroevolutionary investigations. Gould uses two chief arguments: the occurrence of catastrophic extinctions caused by extraterrestrial phenom-ena, such as the meteoric impact of the K/T event; and the hierarchical organization of life. Both arguments, he says, invalidate any efforts of extrapolating genetic and other microevolutionary knowledge to account for macroevolutionary phenomena.

There can be little doubt, in my view, that *macroevolution* and *microevo-lution* investigations are theoretically "decoupled." If the issue is formulated in epistemological language, the matter is obvious.[11] The question of whether macroevolutionary theory can be derived from microevolutionary knowledge can only be answered in the negative, for the reasons I state in the paragraph written more than 20 years ago which Gould quotes and I cite in Note 5. But while the reasons I give are valid (one cannot decide among competing macroevolutionary theories on the basis of microevolutionary knowledge),

Gould's arguments are not compelling. His two chief arguments for autonomy are the *hierarchy* of living systems and the *emergence* of distinctive properties, which cannot be explained as "linear" extrapolations from one level of organization, such as the *gene*, to a higher level, such as the organism.

In Gould's words, the study of *evolution* embodies " a concept of hierarchy— a world constructed not as a smooth and seamless continuum, permitting simple extrapolation from the lowest level to the highest, but as a series of ascending levels, each bound to the one below it in some ways and independent in others... 'emergent' features not implicit in the operation of processes at lower levels, may control events at higher levels."[12] He adds that "the attendant need to reconceptualize trends and stabilities not as optimalities of *selection* upon organisms alone, but as outcomes of interactions among numerous levels of selection, implies an evolutionary world sufficiently at variance from Darwin's own conception that the resulting theory, although still 'selectionist' at its core, must be recognized as substantially different from current orthodoxy... I therefore devote the largest section of this book's second half (Chapters 8 and 9) to defining and defending this hierarchical theory of selection" (Gould, 2002, p. 168). "The hierarchical theory of selection recognizes many kinds of evolutionary individuals, banded together in a rising series of increasingly greater inclusion, one within the next—genes in cells, cells in organisms, organisms in demes, demes in species, species in clades... and we may choose to direct our evolutionary attention to any of the levels" (Gould, 2002, p. 674).

I agree with the thesis that macroevolutionary theories are not reducible to microevolutionary principles, but I argue that it is a mistake to ground this autonomy on the hierarchical organization of life, or purported emergent properties exhibited by higher-level units. The world, and not only the world of life, is hierarchically structured. There is a hierarchy of levels that go from subatomic particles to atoms, through molecules, organelles, cells, tissues, organs, multicellular individuals, and populations, to communities (see Chapter 10). Time adds another dimension of evolutionary hierarchy, with the interesting consequence that transitions from one level to another occur: as time proceeds the descendants of a single *species* may include separate *species*, genera, families, and so forth. But hierarchical differentiation of subject matter is neither necessary nor sufficient for the autonomy of scientific disciplines. It is not necessary because entities of a given hierarchical level can be the subject of diverse disciplines: cells are an appropriate subject of study for cytology, genetics, immunology, and so on. Even a single *event* can be the subject matter of several disciplines. My writing of this paragraph can be studied by a physiologist interested in the workings of muscles and nerves, by a psychologist concerned with thought processes, by a philosopher interested in the epistemological question at issue, and so on. Nor is the hierarchical differentiation of subject matter a *sufficient condition* for the autonomy of scientific disciplines: relativity theory obtains all the way from subatomic particles to planetary motions, and genetic laws apply to multicellular organisms as well as to cellular and even subcellular entities.

One reason alleged by Gould for the theoretical independence of levels within a hierarchy is the appearance of "emergent" properties, which "requires that a trait functioning in *species selection* be emergent at the *species* level" (Gould, 2002, p. 657). The question of *emergence* is an old one, particularly in discussions on the reducibility of biology to the physical sciences. The issue is, for example, whether the functional properties of the kidney are simply the properties of the chemical constituents of that organ. In the context of *macroevolution*, the question is: do *species* exhibit properties different from those of the individual organisms of which they consist? But questions about the *emergence* of properties are ill formed, or at least unproductive, because they can only be solved by definition.[13] The proper way of formulating questions about the relationship between complex systems and their component parts is by asking whether the properties of complex systems can be *inferred* from knowledge of the properties that their components have in isolation. As explained in Chapter 10, the issue of *emergence* cannot be settled by discussion about the "nature" of things or their properties, but is resolvable by reference to our *knowledge* of those objects.

Consider the following simple question raised in Chapter 10. Are the properties of common salt, sodium chloride, simply the properties of sodium and chlorine when they are associated according to the formula NaCl? If among the properties of sodium and chlorine I include their association into table salt and the properties of the latter, the answer is "yes"; otherwise, the answer is "no."[14] But this solution, then, is simply a matter of definition; and resolving the issue by a definitional maneuver contributes little to understanding the relationships between complex systems and their parts.

Is there a rule by which one could reasonably decide whether the properties of complex systems should be listed among the properties of their component parts? I think so. Assume that by studying the components in isolation we can infer the properties they will have when combined with other component parts in certain ways. In such a case, it would seem reasonable to include the "emergent" properties of the whole among the properties of the component parts. (Notice that this solution to the problem implies that a feature which may seem emergent at a certain time might not appear as emergent any longer at a more advanced state of knowledge.) Often, no matter how exhaustively an object (or component part) is studied in isolation, there is no way to ascertain the properties it will have in association with other objects (or component parts). We cannot infer the properties of ethyl alcohol, proteins, or human beings from the study of hydrogen, and thus it makes no good sense to list their properties among those of hydrogen. The important point, however, is that the issue of emergent properties is spurious and it needs to be reformulated in terms of propositions expressing our knowledge, not in terms of the properties of existing entities. It is legitimate to ask whether the *statements* concerning known properties of organisms (but not the properties themselves) can be logically deduced from statements concerning the properties of their physical components.

## THE AUTONOMY OF MACROEVOLUTION

The question of the autonomy of *macroevolution*, like other questions of reduction between scientific disciplines, can only be settled by empirical investigation of the logical consequences of propositions, and not by discussions about the "nature" of things or their properties. What is at issue is not whether the living world is hierarchically organized. It is. Nor is it at issue whether higher-level entities have emergent properties, which is a spurious question. The issue is whether, in a particular case, a set of *propositions* formulated in a defined field of knowledge (eg, *macroevolution*) can be derived from another set of propositions (eg, microevolutionary theory). Scientific theories consist, indeed, of propositions about the natural world. Only the investigation of the logical relations between propositions can establish whether or not one theory or branch of science is reducible to some other theory or branch of science. This implies that a discipline which is autonomous at a given stage of knowledge may become reducible to another discipline at a later time. The reduction of thermodynamics to statistical mechanics became possible (at least in some respects) only after it was discovered that the temperature of a gas bears a simple relationship to the mean kinetic energy of its molecules. The reduction of genetics to chemistry could not take place before the discovery of the chemical nature of the hereditary material. (I am not intimating that genetics can now be fully reduced to chemistry, but only that a partial reduction may be possible now, whereas it was not before the discovery of the structure and mode of replication of *DNA*.)

Microevolutionary processes, as presently known, are compatible with the two models of *macroevolution*—punctualism and *gradualism*. From microevolutionary knowledge, we cannot infer which one of those two macroevolutionary patterns prevails. The conflict between punctualism and *gradualism* is not the only macroevolutionary issue that cannot be decided by logical inference from microevolutionary principles. Many, if not most, macroevolutionary issues, those that distinctively engage the interest of paleontologists, are similarly autonomous—the likes of rates of morphological *evolution*, patterns of *species* extinctions, and historical factors regulating taxonomic diversity. The theories, models, and laws of *macroevolution* cannot be decided by logical inference from microevolutionary principles.

Consider, for example, the question of rates of morphological *evolution*. Three groups of crossopterygian fish flourished during the Devonian. The lungfish (Dipnoi) changed little for hundreds of millions of years and remain as relics. The coelacanths became highly successful in the open ocean until the

Cretaceous, then declined and stagnated, leaving only the relictual *Latimeria*. The rhipidistians, in contrast, evolved into amphibians, reptiles, and, finally, birds and mammals. Models to explain divergent rates of morphological *evolution* must incorporate factors other than microevolutionary principles, including rates of *speciation* and the environmental and biotic conditions that may account for successions of morphological change in some but not other lineages.

Distinctive macroevolutionary theories and models have been advanced concerning such issues as rates of morphological *evolution*, patterns of *species* extinctions, and historical factors regulating taxonomic diversity. The decision as to which one among alternative hypotheses is correct cannot be reached by recourse to microevolutionary principles. Such a decision must rather be based on appropriate tests with the use of macroevolutionary evidence. Thus *macroevolution* is an autonomous field of evolutionary study and, in this epistemologically very important sense, *macroevolution* is decoupled from *microevolution*.

The preceding statements do not imply, however, that macroevolutionary studies cannot be incorporated into the *synthetic theory* of *evolution*. Quite to the contrary, the modern theory of *evolution* is also called the "synthetic" theory of *evolution* precisely because it incorporates knowledge from diverse autonomous disciplines, such as genetics, ecology, *systematics*, and paleontology. The empirical and conceptual discoveries of modern paleontology contribute to the growth of evolutionary theory, much like new branches and incremental growth enlarge and luxuriate a tree, or the baroque-period additions to its Gothic fabric enrich Milan's Duomo, even if at some expense of congruity and simplicity.

Moreover, like the tree's new growth or the cathedral's late ornamentations, the theoretical accretions of Gould and others gain full cogency only as components of the full, preexisting structure. Population-level phenomena are fundamental to long-term *evolution*, because the populations in which macroevolutionary patterns are observed are the same populations that evolve at the microevolutionary level. Moreover, the study of microevolutionary phenomena *is* important to *macroevolution*, because any theory of *macroevolution* that is correct must be compatible with well-established microevolutionary principles and theories. In these two senses—identity at the level of events and compatibility of theories—*macroevolution* cannot be decoupled from *microevolution*.[15]

## Endnotes

1. Gould's language is combative in the extreme. He speaks of a "preemptive strike" (p. 31) against his enemies, their "destruction" (p. 33), their "jealousy" (p. 1021), and how ultimately "we won" (p. 1022). Depressingly frequent throughout the book are such words as "battle," "conflict," "retreat," "victory," and the like.

2. Gould's rich language and elegant metaphors are marred by redundancy, long elaboration, and repetition. The metaphor of a tree with three branches is redundant with that of a coral with three branches, reproduced from a 1670 engraving, and with that of a tripod supported by its three legs. These three metaphors are repeated and elaborated at length. The 1670 coral engraving is reproduced at nearly full-page size twice, identically on pages 18 and 97. Milan's Duomo, introduced as an architectural structure that acquired ornamental and other features centuries after it was built, serves as a metaphor for later elaborations of the fundamental Darwinian logic. The first time this metaphor appears is belabored over five pages (pp. 2–6) and illustrated with two photographs of the cathedral. These two photos, the duplicated coral engraving, plus another 1670 engraving representing two human figures with shells, are the only illustrations for the first 182 pages of the book. The following two paragraphs may serve as examples of Gould's literary style, eloquence, and prolixity:"The specific form of the image—its central metaphorical content, if you will—plays an important role in channeling or misdirecting our thoughts, and therefore also requires careful consideration. In the text of this book, I speak most often of a 'tripod' since central Darwinian logic embodies three major propositions that I have always visualized as supports—perhaps because I have never been utterly confident about this entire project, and I needed some pictorial encouragement to keep me going for 20 years. (And I much prefer tripods, which can hold up elegant objects, to buttresses, which may fly as they preserve great Gothic buildings, but which more often shore up crumbling edifices. Moreover, the image of a tripod suits my major claim particularly well—for I have argued, just above, that we should define the 'essence' of a theory by an absolutely minimal set of truly necessary propositions. No structure, either of human building or of abstract form, captures this principle better than a tripod, based on its absolute minimum of three points for fully stable support in the dimensional world of our physical experience.)" (p. 15)."Galton's Polyhedron, the metaphor and model devised by Darwin's brilliant and eccentric cousin Francis Galton, and then fruitfully used by many evolutionary critics of Darwinism, including St. George Mivart, W.K. Brooks, Hugo de Vries, and Richard Gold-schmidt, clearly expresses the two great, and both logically and historically conjoined, themes of formalist (or structuralist, or internalist, in other terminologies) challenges to functionalist (or adaptationist, or externalist) theories in the Darwinian tradition. This model of *evolution* by facet-flipping to limited possibilities of adjacent planes in inherited structure stresses the two themes—channels set by internal constraint, and evolutionary transition by discontinuous saltation—that structuralist alternatives tend to embrace and that pure Darwinism must combat as challenges to basic components of its essential logic (for channels direct the pathways of evolutionary change from the inside, albeit in potentially positive and adaptive ways, even though some external force, like natural selection, may be required as an initiating impulse; whereas saltational change violates the Darwinian requirement for selection's creativity by vesting the scope and direction of change in the nature and magnitude of internal jumps, and not in sequences of adaptive accumulation mediated by natural selection at each step" (p. 66).

3. A religious architecture metaphor for replacement would have been the splendid Gothic cathedral of Leon, Spain. The 14th-century wealth and exuberant ambition of the citizens of the then capital of Castile moved them to demolish the preexisting 12th-century Romanesque cathedral and build a much larger and taller Gothic cathedral on the same location. This replacement of Romanesque by Gothic cathedrals was common through 14th- and 15th-century Christendom. (How much wiser were the citizens of Salamanca in Spain, who left the Romanesque cathedral standing and built a new Gothic cathedral next to it—which, admirable architectural masterpiece that it is, is nevertheless aesthetically surpassed in my judgment by the earlier Romanesque monument!)

4. Eldredge, N., 1971. The allopatric model and phylogeny in Paleozoic invertebrates. Evolution 25, 156−167; Eldredge, N., Gould, S.J. 1972. Punctuated equilibria: an alternative to phyletic gradualism. In: Schopf, T.J.M. (Ed.), Models in Paleobiology. Freeman, Cooper, Co., pp. 82−115.

5. Gould refers to me, with kind words, as supporting this claim of macroevolutionary autonomy and quotes me at length: "I have particularly appreciated the fairness of severe critics who generally oppose punctuated equilibrium, but who freely acknowledge its legitimacy as a potentially important proposition with interesting implications, and as a testable notion that must be adjudicated in its own macroevolutionary realm. Ayala (1983) has been especially clear and gracious on this point.""If macroevolutionary theory were deducible from microevolutionary principles, it would be possible to decide between competing macroevolutionary models simply by examining the logical implications of microevolutionary theory. But the theory of population genetics is compatible with both punctualism and gradualism; and, hence, logically it entails neither. Whether the tempo and mode of evolution occur predominantly according to the model of punctuated equilibria or according to the model of phyletic gradualism is an issue to be decided by studying macroevolutionary patterns, not by inference from microevolutionary processes. In other words, macroevolutionary theories are not reducible (at least at the present state of knowledge) to microevolution. Hence, macroevolution and microevolution are decoupled in the sense (which is epistemologically most important) that macroevolution is an autonomous field of study that must develop and test its own theories" (Ayala, 1983, as quoted by Gould, 2002, p. 1023).To avoid misunderstanding, I summarize here issues that I have discussed at some length in the paper cited by Gould (p. 1023) and elsewhere. *Macroevolution* and *microevolution* are *not* decoupled in two senses: identity at the level of events and compatibility of theories. First, the populations in which macroevolutionary patterns are observed are the same populations that evolve at the microevolutionary level. Second, macroevolutionary phenomena can be accounted for as the result of known microevolutionary processes. That is, the theory of PE (as well as the theory of *phyletic gradualism*) is consistent with the theory of *population* genetics. Indeed, any theory of *macroevolution* that is correct must be compatible with the theory of *population* genetics, to the extent that this is a well-established theory. The decoupling asserted by me in the quotation cited above by Gould, as well as later in his book, concerns *epistemology*: the logical autonomy of theories.

6. See, for example, Mayr, E., 1963. Animal Species and Evolution. Harvard University Press, Cambridge, MA; Dobzhansky, T., Ayala, F.J., Stebbins, G.L., Valentine, J.W. 1977. Evolution. W.H. Freeman & Co., San Francisco.
7. Mayr, 1963; Dobzhansky, T., 1970. Genetics of the Evolutionary Process. Columbia University Press, New York; Nevo, E., Shaw, C.R., 1972. Genetic variation in a subterranean mammal, Spalax ehrenbergi. Biochemical Genetics 7, 235−241; Dobzhansky et al., 1977; White, M.J.D., 1978. Modes of Speciation. W.H. Freeman, San Francisco; Benado, M., Aguilera, M., Reig, O.A., Ayala, F.J., 1979. Biochemical genetics of Venezuelan spiny rats of the *Proechimys guainae* and *Proechimys trinitatis* superspecies. Genetics 50, 89−97.
The sibling *species* of greatest interest to evolutionists are not recently evolved, but rather *species* that diverged millions of years ago and remain morphologically indistinguishable. For example, several among the closest pairs of sibling *species* of the *Drosophila melanogaster* subgroup diverged from each other about 2 million years ago; other sibling *species* of this subgroup diverged more than 5 million years ago. Sibling *species* exemplify two significant realities of the evolutionary process, namely that *speciation* does not necessarily involve morphological change (the point I am making here, thus contradicting one of the basic claims of PE), and that *species* can persist for millions of years without morphological change—the common and well-known

phenomenon of "*stasis*," which is claimed as the second distinctive component of PE theory. As a long-term student of several groups of sibling *species* of *Drosophila* and other organisms, I must admit to being underwhelmed by both PE claims: the claim of morphological change as a common or even necessary concomitant of *speciation*, because it is false; and the assertion that lineages may remain unchanged for long evolutionary periods, because it was a well-known phenomenon years before the PE theory was formulated.

8. Simpson, G.G., 1944. Tempo and Mode in Evolution. Columbia University Press, New York, p. 197.

9. Selected by the eminent paleontologist James W. Valentine, a supporter of PE, for his Chapter 10 (p. 329) in Dobzhansky et al. (1977).

10. This claim is, however, refuted by the phenomenon commonly observed by *population* geneticists that noticeable morphological change can occur by gradual *gene* substitution impelled by *natural selection*. For an example of what is a ubiquitously observed phenomenon, see Gilchrist, G.W., Huey, R.B., Balanyà, J., Pascual, M., Serra, L., 2004. A time series of evolution in action: a latitudinal cline in wing size in South American *Drosophila subobscura*. Evolution 58, 768−780. Notice also the great morphological diversification of *Drosophila species* in the island of Hawaii over a time span somewhat shorter than a million years.

11. See note 5.

12. Gould, S.J., 1980. Is a new general theory of evolution emerging? Paleobiology 6, p. 121.

13. See Chapter 10 and Ayala, F.J. 1983. Beyond Darwinism? The challenge of macroevolution to the synthetic theory of evolution. In: Asquith. P.D., Nickles, T. (Eds.), PSA 1982, vol. 2, Philosophy of Science Association: East Lansing, Michigan, pp. 275−291.

14. I will return to the culinary *analogy* formulated in Chapter 10 about the Spanish gazpacho soup. The question I have just asked would be: are the flavors of gazpacho the same as the flavors of its components? As in the corresponding cases of table salt and *macroevolution*, the answer would be "yes" if among the components' flavors we include the flavors they yield when suitably combined with the other components in gazpacho soup. But if we cannot predict gazpacho's flavors from what we know by tasting each component separately, the appropriate answer would be "no." To say that table salt is *nothing else* than sodium and chlorine (or similarly for macroevolutionary processes or gazpacho) is to commit the *nothing-but* fallacy.

15. A fitting architectural metaphor of *The Structure of Evolutionary Theory* and, more generally, Gould's contribution to evolutionary theory is the gorgeous Portada del Obradoiro, the western façade of the magnificent cathedral of Santiago de Compostela (*campus stellae*, "meadow of the stars") in the northwest corner of Spain, one of the largest and most beautiful Romanesque cathedrals ever built. Santiago's cathedral was built between 1075 and 1128, under the successive direction of Maestro Bernardo "the older," Maestro Roberto, and Maestro Bernardo "the younger," of huge dimensions, suitable to accommodate the thousands of pilgrims who attended mass and other religious services after their months-long pilgrimage from all parts of Europe. The cathedral was built over the popularly believed burial place of the apostle Santiago, Jesus' cousin and disciple. Santiago's tomb was the most important destination of Christian pilgrimage through much of the Middle Ages, when Jerusalem and the Holy Land were not accessible to Christians owing to Muslim occupation. The huge Obradoiro façade, a splendid example of Spanish baroque, was built around 1740 under the direction of the architect Casas y Nóvoa. The heavily ornamental Portada del Obradoiro dominates the large Plaza del Obradoiro where the pilgrims gathered, some newly arrived, others emerging from the elegant hospital that Ferdinand and Isabella had donated to attend pilgrims in need, which dominates the north side of the square. After the mid-18th century the pilgrims entered the cathedral, as they do now, through the Obradoiro gates and found themselves facing another

façade, also of magnificent scale if somewhat smaller, the Pórtico de la Gloria, which was completed around 1188 under the direction of the great sculptor Maestro Mateo. This Romanesque façade, of arresting beauty, consists of three pointed arches framed by splendid sculptures of Santiago and other apostles, prophets, and saints. Now, as in past centuries, pilgrims, after crossing the Obradoiro façade, pause in the atrium behind it, kiss the feet of the saint at the center of the Pórtico de la Gloria, and enter the Romanesque cathedral, which was not altered during the construction of the Obradoiro façade or later. *The Structure of Evolutionary Theory* is, like the Obradoiro façade, an enormous construction of considerable beauty, behind which stands the theory of *evolution*, which, like the Romanesque cathedral, has lost nothing of its magnificence in spite of the façade in front of it. The Romanesque cathedral of Santiago de Compostela and its baroque façade added centuries later are a more apposite metaphor of the theory of *evolution* and Gould's theoretical constructs than the Gothic Duomo of Milan and its baroque accretions.

# Part III

# Ethics, Aesthetics, and Religion

# Chapter 12

# Ethics

## INTRODUCTION

*Ethics* is a human universal. People have moral values: that is, they accept
standards according to which their conduct is judged either right or wrong,
good or evil. The particular norms by which moral actions are judged vary to
some extent from individual to individual and from culture to culture (although
some norms, like do not kill, do not steal, and honor one's parents, are
widespread and perhaps universal), but value judgments concerning human
behavior are made in all cultures. This universality raises the questions of
whether the moral sense is part of human nature, one more dimension of our
biological makeup, and whether ethical values may be the product of biological *evolution* rather than being prescribed by religious and cultural
traditions.

I define moral behavior for the present purposes as the actions of a person
when taking into account in a sympathetic way the impact the actions have on
others. A related definition is advanced by David Copp in *The Oxford
Handbook of Ethical Theory* (2006, p. 4): "we can take a person's moral beliefs to be the beliefs she has about how to live her life when she takes into
account in a sympathetic way the impact of her life and decisions on others."
The related concept of *altruism* may be defined as "unselfish regard for or
devotion to the welfare of others" (*Webster's New Collegiate Dictionary*, tenth
ed., 1998). *Altruism*, however, is usually taken to imply some cost to the
altruist for the benefit of others, and this is the sense in which "altruism" is
used here. I use the term "ethical behavior" as a synonym for "moral
behavior," and "morality" and "ethics" as synonyms of each other, except
when explicitly noted or contextually obvious that they are used with a
somewhat different meaning. Some authors use "morality" or "virtue ethics"
in a broader sense that includes good feelings in regard to others and excludes
inappropriate thoughts or desires, such as entertaining sexual desires for
somebody else's wife or wishing that something harmful would happen to
others. So long as these thoughts or desires are not transformed into actions,
they are not included in the sense of "morality" used here. Actions that may be
thought to be evil or sinful in some moral systems, such as masturbation, are

Evolution, Explanation, Ethics, and Aesthetics. http://dx.doi.org/10.1016/B978-0-12-803693-8.00012-8

not included either as moral behaviors in the sense used here, so long as the actions have no consequences for others.

There are many theories concerned with the rational grounds for *morality*, such as deductive theories seeking to discover the axioms or fundamental principles that determine what is morally correct on the basis of direct moral intuition, and theories like logical positivism or existentialism that negate rational foundations of *morality*, reducing moral principles to social decisions or emotional and other irrational grounds. After the publication of Darwin's theory of evolution by natural selection, several philosophers as well as scientists attempted to find in the evolutionary process the justification for moral behavior.

Aristotle and other philosophers of classical Greece and Rome, as well as many other philosophers throughout the centuries, held that humans hold moral values by nature. A human is not only *Homo sapiens*, but also *Homo moralis*. But biological *evolution* brings about two important issues: timing and causation. We do not attribute ethical behavior to animals (surely, not to all animals and not to the same extent as to humans, in any case). Thus *evolution* raises distinct questions about the origins and tenets of moral behavior. When did ethical behavior come about in human evolution? Did modern humans have an ethical sense from the beginning? Did Neanderthals hold moral values? What about *Homo erectus* and *Homo habilis*? And how did the moral sense evolve? Was it directly promoted by natural selection? Or did it come about as a by-product of some other attribute (such as rationality) that was the direct target of selection? Alternatively, is the moral sense an outcome of *cultural evolution* rather than of biological evolution?

## DARWIN AND THE MORAL SENSE

Two years after returning from his trip in HMS *Beagle* (1826–1831), Darwin began gathering contemporaneous literature on human moral behavior, including such works as William Paley's *The Principles of Moral and Political Philosophy* (1785), which he had encountered earlier while a student at the University of Cambridge, and the multivolume *Illustrations of Political Economy* by Harriet Martineau, published in 1832–1834. These two authors, like other philosophers of the time, maintained that *morality* was a conventional attribute of humankind rather than a naturally determined human attribute, using an argument often exploited in our days: the diversity of moral codes.

The proliferation of ethnographic voyages had brought to light the great variety of moral customs and rules. This variation is something Darwin had observed in South America among the Indian populations. But this apparent dispersion had not distracted him. He eventually developed a more complex and subtle theory of the moral sense than those contemporaneous authors, a theory that, implicitly at least, recognized moral behavior as a biologically determined human universal but with culturally evolved differences. For Darwin, the ethnographic diversity of moral customs and rules came about as an adaptive response to the environmental and historical conditions, unique in every different place, without necessarily implying that *morality* was an acquired, rather than natural, human trait.

A variable adaptive response could very well derive from some fundamental capacity, a common substrate, unique for the whole human race, but capable of becoming expressed in diverse directions. Darwin did not attribute the universality of *morality* to supernatural origin, but rather saw it as a product of *evolution* by *natural selection*. The presence of a universal and common foundation, endowing humans with an ethical capacity, was for Darwin compatible with different cultures manifesting different stages of moral *evolution* and with different sets of moral norms.

Darwin's first sustained discussion of *morality* is in *The Descent of Man* (1871, pp. 67–102). His two most significant points concerning the *evolution* of *morality* are stated early in Chapter 3 of the book: that moral behavior is a necessary attribute of advanced intelligence, as it occurs in humans, and thus moral behavior is biologically determined; and that the norms of *morality* are not biologically determined but a result of human collective experience, or human culture as we would now call it.

Darwin asserts that the moral sense is the most important difference "between man and the lower animals": "I fully subscribe to the judgment of those writers who maintain that of all the differences between man and the lower animals, the moral sense or conscience is by far the most important" (Darwin, 1871, p. 62). Darwin belongs to an intellectual tradition, originating in the Scottish Enlightenment of the 18th and 19th centuries, which identifies the moral sense as a behavior based on sympathy that leads human behavior.

Darwin then proceeds to assert his view that moral behavior is strictly associated with advanced intelligence: "The following proposition seems to me in a high degree probable—namely, that any animal whatever, endowed with well-marked social instincts, would inevitably acquire a moral sense or conscience, as soon as its intellectual powers had become as well developed, or nearly as well developed, as in man" (Darwin, 1871, pp. 68–69). This is a hypothetical issue, because no other animal has ever reached the level of human mental faculties, language included. But it is an important statement, because Darwin is affirming that the moral sense, or conscience, is a necessary consequence of high intellectual powers, such as exist in modern humans. Therefore, if our intelligence is an outcome of *natural selection*, the moral sense would also be an outcome of *natural selection*. Darwin's statement further implies that the moral sense is not by itself directly promoted by *natural selection* but only indirectly as a necessary consequence of high intellectual powers, which are the attributes that *natural selection* is directly promoting.

## MORAL BEHAVIOR VERSUS MORAL NORMS

Darwin states that even if some animal could achieve a human-equivalent degree of *development* of its intellectual faculties, we cannot conclude that it would also acquire exactly the same moral sense as ours. "I do not wish to maintain that any strictly social animal, if its intellectual faculties were to become as active and as highly developed as in man, would acquire the same moral sense as ours... they might have a sense of right and wrong, though led by it to follow widely different lines of conduct" (Darwin, 1871, p. 70). These statements imply that, according to Darwin, having a moral sense does not by itself determine what the moral norms would be: which sorts of actions might by sanctioned by the norms and which would be condemned.

This distinction is important—indeed, it is central to the theory I advance herein. Much of the historical controversy, particularly between scientists and philosophers, as to whether the moral sense is or is not biologically determined has arisen owing to a failure to make the distinction. Scientists often affirm that *morality* is a human biological attribute because they are thinking of the predisposition to pass moral judgment: that is, to judge some actions as good and others as evil. Some philosophers, such as William Paley and Harriet Martineau, cited earlier, as well as many contemporary philosophers argue that *morality* is not biologically determined, but rather comes from cultural traditions or religious beliefs, because they are thinking about moral codes, the sets of norms that determine which actions are judged to be good and which are evil. They point out that moral codes vary from culture to culture, and therefore are not biologically predetermined.

I consider this distinction fundamental (Ayala, 1987b, 2007, 2010). Thus, I argue that the question of whether ethical behavior is biologically determined may refer to either one of the two issues. First, is the capacity for *ethics*—the proclivity to judge human actions as either right or wrong—determined by the biological nature of human beings? Second, are the systems or codes of ethical norms accepted by humans biologically determined? A similar distinction can be made with respect to language. The question of whether the capacity for symbolic creative language is determined by our biological nature is different from the question of whether the particular language we speak—English, Spanish, Chinese, etc.—is biologically determined, which in the case of language obviously is not.

The distinction between the *predisposition* to judge certain sorts of actions as morally either good or evil and the *norms* according to which we determine which actions are good and which are evil has affinity with the distinction made by moral philosophers between *metaethics* and *normative ethics*. The subject of metaethics is why we ought to do what we ought to do, while normative *ethics* tells us what we ought to do. I propose that the moral evaluation of actions emerges from human rationality, or, in Darwin's terms, from our highly developed intellectual powers. Our high intelligence allows us to anticipate the consequences of our actions with respect to other people, and thus to judge them as good or evil in terms of their consequences for others. But I propose that the norms according to which we decide which actions are good and which are evil are largely culturally determined, although conditioned by biological predispositions.

## DARWINIAN AFTERMATH

Herbert Spencer (1820–1903) was among the first philosophers to seek the grounds of *morality* in biological *evolution*. In *The Principles of Ethics* (1893), Spencer seeks to discover values that have a natural foundation. He argues that the theory of organic *evolution* implies certain ethical principles. Human conduct must be evaluated, like any biological activity whatsoever, according to whether it conforms to the life process; therefore, any acceptable moral code must be based on *natural selection*, the law of struggle for existence. According to Spencer, the most exalted form of conduct is that which leads to a greater duration, extension, and perfection of life; the *morality* of all human actions must be measured by that standard. Spencer proposes that, although exceptions exist, the general rule is that pleasure goes with that which is biologically useful, whereas pain marks what is biologically harmful. This is an outcome of *natural selection*; thus, while doing what brings them pleasure and avoiding what is painful, organisms improve their chances for survival. With respect to human behavior, we see that we derive pleasure from virtuous behavior and pain from evil actions—associations which indicate that the *morality* of human actions is also founded on biological nature.

Spencer proposes as the general rule of human behavior that anyone should be free to do anything they want, so long as it does not interfere with the similar freedom to which others are entitled. The justification of this rule is found in organic evolution: the success of an individual, plant or animal, depends on its ability to obtain that which it needs. Consequently, Spencer reduces the role of the state to protecting the collective freedom of individuals so that they can do as they please. This *laissez-faire* form of government may seem ruthless, because individuals would seek their own welfare without any consideration for others (except for respecting their freedom), but Spencer believes that it is consistent with traditional Christian values. It may be added that although Spencer sets the grounds of *morality* on biological nature and nothing else, he admits that certain moral norms go beyond that which is biologically determined; these are rules formulated by society and accepted by tradition.

Social Darwinism, in Spencer's version or some variant form, was fashionable in European and American circles during the latter part of the 19th century and the early years of the 20th century, but it has few or no distinguished intellectual followers at present. Spencer's critics include the evolutionists Julian Huxley and C.H. Waddington, who nevertheless maintain that organic *evolution* provides grounds for a rational justification of ethical codes. For Julian Huxley (1953; Huxley and Huxley, 1947), the standard of *morality* is the contribution that actions make to evolutionary *progress*, which goes from less to more "advanced" organisms. For Waddington (1960), the *morality* of actions must be evaluated by their contribution to human *evolution*.

Huxley's and Waddington's views are based on value judgments about what is or is not progressive in *evolution*. But, contrary to Huxley's claim, there is nothing objective in the evolutionary process itself (ie, outside human considerations; see Ayala, 1982, 1988, 2007) that makes the success of bacteria, which have persisted as such for more than 2 billion years and consist of a huge diversity of *species* and astronomic numbers of individuals, less desirable than that of the vertebrates, even though the latter are more complex. The same objection can be raised against Waddington's human *evolution* standard of biological *progress*. Are insects, of which more than 1 million *species* exist, less desirable or less successful from a purely biological perspective than humans or any other mammal species? Waddington fails to demonstrate why the promotion of human biological *evolution* by itself should be the standard to measure what is morally good.

More recently, numerous philosophers as well as scientists have sought to give accounts of moral behavior as an evolutionary outcome (eg, Ayala, 1974, 2010, 2016; Blackmore, 1999; Hauser, 2006; Maienschein and Ruse, 1999; Ruse, 1995; Sober and Wilson, 1998; Wilson, 2012). Particularly notable are the contributions of Edward O. Wilson (1975, 1978, 1998), founder of *sociobiology* as an independent discipline engaged in discovering the biological foundations of all social behavior. Wilson and other sociobiologists, as well as the derivative subdisciplines of evolutionary psychology (eg, Barkow et al., 1992) and memetics (Blackmore, 1999), have sought to solve the naturalistic fallacy by turning it on its head. They assert that moral behavior does not exist as something distinct from biological, or biologically determined, behavior. As Ruse and Wilson (1985) asserted, "Ethics is an *illusion* put in place by natural selection to make us good cooperators" (p. 50). I return later to these sociobiological and related proposals.

## MORAL BEHAVIOR AS RATIONAL BEHAVIOR

The first proposition I advance here, fully consistent with Darwin's ideas, is that humans, because of their high intellectual powers, are necessarily inclined to make moral judgments and accept ethical values; that is, to evaluate certain kinds of actions as either right or wrong. The claim made is that moral behavior is a necessary outcome of the biological makeup of humans, a product of their *evolution*. This view falls within the metaethical theories known as deontological or rational. It is the exalted degree of rationality that we humans have achieved that makes us moral beings. Humans are *H. moralis* because they are *Homo rationalis*.

This thesis does not imply that the norms of *morality* are also biologically determined or are unambiguous consequences of our rationality. Independent of whether or not humans have a biologically determined moral sense, it remains to be ascertained whether particular moral prescriptions are in fact determined by the biological nature of humans, or whether they are products of *cultural evolution*, be these chosen by a society or established by religious beliefs, or even selected according to individual preferences. Even if we were to conclude that people cannot avoid having moral standards of conduct, it might be that the choice of the particular standards used for judgment would be arbitrary or a product of *cultural evolution*. The need for having moral values does not necessarily tell us what the moral values should be, just as having the capacity for language does not determine which language we shall speak.

I first argue that humans are ethical beings by their biological nature: that humans evaluate their behavior as either right or wrong, moral or immoral, as a consequence of their eminent intellectual capacities, which include self-awareness and abstract thinking. These intellectual capacities are products of the evolutionary process, but they are distinctively human. Thus I assert that ethical behavior is not causally related to the social behavior of animals, including *kin selection* and the so-called "reciprocal altruism."

A second argument put forward here is that the moral norms according to which we evaluate particular actions as either morally good or morally bad (as well as the grounds that may be used to justify these norms) are products of cultural, not biological, *evolution*. The norms of *morality* belong, in this respect, to the same *category* of phenomena as political and religious institutions, or the arts, sciences, and technology, as well as the particular languages we speak. The moral codes, like these other products of human culture, are often consistent with the biological predispositions of the human *species*. But many moral norms are formulated independently of biological necessity or predisposition, simply because they do not have necessary biological consequences. Biological welfare (survival and reproduction) is not obviously a determinant of all ethical norms in any given society or culture.

Moral codes, like any other cultural system, depend on the existence of human biological nature and must be consistent with it, in the sense that they could not counteract it without promoting their own demise. Moreover, the acceptance and persistence of moral norms are facilitated when they are consistent with biologically conditioned human behaviors. But moral norms are independent of such behaviors in the sense that some norms may not favor, and may hinder, the survival and reproduction of the individual and its genes, which processes are the targets of biological *evolution*. Discrepancies between accepted moral rules and biological survival are, however, necessarily limited in scope or they would otherwise lead to the extinction of the groups accepting such discrepant rules.

## CONDITIONS FOR ETHICAL BEHAVIOR

We may refer to the *moral sense* in its strict meaning as the evaluation of some actions as virtuous, or morally good, and others as evil, or morally bad. *Morality* in this sense is the urge or predisposition to judge actions as either right or wrong in terms of their consequences for other human beings. In this sense, humans are moral beings by nature because their biological constitution determines the presence in them of the three necessary conditions for ethical behavior: the ability to anticipate the consequences of one's own actions, the ability to make value judgments, and the ability to choose between alternative courses of action. These abilities exist as a consequence of the eminent intellectual capacity of human beings. Notice, as discussed later, that the position taken here is not the one known as "utilitarianism," because it has not been claimed that maximizing the benefits to others, and to as many others as possible, is the ultimate standard by which the *morality* of actions should be determined.

The ability to anticipate the consequences of one's own actions is the most fundamental of the three conditions required for ethical behavior. Only if I can anticipate that pulling the trigger will fire the bullet, which in turn will strike and kill my enemy, can the action of pulling the trigger be evaluated as nefarious. Pulling a trigger is not in itself a moral action; it becomes so by virtue of its relevant consequences. My action has an ethical dimension only if I do anticipate these consequences.

The ability to anticipate the consequences of one's actions is closely related to the ability to establish the connection between means and ends: that is, of seeing a means precisely as a means, as something that serves a particular end or purpose. This ability to establish the connection between means and their ends requires the ability to anticipate the future and form mental images of realities not present or not yet in existence.

The ability to establish the connection between means and ends happens to be the fundamental intellectual capacity that has made possible the *development* of human culture and technology. An evolutionary scenario, seemingly the best hypothesis available, proposes that the remote evolutionary roots of this capacity to connect means with ends may be found in the *evolution* of bipedalism, which transformed the anterior limbs of our ancestors from organs of locomotion into organs of manipulation. The hands thereby gradually became organs adept for the construction and handling of objects for hunting and other activities that improved survival and reproduction: that is, which increased the reproductive *fitness* of their carriers. Eventually, our ancestors of about 2 million years ago advanced from just using as tools existing objects, such as a stone or a wooden stick, to make themselves tools that they could use. The construction of tools depends not only on manual dexterity, but on perceiving them precisely as tools, as objects that help to perform certain actions: that is, as means that serve certain ends or purposes—a knife for cutting, an arrow for hunting, an animal skin for protecting the body from the cold. According to this evolutionary scenario, *natural selection* promoted the intellectual capacity of our bipedal ancestors because increased intelligence facilitated the perception of tools as tools, and therefore their construction and use, with the ensuing amelioration of biological survival and reproduction.

The *development* of the intellectual abilities of our ancestors took place over several million years, gradually increasing the ability to connect means with their ends and, hence, the possibility of making ever more complex tools serving more remote purposes. According to the theory proposed here, the ability to anticipate the future, essential for ethical behavior, is therefore closely associated with the *development* of the ability to construct tools—an ability that has produced the advanced technologies of modern societies and is largely responsible for the success of humans as a biological *species*.

The second condition for the existence of ethical behavior is the ability to make value judgments, to perceive certain objects or deeds as more desirable than others. Only if I can see the death of my enemy as preferable to his survival (or vice versa) can the action leading to his demise be thought of as moral. If the consequences of alternative actions are neutral with respect to value, an action cannot be characterized as ethical. Values are of many sorts: not only ethical, but also aesthetic, economic, gastronomic, political, and so on. But in all cases, the ability to make value judgments depends on the capacity for abstraction: that is, on the capacity to perceive actions or objects as members of general classes. This makes it possible to compare objects or actions with one another and to perceive some as more desirable than others. The capacity for abstraction requires an advanced intelligence such as exists in humans, and in them alone.

As noted earlier, the model advanced here does not necessarily imply the ethical theory known as *utilitarianism* (or, more generally, consequentialism). According to so-called "act consequentialism," the rightness of an action is determined by the value of its consequences, so the morally best action in a particular situation is the one the consequences of which would have the most benefit to others. I propose that the *morality* of an action depends on our ability to anticipate the consequences of our actions, and make value judgments about such consequences; but it is not asserted that the *morality* of actions is exclusively measured in terms of how beneficial their consequences will be to others.

The third condition necessary for ethical behavior is the ability to choose between alternative courses of actions. Pulling the trigger can be a moral action only if you have the option not to pull it. A necessary action beyond conscious control is not a moral action: the circulation of the blood or the process of food digestion is not a moral action. Whether there is free will is a question much discussed by philosophers, and the arguments are long and involved (for example, Ruse, 2006). Here, I advance two considerations that are commonsense evidence of the existence of free will. One is personal experience, which indicates that the possibility to choose between alternatives is genuine rather than only apparent. The second consideration is that when we confront a given situation that requires action on our part, we are mentally able to explore alternative courses of action, thereby extending the field within which we can exercise our free will.

In any case, if there were no free will, there would be no ethical behavior; *morality* would only be an illusion. A point to be made, however, is that free will is dependent on the existence of a well-developed intelligence, which makes it possible to explore alternative courses of action and to choose one or another in view of the anticipated consequences.

## ADAPTATION OR EXAPTATION?

We can now consider explicitly two issues that are largely implicit in the previous section. I earlier proposed that the moral sense emerges as a necessary implication of our high intellectual powers, which allow us to anticipate the consequences of our actions and evaluate such consequences. But is it the case that the moral sense may have been promoted by *natural selection* in itself, and not only indirectly as a necessary consequence of our exalted intelligence? The question in evolutionary terms is whether the moral sense is an *adaptation* or, rather, an *exaptation*. Evolutionary biologists define exaptations as features of organisms that evolve because they serve some particular function, but are later coopted to serve a different function which was not originally the target of *natural selection*. The new function may replace the older function or coexist with it. Feathers seem to have evolved first for conserving temperature, but were later coopted in birds for flying. The beating of the human heart is an *exaptation* used by doctors to diagnose the state of health, although this is not why it evolved in our ancestors. The issue at hand is whether moral behavior was directly promoted by *natural selection* or rather whether it is a consequence of our exalted intelligence, which was the target of *natural selection* because it made possible the construction of tools. Art, literature, religion, and many other human cultural activities might also be seen as exaptations that came about as consequences of high intelligence and tool making.

The second issue is whether some animals, such as apes or other nonhuman primates, may have a moral sense, however incipient, either as directly promoted by *natural selection* or as a consequence of their own intelligence.

The position I argue here is that the human moral sense is an *exaptation*, not an *adaptation*. The moral sense consists of *judging* certain actions as either right or wrong, not of choosing and carrying out some actions rather than others, or evaluating them with respect to their practical consequences. It seems unlikely that making moral judgments would promote the reproductive *fitness* of those judging an action as good or evil. Nor does it seem likely that there might be some form of "incipient" ethical behavior that would then be further promoted by *natural selection*. The three necessary conditions for there being ethical behavior are manifestations of advanced intellectual abilities.

Indeed, it rather seems that the target of *natural selection* was the *development* of advanced intellectual capacities. This was favored by *natural selection* because the construction and use of tools improved the strategic position of our biped ancestors. In the account I advance here, once bipedalism evolved and after tool making and using became practical, those individuals more effective in these functions had a greater probability of biological success. The biological advantage provided by the *design* and use of tools persisted long enough so that intellectual abilities continued to increase, eventually yielding the eminent *development* of intelligence that is characteristic of *H. sapiens*.

## ALTRUISM AND GROUP SELECTION

A related question is whether *morality* would benefit a social group within which it is practiced, and, indirectly, individuals as members of this group. This seems likely to be the case if indeed moral judgment would influence individuals to behave in ways that increase cooperation or benefit the welfare of the social group in some way: for example, by reducing crime or protecting private property. This brings about the issue of whether there is in humans "group selection," and the related issues of *kin selection* and inclusive *fitness*, which I discuss later.

Altruistic behavior is generally not favored within a particular animal *population* because mutations that adopt selfish over altruistic behavior will be favored by *natural selection* within the *population*, so selfish alleles may drive out altruistic alleles. An altruist incurs a cost to benefit others; a selfish individual benefits from the altruist's behavior without incurring any cost (see later). However, it might be the case that populations with a preponderance of altruistic alleles will survive and spread better than populations consisting of selfish alleles. This would be *group selection*. But typically there are many more individual organisms than there are populations, and individuals are born, procreate, and die at rates much higher than populations. Thus the rate of multiplication of selfish individuals over altruists is likely to be much higher than the rate at which altruistic populations multiply relative to predominantly selfish populations.

There is, however, an important difference between animals and humans that is relevant in this respect. Namely, the *fitness* advantage of selfish over altruistic behavior does not apply to humans, because humans can *understand* the benefits of altruistic behavior (to the group and indirectly to themselves), and thus adopt *altruism* and protect it, by laws or otherwise, against selfish behavior that harms the social group. As Darwin wrote in *The Descent of Man*: "It must not be forgotten that, although a high standard of morality gives but a slight or no advantage to each individual man and his children over the other men of the same tribe, yet that an advancement in the standard of morality and an increase in the number of well-endowed men will certainly give an immense advantage to one tribe over another" (Darwin, 1871, p. 159).

The theory of *sociobiology* advances a ready answer to the second question raised earlier, of whether *morality* occurs in other animals, even if only as a rudiment. The theory of *kin selection*, sociobiologists argue, explains altruistic behavior, to the extent that it exists in other animals as well as in humans. I propose, however, that properly moral behavior does not exist, even incipiently, in nonhuman animals. The reason is that the three conditions required for ethical behavior depend on an advanced intelligence—which includes the capacities for free will, abstract thought, and anticipation of the future—such as exists in *H. sapiens* and in no other living *species*. It is the case that certain animals exhibit behaviors analogous with those resulting from ethical actions in humans, such as the loyalty of dogs or the appearance of compunction when they are punished. But such behaviors are either genetically determined or elicited by training (conditioned responses). Genetic determination and not moral evaluation is also what is involved in the altruistic behavior of social insects and other animals. I argue later that biological *altruism* (altruism$_b$) and moral *altruism* (altruism$_m$) have disparate causes: *kin selection* in altruism$_b$, and regard for others in altruism$_m$.

The capacity for *ethics* is an outcome of gradual *evolution*, but it is an attribute that only exists when the underlying attributes (ie, intellectual capacities) reach an advanced degree. The necessary conditions for ethical behavior only come about after the crossing of an evolutionary threshold. The approach is gradual, but the conditions only appear when a degree of intelligence is reached such that the formation of abstract concepts and the anticipation of the future are possible, even though we may not be able to determine when the threshold was crossed. Thresholds occur in other evolutionary developments—for example, in the origins of life, multicellularity, and sexual reproduction—as well as in the *evolution* of abstract thinking and self-awareness. Thresholds also occur in the physical world: for example, water heats gradually, but at 100°C boiling begins and the transition from liquid to gas starts suddenly. Surely, human intellectual capacities came about by gradual *evolution*. But when looking at the world of life as it exists today, it would seem that there is a radical breach between human intelligence and that of other animals. The rudimentary cultures that exist in chimpanzees (Whiten et al., 1999, 2005) do not imply the advanced intelligence required for moral behavior.

The question remains: when did *morality* emerge in the human lineage? Did *H. habilis* or *H. erectus* have morality? What about the Neanderthals, *Homo neanderthalensis*? When in hominid *evolution morality* emerged is difficult to determine. The advanced degree of rationality required for moral behavior may only have been reached at the time when creative language came about, and perhaps dependent on the *development* of creative language. When creative language came about in human *evolution* is not, however, a question discussed here.

## WHENCE MORAL CODES?

Moral behavior, I have proposed, is a biological attribute of *H. sapiens*, because it is a necessary consequence of our biological makeup, namely our high intelligence. But moral codes, I argue, are products not of biological *evolution*, but of *cultural evolution*.

It must first be stated that moral codes, like any other cultural systems, cannot survive for long if they are outright contrary to our biology. The norms of *morality* must be consistent with biological nature, because *ethics* can only exist in human individuals and human societies. One might therefore also expect, and it is the case, that accepted norms of *morality* will promote behaviors that increase the biological *fitness* of those who adopt them, such as caring for children. But the correlation between moral norms and biological *fitness* is neither necessary nor indeed always the case: some moral precepts common in human societies have little or nothing to do with biological *fitness*, and some are contrary to the interests of *fitness*.

Before going any further, it seems worthwhile to consider briefly the proposition that the justification of the codes of *morality* derives from religious convictions and only from them. There is no necessary, or *logical*, connection between religious faith and moral values, although there usually is a motivational or psychological connection. Religious beliefs do explain why people accept particular ethical norms, because people are motivated to do so by their religious convictions. But in following the moral and other dictates of one's religion, one is not rationally justifying the moral norms that one accepts. It may, of course, be possible to develop such rational justification: for example, when a set of religious beliefs contains propositions about human nature and the world, from which a variety of ethical norms can be logically derived. Indeed, religious authors, including, for example, Christian theologians, do often propose to justify their *ethics* on rational foundations concerning human nature. But in this case, the logical justification of the ethical norms does not come from religious faith as such, but from a particular (religious) conception of the world; it is the result of philosophical analysis grounded on religious premises.

It may well be that the motivational connection between religious beliefs and ethical norms and other values is the decisive one for the religious believer. But this is true in general: most people, religious or not, accept a particular set of values for social reasons, without trying to justify them rationally by means of a theory from which the moral norms can be logically derived. They accept the values that prevail in their societies, because they have learned such norms from parents, school, or religious and other authorities. The question therefore remains: how do moral codes come about?

The short answer is, as already stated, that moral codes are products of *cultural evolution*, a distinctive human mode of *evolution* that has surpassed the biological mode because it is a more effective form of *adaptation*; it is faster than biological *evolution*, and it can be directed. *Cultural evolution* is based on cultural heredity, which is Lamarckian rather than Mendelian, so acquired characteristics are transmitted. Most important, cultural heredity does not depend on biological inheritance, from parents to children, but is transmitted among individuals of the same or different generations without biological bounds. A cultural *mutation*, an invention (think of the laptop computer, the cell phone, or rock music), can be extended to millions and millions of individuals in less than one generation.

Darwin's Chapter V of *The Descent of Man* (1871) is entitled, "On the Development of the Intellectual and Moral Faculties during Primeval and Civilized Times." There he writes: "There can be no doubt that a tribe including many members who, from possessing in a high degree the spirit of patriotism, fidelity, obedience, courage, and sympathy, were always ready to give aid to each other and to sacrifice themselves for the common good, would be victorious over most other tribes; and this would be natural selection. At all times throughout the world tribes have supplanted other tribes; and as morality is one element in their success, the standard of morality and the number of well-endowed men will thus everywhere tend to rise and increase" (pp. 159—160).

Darwin is making two important assertions. First, *morality* may contribute to the success of some tribes over others; moral behavior amounts to *natural selection* in the form of *group selection*. Second, the standards of *morality* will tend to improve over human history, because the higher the standards of a tribe, the more likely the success of the tribe. This assertion depends on which standards are thought to be "higher" than others. If the higher standards are defined by their contribution to the success of the tribe, then the assertion is circular. But Darwin asserts that there are some particular standards that, in his view, would contribute to tribal success: patriotism, fidelity, obedience, courage, and sympathy.

## SOCIOBIOLOGY'S ACCOUNT OF MORAL BEHAVIOR

Darwin was puzzled by the social organization and behavior of hymenopterans: bees, wasps, ants, and termites. Consider Meliponinae bees, with hundreds of *species* across the tropics. These stingless bees have typically single-queen colonies with hundreds to thousands of workers. The queen generally mates only once. The worker bees toil, building the hive and feeding and caring for the eggs and larvae, even though they themselves are sterile and only the queen produces progeny. Assume that in some ancestral hive a *gene* arose that prompts worker bees to behave as they now do. It would seem that such a *gene* would not be passed on to the following generation because worker bees do not reproduce. But such inference would be erroneous.

Meliponinae bees, like other hymenopterans, have a haplo—diploid system of reproduction. Queen bees produce two kinds of eggs: some are unfertilized and develop into males (which are therefore *haploid*, ie, they carry only one set of genes); others are fertilized (hence are diploid, ie, they carry two sets of genes) and develop into worker bees and occasionally into a queen. W.D. Hamilton (1964) demonstrated that with such a reproductive system the queen's daughters share three-quarters of their genes among them, whereas the queen and her daughters share only one-half of their genes. Hence the worker-bee genes are more effectively propagated by workers caring for their sisters than if they produced and cared for their own daughters. *Natural selection* can thus explain the existence in social insects of sterile castes, which exhibit a most extreme form of apparently altruistic behavior by dedicating their life to caring for the progeny of another individual, the queen.

Hamilton's discovery solved the mystery that had puzzled Darwin and continued puzzling specialists in hymenopteran biology and other evolution-ists for somewhat more than a century. In 1975 the notable ant specialist Edward O. Wilson published *Sociobiology: The New Synthesis*, a treatise appropriately considered the founding document of the new discipline of *sociobiology*. His last chapter was concerned with the social organization of human societies, with the telling title "Man: From Sociobiology to Sociology" and sections dedicated to "Culture, Ritual, and Religion" and "Ethics." The first sentence of the "Ethics" section startled many readers: "Scientists and humanists should consider together the possibility that the time has come for ethics to be removed temporarily from the hands of the philosophers and biologicized" (p. 562). Wilson (1975, 1978, 1998), like other sociobiologists (Alexander, 1979; Barash, 1977; see also Ruse, 2000, 2006, 2010, 2012; Sober, 1993; Sober and Wilson, 1998), sees that *sociobiology* may provide the key for finding a naturalistic basis for *ethics*.

According to Wilson (1975), "The requirement for an evolutionary approach to ethics is self-evident. It should also be clear, for example, that no single set of moral standards can be applied to all human populations, let alone all sex-age classes within each population. To impose a uniform code is therefore to create complex, intractable moral dilemmas" (p. 564). Moral pluralism is, for Wilson, "innate." It seems therefore, according to Wilson, that biology helps us at the very least to decide that certain moral codes (eg, all those pretending to be universally applicable) are incompatible with human nature and therefore unacceptable.

However, Wilson (1978) goes further when he writes: "Human behavior—like the deepest capacities for emotional response which drive and guide it—is the circuitous technique by which human genetic material has been and will be kept intact. *Morality has no other demonstratable ultimate function*" (p. 167). How is one to interpret this statement? It is possible that Wilson is simply giving the reason why ethical behavior exists at all? His proposition would be that humans are prompted to evaluate their actions morally as a means to preserve their genes, their biological nature. But this proposition is, in my view, erroneous. Human beings are by nature ethical beings in the sense I have expounded: they judge their actions morally because of their innate ability to anticipate the consequences of their actions, to formulating value judgments, and to have free choice. Human beings exhibit ethical behavior by nature and necessity, rather than because such behavior would help to preserve their genes or serve any other purpose.

Wilson's statement may alternatively be read as a justification of human moral codes: their function would be to preserve human genes. But this would entail the naturalistic fallacy and, worse yet, would seem to justify a *morality* that most people detest. The "naturalistic fallacy" (Moore, 1903) consists in identifying what "is" with what "ought to be." This error was pointed out by Hume (1740 [1978]): "In every system of morality which I have hitherto met with I have always remarked that the author proceeds for some time in the ordinary way of reasoning... when of a sudden I am surprised to find, that instead of the usual copulations of propositions, is and is not, I meet with no proposition that is not connected with an ought or ought not. This change is imperceptible; but is, however, of the last consequence. For as this ought or ought not express some new relation or affirmation, it is necessary that it should be observed and explained; and at the same time a reason should be given, for what seems altogether inconceivable, how this new relation can be a deduction from others, which are entirely different from it." (p. 469). The naturalistic fallacy occurs whenever inferences using the terms "ought" or "ought not" are derived from premises that do not include such terms but are rather formulated using the connections "is" or "is not." An argument cannot be logically valid unless the conclusions only contain terms that are also present in the premises. To proceed logically from that which "is" to what "ought to be," it is necessary to include a premise that justifies the transition between the two expressions. But this transition is what is at stake, and one would need a previous premise to justify the validity of the one making the transition, and so on in a regression *ad infinitum*. In other words, from the fact that something is the case, it does not follow that it ought to be so in the ethical sense; *is* and *ought* belong to disparate logical categories. Because *evolution* has proceeded in a particular way, it does not follow that this course is morally right or desirable. The justification of ethical norms using biological *evolution*, or any other natural process, can only be achieved by introducing value judgments, human choices that prefer one rather than another object or process. Biological nature is in itself morally neutral.

Moreover, if the preservation of human genes (of the individual or of the species) is the purpose that moral norms serve, Spencer's social Darwinism would seem right: racism or even genocide could be justified as morally correct if they were perceived as the means to preserve those genes thought to be good or desirable and eliminate those thought to be bad or undesirable. Surely, however, Wilson is not intending to justify racism or genocide.

I believe that what Wilson and other sociobiologists are saying is something else, something of great philosophical import that has been stated, with characteristic verve and clarity, by Michael Ruse (Ruse, 2010; Ruse and Wilson, 1985; see also Ruse, 2012): "To be blunt, my Darwinism says that substantive morality is a kind of *illusion* [italics added], put in place by our genes, in order to make us good social cooperators." Ruse proceeds to explain why the illusion of *ethics* is a powerful adaptation: "I would add that the reason why the illusion is such a successful *adaptation* is that not only do we believe in substantive *morality*, but we also believe that substantive *morality* does have an objective foundation. An important part of the phenomenological experience of substantive *ethics* is not just that we feel that we ought to do the right and proper thing, but that we feel that we ought to do the right and proper thing because it truly is the right and proper thing" (Ruse, 2010, p. 309).

The deceit perpetrated on us by our genes is complete: "There are in fact no foundations, but we believe that in some sense there are" (Ruse, 2010, p. 309).

How come "selfish genes" move us to act altruistically and otherwise behave in ways that seem morally right? The answer comes, according to sociobiologists, from the theory of *kin selection* that explains the *altruism* of haplo–diploid insects and much more, as well as from other related theoretical constructs such as inclusive *fitness* and reciprocal *altruism*. The sociobiologists' argument concerning normative *ethics* is not that the norms of *morality* can be grounded in biological *evolution*, but rather that *evolution* predisposes us to accept certain moral norms, namely those that are consistent with the "objectives" of *natural selection*. It is because of this predisposition that human moral codes sanction patterns of behavior similar to those encountered in the social behavior of animals. According to sociobiologists, the commandment to honor one's parents, the incest taboo, the greater blame usually attributed to the wife's adultery than to the husband's, and the banning or restriction of divorce are among the numerous ethical precepts and practices that endorse behaviors promoted by *natural selection*. Sociobiologists reiterate their conviction that science and *ethics* belong to separate logical realms; that one may not infer what is morally right or wrong from a determination of how things are or are not in nature. The sociobiologists avoid the naturalistic fallacy by the drastic move of denying that ethical behavior exists as an activity with different causation than any other activities or traits simply determined by our genes. Ethical behavior is simply an expression of our genes and a direct consequence of *natural selection*, as it adapts humans, as well as other organisms, to their environments.

## ALTRUISM: BIOLOGICAL AND MORAL

Evolutionists for years struggled to find an explanation for the apparently altruistic behavior of animals. When a predator attacks a herd of zebras, adult males attempt to protect the young in the herd, even if they are not their progeny, rather than fleeing. When a prairie dog sights a coyote, it will warn other members of the colony with an alarm call, even though by drawing attention to itself this increases its own risk. Darwin tells the story of adult baboons protecting the young. "Brehm encountered in Abyssinia a great troop of baboons who were crossing a valley: some had ascended the opposite mountain, and some were still on the valley: the latter were attacked by the dogs, but the old males immediately hurried down from the rocks, and with mouths widely opened, roared so fearfully, that the dogs precipitately retreated. They were again encouraged to the attack; but by this time all the baboons had reascended the heights, excepting a young one, about six months old, who, loudly calling for aid, climbed on a block of rock and was surrounded. Now one of the largest males, a true hero, came down again from the mountain, slowly went to the young one, coaxed him, and triumphantly led him away—the dogs being too much astonished to make an attack" (Darwin, 1871, p. 124).

Examples of altruistic behaviors of this kind are multiple. But to speak of animal *altruism* is not to claim that explicit feelings of devotion or regard are present in animals, but rather that they act for the welfare of others at their own risk just as humans are expected to do when behaving altruistically. However, in this particular case Darwin uses a word that deserves attention: the baboon that comes down the mountain is called "a true hero." Heroism is an ethical concept. Is Darwin using it in this sense or only metaphorically? Darwin's views are clearly stated, as noted earlier: "of all the differences between man and the lower animals, the moral sense or conscience is by far the most important."

The problem is precisely how to justify altruistic behavior in terms of *natural selection*. Assume, for illustration, that in a certain *species* there are two alternative forms of a *gene* (two alleles), of which one but not the other promotes altruistic behavior. Individuals possessing the altruistic *allele* will risk their lives for the benefit of others, whereas those possessing the non-altruistic *allele* will benefit from the altruistic behavior of their partners without risking themselves. Possessors of the altruistic *allele* will be more likely to die or fail to reproduce, and the *allele* for *altruism* will therefore be eliminated more often than the nonaltruistic *allele*. Eventually, after some generations, the altruistic *allele* will be completely replaced by the non-altruistic one. So how is it that altruistic behaviors are common in animals without the benefit of ethical motivation? The explanation comes from the theory of *kin selection*.

To ascertain the consequences of *natural selection* it is necessary to take into account a *gene*'s effects not only on a particular individual but also on all individuals possessing that *gene*, as in the explanation of the social organization of bees and other hymenopterans. When considering altruistic behavior, one must take into account not only the costs for the altruistic individual but also the benefits for other possessors of the same *allele*. Zebras live in herds where individuals are blood relatives. This is also the case for baboon troops. A *gene* prompting adults to protect the defenseless young would be favored by *natural selection* if the benefit (in terms of saved individuals that are carriers of that gene) is greater than the cost (due to the increased risk or other costs of the protectors). An individual lacking the altruistic *gene* and carrying instead a nonaltruistic one will not incur a cost or risk its life, but the nonaltruistic *gene* is partially eradicated with the death of each defenseless relative.

It follows from this line of reasoning that the more closely related the members of a herd, troop, or animal group are, the more altruistic behavior should be present. This seems to be generally the case. Consider parental care, which is most obvious in the genetic benefits it entails. Parents feed and protect their young because each child has half the genes of each parent: the genes are protecting themselves, as it were, when they prompt a parent to care for its young. That is why parental care is widespread among animals.

Sociobiologists point out that many of the moral norms commonly accepted in human societies sanction behaviors also promoted by *natural selection*, such as the commandment to honor one's parents and the incest taboo, as pointed out earlier. Once again, the sociobiologists' argument is that human ethical norms are sociocultural correlates of behaviors fostered by biological *evolution*. Ethical norms protect such evolution-determined behaviors as well as being specified by them.

The sociobiologists' arguments, however, are misguided (Ayala, 2010). Consider *altruism* as an example. *Altruism* in the biological sense (altruism$_b$) is defined in terms of the *population* genetic consequences of a certain behavior. Altruism$_b$ is explained by the fact that genes prompting such behavior are actually favored by *natural selection* (when inclusive *fitness* is taken into account), even though the *fitness* of the behaving individual is decreased. But *altruism* in the moral sense (altruism$_m$) is explained in terms of motivations: a person chooses to risk his/her own life (or incur some cost) for the benefit of somebody else. The similarity between altruism$_b$ and altruism$_m$ is only with respect to the consequences: an individual's chances are improved by the behavior of another individual who incurs a risk or cost. The underlying causations are completely disparate: the ensuing genetic benefits in altruism$_b$, and regard for others in altruism$_m$. As Darwin put it, the altruistic behaviors of a baboon and a human are similar in that they both benefit other individuals, but differ in that humans carry out an assessment, which baboons do not. Humans make moral judgments as a necessary consequence of their eminent intellectual abilities. Their judgments, as well as the moral norms on which they are based, are not always accompanied by biological gain.

Parental care is a behavior generally favored by *natural selection* and is present in virtually all codes of *morality*, from primitive to more advanced societies. There are other human behaviors sanctioned by moral norms that have biological correlates favored by *natural selection*. One example is monogamy, which occurs in some animal *species* but not in many others. It is also sanctioned in many human cultures, but not all. Polygamy is accepted in some current human cultures, and surely was more so in the past. Food sharing outside the mother—offspring unit rarely occurs in primates, with the exception of chimpanzees and capuchin monkeys, although even in chimpanzees food sharing is highly selective and often associated with reciprocity. A more common form of mutual aid among primates is coalition formation: alliances are formed in fighting other conspecifics, although these alliances are labile, with partners readily changing partners.

One interesting behavior associated with a sense of justice, or equal pay for equal work, is described by Sarah Brosnan and Frans de Waal (2003; see also de Waal, 1996) in the brown capuchin monkey, *Cebus paella*. Monkeys responded negatively to unequal rewards in exchanges with a human experimenter. Monkeys refused to participate in an exchange when they witnessed that a conspecific had obtained a more attractive reward for equal effort.

Is the capuchin behavior phylogenetically related to the human virtue of justice? This seems unlikely, since similar behavioral patterns have not been observed in other primates, including apes, phylogenetically closer to humans. Moreover, it was later shown that capuchin monkeys reject a food reward when a more attractive reward is visible, whether this more attractive reward is offered to other monkeys or is simply present within their sight. Cannibalism is practiced by chimps, as well as by human cultures of the past. Do we have a phylogenetically acquired predisposition to cannibalism as a morally acceptable behavior? This seems unlikely. Moral codes arise in human societies by *cultural evolution*. The moral codes that tend to be widespread are those that lead to successful societies.

Since time immemorial, human societies have experimented with moral systems. Some have succeeded and spread widely through humankind, like the Ten Commandments, although other moral systems persist in different human societies. Many moral systems of the past have surely become extinct because they were replaced or the societies that held them became extinct. The moral systems that currently prevail in humankind are those that were favored by *cultural evolution*. They were propagated within particular societies for reasons that might be difficult to fathom, but that surely must have included the perception by individuals that a particular moral system was beneficial for them, at least to the extent that it was beneficial for their society by promoting social stability and success. Acceptance of some precepts in many societies is reinforced by civil authorities (eg, those who kill or commit adultery will be punished) and religious beliefs (God is watching, and you will go to hell if you misbehave). Religious, legal, and political systems, as well as belief systems, are themselves outcomes of *cultural evolution*.

## GENE–CULTURE COEVOLUTION

A different explanation of the *evolution* of the moral sense has been advanced by proponents of the theory of "gene–culture coevolution" (Richerson and Boyd, 2005; Strimling et al., 2009; Richerson et al., 2010; see also Greene et al., 2001; Haidt, 2007). It is assumed that cultural variation among tribes in terms of patriotism, fidelity, sympathy, and other moralizing behaviors may have occurred incipiently in early hominid populations, starting at least with *H. habilis*. This cultural variation may have selected for genes that endowed early humans with primitive moral emotions. Primitive moral emotions would in turn have facilitated the *evolution* of more advanced cultural codes of *morality*. Repeated rounds of *gene–culture* coevolution would gradually increase both the moral sense itself and the systems of moral norms. That is, the *evolution* of *morality* would have been directly promoted by *natural selection* in a process where the moral sense and moral norms would coevolve.

The gene–culture coevolution account of the *evolution* of *morality* is, of course, radically different from the theory I advanced earlier, in which moral behavior evolved not because it increased *fitness* but as a consequence of advanced intelligence, which allowed humans to see the benefits that adherence to moral norms brings to society and its members. The extreme variation in moral codes among recent human populations and the rapid *evolution* of moral norms over short time spans (see later) would seem to favor the explanation I propose. *Gene–*culture coevolution would rather lead to a more nearly universal system of *morality*, which would have come about gradually as our hominid ancestors evolved toward becoming *H. sapiens*.

Empathy, or the predisposition to assimilate mentally the feelings of other individuals, has recently been extensively discussed in the context of altruistic or moral behavior. Incipient forms of empathy seem to be present in other animals. In humans, increasing evidence indicates that we automatically simulate the experiences of other humans (Gazzaniga, 2005, pp. 158–199, 2008). Empathy is a common human phenomenon, surely associated with our advanced intelligence, which allows us to understand the harms or benefits that impact other humans, as well as their associated feelings. Empathic humans may consequently choose to behave according to how their behavior will affect those for whom they feel empathy. That is, human empathy occurs because of our advanced intelligence. Humans may choose to behave altruistically or not—that is, morally or not—in terms of the anticipated consequences of their actions for others.

The question remains: when did *morality* emerge in the human lineage? When in hominid *evolution morality* emerged is difficult to determine. As pointed out earlier, the capacity for *ethics* is an outcome of gradual *evolution*, but it only comes about when the underlying attributes (ie, intellectual capacities) reach an advanced degree. When in human evolutionary history this threshold was reached would seem impossible to determine, at least at the current state of knowledge.

One additional observation is that the norms of *morality*, as they exist in any particular culture, are felt to be universal within that culture. Yet, similarly to other elements of culture, they are continuously evolving, often within a single generation. As Steven Pinker pointed out, Western societies have recently experienced the moralization and amoralization of diverse behaviors. Thus "smoking has become moralized... now treated as immoral... At the same time many behaviors have become amoralized, switched from moral failings to lifestyle choices. They include divorce, illegitimacy, working mothers, marijuana use and homosexuality" (Pinker, 2002, p. 34). But it is worth noticing, once again, that the legal and political systems that govern human societies, as well as the belief systems held by religion, are themselves outcomes of *cultural evolution* as it has eventuated over human history, particularly over the last few millennia.

# Chapter 13

# Aesthetics

## INTRODUCTION

"Aesthetics" is a term forged by Alexander Gottlieb Baumgarten (1714–1762) to refer to a philosophy of taste and beauty, but the appreciation of beauty transcends the German Enlightenment. Indeed, the capacity for appreciating beauty, the aesthetic qualities of objects, motions, and sounds, is a human universal; all human groups possess this competence. The capacity for producing aesthetic items is also universal: sculptors, painters, dancers, and musicians are pervasive in all human cultures and historical epochs. However, *appreciating* aesthetic attributes goes beyond producing them in two respects: First, "artists" or producers of beauty make up a small fraction of individuals in human groups; in contrast, "spectators" are numerous. This difference becomes, nevertheless, reduced or even eliminated if we extend the production of beauty beyond those who might be considered professional artists to the decoration or furnishing of a home, the preparation and presentation of food for a meal, and singing or dancing in the course of ordinary life. A second difference is that it is possible to appreciate aesthetic qualities in natural objects and events, such as a sunset or sunrise, a snow-covered mountain, a forest, or bird songs (Cela-Conde and Ayala, 2007, 2016).

An important difference between *ethics* and aesthetics is that ethical values are considered, by and large, "objective," while aesthetic values are considered much more "subjective." Individuals within a given culture, and largely across cultures, usually agree about what is necessarily wrong and what is absolutely right, while qualifying an object or song as beautiful tolerates considerably more diversity of opinion: *de gustibus non est disputandum* (Bernardini, 2015).

## ORIGINS

It does not seem possible to determine the phylogenetic origin of the human capacity for appreciating beauty. Neither *fossil* nor archeological records provide convincing evidence to ascertain the appearance of this competence. It is not possible to determine whether spectators with the ability to appreciate landscapes, motions, or songs existed in previous human *species*, or when in the *evolution* of *Homo sapiens* or our closely related *species Homo*

Evolution, Explanation, Ethics, and Aesthetics. http://dx.doi.org/10.1016/B978-0-12-803693-8.00013-X
261

*neanderthalensis* the capacity to appreciate beauty came about. It seems the best we might be able to do is determine when human artisans first started to produce objects or activities with aesthetic qualities, and conclude that the appreciation of beauty would already be present among their human partners.

Ascertaining when the production of objects with aesthetic attributes first came about in human *evolution* is less elusive than ascertaining when the appreciation of beauty first appeared, but even so it is far from obvious. The *Paleolithic* polychromies representing bisons and horses in the caves of Altamira or Lascaux are doubtless art works, but what about the bifacial symmetric handaxes made by *Homo erectus* 1 million years ago and earlier, and even the first recognizable stone tools produced by *Homo habilis* as early as 2.5 million years ago? I first explore the production of objects doubtless produced to serve useful purposes, such as handaxes, and leave unanswered for the earliest such useful objects whether aesthetic considerations may have played a role in their *design*.

The start of the *Paleolithic*, or "Old Stone Age," is defined precisely by the appearance of the first stone tools, around 2.6 Mya (million years ago), associated in turn with *H. habilis*, which is the first *species* of *Homo*, named *habilis* precisely for its ability, the first in human *evolution*, to construct tools. The *Paleolithic* lasted until the Neolithic, 12—10 kya (thousand years ago), and the "Modern Stone Age" or Holocene that lasted until about 2000 years BCE, when modern history starts.

*Paleolithic* history is often recognized by a succession of stages or "cultures," characterized by distinct tools and other products of human activity and named after localities where the characteristic human objects were found. Successively, these cultures are known as Oldowan (named after the Olduvai site in Tanzania), from about 2.5 Mya until about 1.5 Mya; Acheulan (named after the St. Acheul site in France), from about 1.5 Mya until 150—120 kya, encompassing African and European *H. erectus* and archaic *H. sapiens*; Mousterian, 150—120 kya until about 40 kya, with *H. neanderthalensis* and *H. sapiens* in Europe; succeeded by the Aurignacian culture, lasting from about 40 to about 18 kya, during which clearly defined artistic expressions appear as decorations on cave walls; and then the Magdalenian, with the highest artistic expression of the *Paleolithic*, from 18 to 10 kya, coinciding with the peak of the Würm glaciation, the last and most severe impacting our planet. The last three cultures are designated after the French sites of Le Moustier, Aurignac, and La Magdalene.

## OLDOWAN AND ACHEULEAN CULTURES

Olduvai Gorge (Tanzania) has been a source of numerous *hominin* fossils since the middle of the 20th century. Although Olduvai Gorge is not the first place in which stone tools were found, it gave its name to the earliest known lithic industry: Oldowan culture. A lithic industry can be described as a set of diverse stones manipulated by hominins to obtain tools to cut, scrape, or hit. They are diverse tools obtained by hitting pebbles of different hard materials. Silex, quartz, flint, granite, and basalt are some of the materials used for tool making. In the Oldowan culture the size of the round-shaped cores is variable, but they usually fit comfortably in the hand; they are tennis-ball-sized stones. Many tools belonging to different traditions fit within these generic characteristics. What specifically identifies Oldowan culture is that its tools are obtained with very few knocks, sometimes only one. The resultant tools are misleadingly crude in appearance, although it is not easy to hit the stones with enough precision to obtain cutting edges and efficient flakes (Fig. 13.1).

The idea we have of their function depends on the way we interpret the *adaptation* of the hominins who used them. The most important tools would be the handaxes that allow hitting a *cranium* or femur hard enough to break it. But for butchering almost whole animals, flakes would be the essential tools.

The Oldowan culture was not restricted to Olduvai. Stone tools have also been found at Kenyan and Ethiopian sites, some older than Olduvai, such as Gona in the Middle Awash region (Ethiopia), where some findings extend back to 2.6–2.5 Mya. Other Ethiopian sites include Hadar and Ono and, just south of them in Kenya, the Lokalalei site in west Turkana—these sites, dated 2.0–2.3 Mya, collectively have yielded thousands of stone tools. The dates for these sites are compatible with the attribution of their stone tools to *H. habilis*. The oldest *H. habilis* fossils from the Afar region in Ethiopia have recently been dated between 2.75 and 2.80 Mya (Villmoare et al., 2015; DiMaggio et al., 2015).

The issue arises as to what sort of knowledge was possessed by the *H. habilis* hominins who produced these stone tools. There can be little doubt that they had acquired some capacity for abstract knowledge and forming images of realities not present, manifest in the construction of tools that would be used at a later time, for purposes not immediately present. However, the tools were crude, showing no evidence of any aesthetic intention (Fig. 13.2). This changed later, during the ensuing Acheulan culture, when symmetric, carefully designed handaxes were produced that betray an appreciation of beauty. Their creators were other more-developed *hominin species*, which include *H. erectus*, *H. neanderthalensis*, and perhaps archaic *H. sapiens*.

The relatively advanced techniques and diversity of tools found in sites of the Oldowan culture have motivated some paleontologists to suggest that there must have been a pre-Oldowan lithic culture, even though no *hominin* fossils have been found at the sites where the earlier tools rest. Recent findings suggest that indeed this may be the case. The journal *Nature* (Callaway, 2015a) reported the finding of stone cores and flakes, likely intentionally crafted, at the Lomekwi site, west of Kenya's Lake Turkana, in sediments dated 3.3 Mya, much older than *H. habilis*. Sonia Harmand of Stony Brook University in New York reported the findings at a meeting of the Paleoanthropology Society in San Francisco on April 14, 2015. Harmand's team concluded that the tools represent a distinct culture, which they named the Lomekwian. Harmand

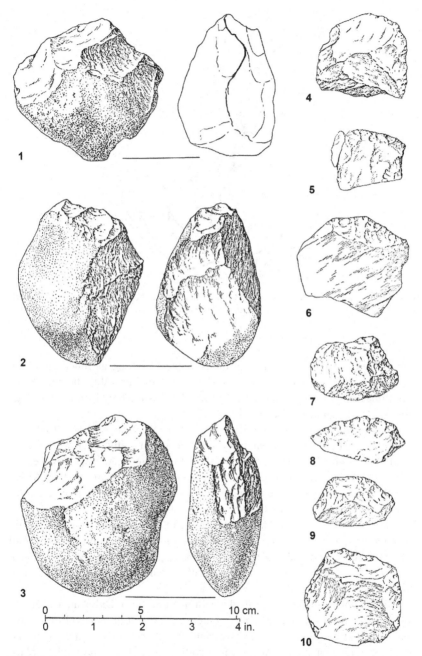

**FIGURE 13.1**    Oldowan tools. 1—3: lava choppers; 4—10, quarcite flakes. *Drawing by Leakey, M.D., 1971. Olduvai Gorge. In: Excavations in Beds I and II 1960—1963, vol. 3. Cambridge University Press, Cambridge.*

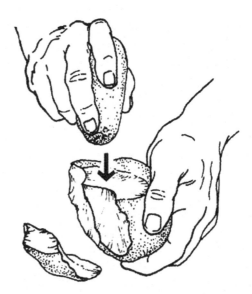

**FIGURE 13.2**   Flake production by the Oldowan technique. *Picture from Plummer, T., 2004. Flaked stones and old bones: biological and cultural evolution at the dawn of technology. Yearbook of Physical Anthropology 47, 118–164.*

pointed out that the cores are enormous, some weighing as much as 15 kg, which is surprising considering the small size of the australopithecines (*Australopithecus species*, predating *H. sapiens*). How could they handle such large stones? And what were they used for?

The transition, described by Mary Leakey (1975), from Oldowan tools to a different and more advanced industry, the Acheulean culture, can be observed in Olduvai. Acheulean tools made with great care were identified for the first time at the St. Acheul site (France). Acheulean culture appeared in East Africa slightly over 1.5 Mya, and extended to the rest of the Old World to a greater or lesser degree until around 0.3 Mya. Its most characteristic element is the biface, "teardrop shaped in outline, biconvex in cross-section, and commonly manufactured on large (more than 10 cm) unifacially or bifacially flaked cobbles, flakes, and slabs" (Noll and Petraglia, 2003). But the term *biface* corresponds to a form of manufacture rather than to a tool. Bifaces led to different utensils, such as those shown in Fig. 13.3.

Mary Leakey (1975) described the transition from Oldowan culture to Acheulean culture as gradual. Other studies, as argued by Isaac (1969), have convincingly shown that the improvement of the necessary techniques to go from the Oldowan to the Acheulean traditions could not have taken place gradually. A completely new type of manipulation would have appeared with Acheulean culture, a true change in the way of carrying out the operations involved in tool making.

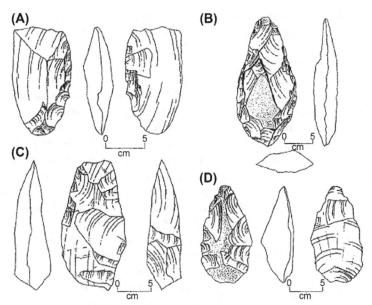

**FIGURE 13.3**    Bifaces destined to different uses: (A) cleaver; (B) handaxe; (C) knife; (D) pick. *From Noll, M.P., Petraglia, M.D., 2003. Acheulean bifaces and early human behavioral patterns in East Africa and South India. In: Soressi, M., Dibble, H.L. (Eds.), Multiple Approaches to the Study of Bifacial Technologies. University of Pennsylvania Museum of Archaeology and Anthropology, Philadelphia, pp. 31–53.*

The cultural sequence identified by Mary Leakey (1975) involved a three-stage transition (Table 13.1): first, the *evolution* of progressively more sophisticated techniques within the Oldowan culture itself; second, the coexistence of Oldowan and Acheulean tools; and third, the disappearance of the

**TABLE 13.1** Cultural Sequence at Olduvai Established by Mary Leakey

| Beds | Age (Mya) | Number of Pieces | Industries |
|---|---|---|---|
| Masek | 0.2 | 187 | Acheulean |
| IV | 0.7–0.2 | 686 | Acheulean |
| | | 979 | Developed Oldowan C |
| Middle part of III | 1.5–0.7 | 99 | Acheulean |
| Middle part of II | 1.7–1.5 | 683 | Developed Oldowan A |
| I and lower part of II | 1.9–1.7 | 537 | Oldowan |

Modified from Leakey, M.D., 1975. Cultural patterns in the Olduvai sequence. In: Butzer, K.W., Isaac, G.L. (Eds.), After the Australopithecines. Mouton, The Hague, pp. 477–493.

former, and further *development* of Acheulean techniques. Isaac (1969) described that sequence in terms of four cultural–stratigraphical associations, from the oldest to the youngest, as Oldowan, developed Oldowan, lower Acheulean, and upper Acheulean.

To what extent can the Acheulean tradition be considered a continuation or a rupture regarding Oldowan? Was developed Oldowan a transition phase toward subsequent cultures? The required technique to execute the Acheulean bifaces is different from Oldowan in several features. The first difference is the succession of strikes required to produce a handaxe, which contrasts with the few and unorganized strikes required to manufacture a protobiface. The production of very long oval flakes (more than 10 cm), characteristic of Acheulean techniques, is its key difference from the advanced Oldowan traditions. The shape of those long flakes is not very different from the bifaces themselves. This is why Isaac (1969) suggested that they could be transformed into handaxes without too much effort. The appearance of the technique for producing flakes suddenly changed the possibilities for tool manufacture.

Given that the production of those flakes involves starting from large cores, the availability of quarries with such raw materials can determine important differences in the cultural content of different sites. The manipulation of large blocks of material (mostly lava and quartzite) to produce long flakes seems to have been the turning point for the *development* of the Acheulean culture. It would also have involved risk for those who had to manipulate stones of large size (Schick and Toth, 1993).

The most advanced Acheulean stage includes handaxes with such symmetrical and carefully elaborated edges that they must have required the so-called soft-hammer technique (Fig. 13.4), which uses softer hammers than the cores themselves, such as wood or bone. Knapping with such a tool allowed more precise control, and certainly required more time. Schick and Toth (1993) provided a detailed description of the process.

Three events are usually considered to have taken place together during the *evolution* of early and mid-Pleistocene hominins: the appearance of *H. erectus*, Acheulean culture, and the first migration of hominins out of the African continent. The usual interpretation suggests that these three events are related. Leaving Africa confronted hominins with climates colder than that in the African Rift Valley. *Adaptation* to those extreme conditions was made possible by cultural novelties, such as the control of fire. The *adaptation* was achieved by *erectus*-grade hominins, associated with Acheulean culture.

The central element of Acheulean culture, the handaxe, is absent in many early European sites with signs of human presence (Italy, France, Germany, Czech Republic, and Spain). It was not until a second colonizing wave, which took place about 0.5 Mya, that Acheulean handaxes were introduced. Sites corresponding to this time interval include Torralba and Ambrona (Spain), St. Acheul and Abbeville (France), Swanscombe, Boxgrove, and Hoxne (England), and Torre di Pietra and Venosa-Notarchirico (Italy).

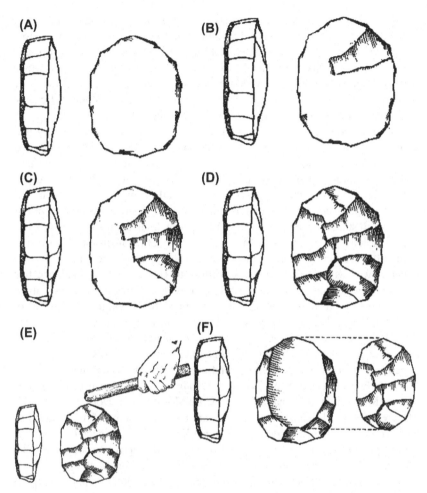

**FIGURE 13.4**   Levallois technique. Phases in the construction of a tool by flake removal. (A) Preparation of an adequately shaped core. (B and C) Removal of flakes. (D) The prepared platform is obtained. (E) A last blow with a soft hammer separates the tool from the core (F). *Drawings from www.hf.uio.no/iakh/forskning/sarc/iakh/lithic/LEV/Lev.htm.*

## MOUSTERIAN CULTURE AND NEANDERTHALS

Mousterian culture (c.100—40 kya) is the lithic tool tradition that evolved from Acheulean culture during the middle *Paleolithic*. It was followed by the Aurignacian technical and artistic explosion (40—18 kya), with tools and decorated objects that contrast sharply with the Mousterian and earlier cultures. Mousterian culture has controversial features, including objects clearly created with a decorative intention. Participants were *H. neanderthalensis*, as well as modern *H. sapiens*.

Mousterian tool-making techniques produced tools that were much more specialized than Acheulean ones: the Mousterian tools were given a form before sharpening their edges. The most typical Mousterian tools found in Europe and the Near East are flakes produced by means of the Levallois technique, which were subsequently modified to produce diverse and shaper edges. Objects made from bone are less frequent, but up to 60 types of flake and stone foil can be identified, which served different functions (Bordes, 1979).

The Levallois technique appeared during the Acheulean period, and was used thereafter. Its pinnacle was reached during the Mousterian culture. The purpose of this technique is to produce flakes or foils with a very precise shape from stone cores that serve as raw material. The cores must first be carefully prepared by trimming their edges, removing small flakes until the core has the correct shape. Thereafter, with the last blow, the desired flake, a Levallois point for instance, is obtained (Fig. 13.4). The final results of the process, which include points and scrapers among many other instruments, are subsequently modified to sharpen their edges. The amazing care with which the material was worked constitutes, according to Bordes (1953), evidence that these tools were intended to last for a long time in a permanent living location. They also seem to reflect an aesthetic intention. Tools obtained by means of the Levallois technique are typical of European and Near East Mousterian sites. Bifaces, in contrast—so abundant in Acheulean sites—are scarce. The difference has to do mostly with the manipulation of the tools; scrapers were already produced using Acheulean, and even Oldowan, techniques.

Most European and Near East sites belonging to the Würm glacial period contain Mousterian tools (named after the Le Moustier site, Dordogne, France). The archaeological richness and sedimentary breadth of some of these sites, like La Ferrassie, La Quina, and Combe-Grenal, grants them a special interest for studying the interaction between cultural utensils and adaptive responses. Similar Mousterian utensils have appeared in the Near East at Tabun, Skuhl, and Qafzeh.

The identification between the Mousterian culture and *H. neanderthalensis* has been considered so consistent that, repeatedly, European sites yielding no human specimens, or with scarce and fragmented remains, were attributed to Neanderthals on the sole basis of the presence of Mousterian utensils. Despite the difficulties inherent in associating a given *species* with a cultural tradition, it was thought to be beyond doubt that Mousterian culture was part of the Neanderthal identity. Scrapers and Levallois points, which were very similar to the typical European ones, turned up also in Near East sites (Bar-Yosef and Vandermeersch, 1993). Neanderthals also lived there, of course (Fig. 13.5), but, in contrast with European sites, a distinction could not be drawn between localities that had housed Neanderthals and anatomically modern humans solely on the grounds of the cultural traditions. The more or less systematic distinction between Neanderthal—Mousterian and Cro-Magnon—Châtelperronian (or Aurignacian, or Magdalenian) helped to clarify the situation in Europe. But it could not be transferred to the Near East, where sites occupied by Neanderthals

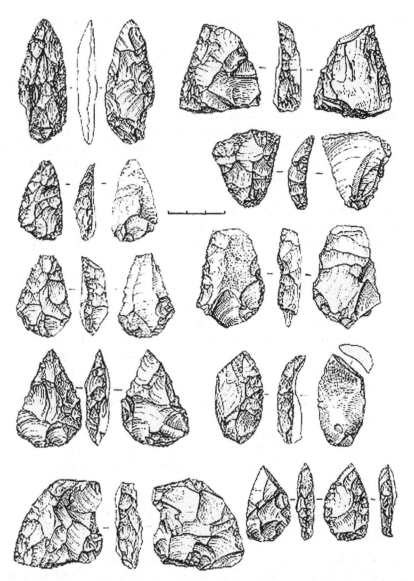

**FIGURE 13.5**    Mousterian handaxes from Mezmaiskaya cave (Caucasus). *From Doronichev, V., Golovanova, L., 2003. Bifacial tools in the lower and middle Paleolithic of the Caucasus and their contexts. In: Soresi, M., Dibble, H.L. (Eds.), Multiple Approaches to the Study of Bifacial Technologies. Museum of Archaeology and Anthropology, Philadelphia, PA, pp. 77—107.*

and those inhabited by anatomically modern humans, proto-Cro-Magnons, yielded the same utensils of the Mousterian tradition.

This implies several things. First, cultural sharing was common during the middle *Paleolithic*, at least in Levant sites. Second, during the initial stages of their occupation of the eastern shore of the Mediterranean, anatomically modern humans made use of the same utensils as Neanderthals. Hence, it seems that at the time there was no technical superiority of modern humans over Neanderthals. The third and most important implication has to do with the inferences that can be made because Neanderthals and *H. sapiens* shared identical tool-making techniques. The production of Mousterian tools required complex mental capabilities. Were Neanderthal cognitive abilities as complex as those currently characteristic of our own species? Some authors have argued in favor of high cognitive capacities in Neanderthals, seeing other kinds of items as indications of Neanderthal aesthetic, religious, symbolic, and even maybe linguistic capacities. Numerous Neanderthal burials can be interpreted as a functional response to the need to dispose of the bodies, even if only for hygienic reasons. But they could also be understood as the reflection of transcendent thinking, beyond the simple human motivation of preserving the bodies of deceased loved ones, probably reflecting symbolic and even religious intentions.

A possible key to the symbolic thought of *H. neanderthalensis* could come from stone and bone objects belonging to the Mousterian tradition. Making tools to use them in one way or another requires a capacity to formulate objectives and anticipate behaviors. Beyond their utility, Acheulean handaxes are beautiful objects (Fig. 13.6). Were these tools created with the *intention* of being beautiful? Bifaces could be an early manifestation of the *evolution* of preferences for lateral symmetry. It is true that the Acheulean symmetrical tools and, more so, Mousterian tools turned into beautiful objects, artistic representations for us, who live hundreds of thousands of years after the objects were manufactured. But were they also so perceived, at least to some extent, by those who manufactured them? Can a gradual and slow *evolution* toward more advanced symbolic objects be documented? Or, rather, did symbolic expression and perception come about relatively suddenly, late in human evolution?

## SYMBOLISM AND AESTHETICS: AURIGNACIAN AND MAGDALENIAN CULTURES

The symmetry of Acheulean handaxes could be considered as a possible indication of *symbolism*. The earliest of these tools are about 1.5 million years old (Leakey, 1975). Realistic representations, such as the Altamira, Lascaux, Chauvet, and other cave paintings in Spain and France were made toward the end of the upper *Pleistocene*, about 14 kya. There is an enormous time gap between these two cultural manifestations (symmetry and realistic painting). Within this gap comes the Mousterian culture characteristic of the European middle *Paleolithic*, ranging from about 100 to 40 kya. The ensuing Aurignacian (c.40–18 kya) and Magdalenian (c.18–10 kya) cultures include drawings, engravings, and paintings with unquestionable intentional aesthetic and

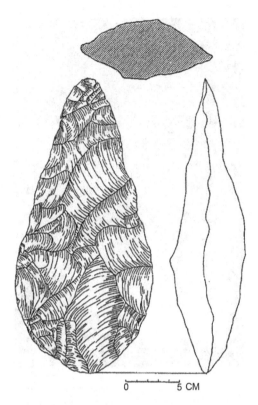

0 ————————— 5 CM

**FIGURE 13.6**   Handaxe from Isimila (Tanzania, c.300,000 year). As Tomas Wynn observed, this artifact has congruent symmetry in three dimensions. *Illustration from Wynn, T., 2007. Archaeology and cognitive evolution. Behavioral and Brain Sciences 25 (3), 389–402.*

symbolic characteristics, much beyond the *symbolism* and beauty that might be attributed to the Mousterian and earlier cultures.

There are two mutually exclusive hypotheses about the process that led to the massive production of artistic representations unquestionably charged with *symbolism*: the gradual and explosive models. The former argues that the capacity to appreciate Acheulean "beautiful forms" evolved gradually and continuously, leading to the great abundance of late-Paleolithic artistic objects. This gradual model does not refer to an origin of art, which is thought to be fuzzy, widespread in space, and continuous in time. According to this model, the initial manifestations of that origin were scarce; slowly, over a long period of time, they became progressively generalized. In contrast, the explosive model of the *symbolism* characteristic of art argues that it appeared fairly suddenly during the late *Paleolithic* and is exclusively an attribute of modern humans. The great cognitive transformation evinced in the upper-Paleolithic artistic explosion must have included different capacities for *adaptation* in our ancestors. Several authors have suggested coevolutionary sequences of

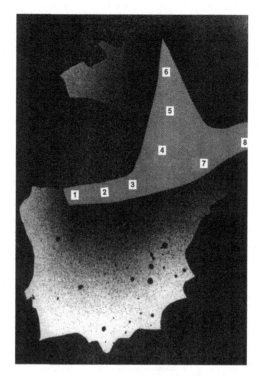

**FIGURE 13.7**  The seven central regions of Paleolithic cave art: 1, Asturias; 2, Cantabria; 3, País Vasco; 4, Dordogne (Lot and Tarn); 5, Dordogne (Gironde); 6, Dordogne (Charente and Yonne); 7, Aragon and Eastern Pyrenees; 8, Garde, Ardèche, Ain. The caves of Altamira and El Castillo are in region 2; Lascaux is in region 5; Chauvet is in region 8 (Vaquero Turcios, 1995, p. 21).

cultural manufactures and communicative abilities. Some of these models (Davidson and Noble, 1989) put forward a hypothesis that relates the origin of art itself, and not just the general cultural sequence, with the origin of language. They believe that drawing requires prior communication. Drawing later transformed communication into language. Davidson and Noble (1989) argue that pictorial representations are halfway between reality and language. There are plenty of hypothetical speculations and models of the *evolution* of language, but this is a matter that is not pursued here.

The appearance of cave art early in the Aurignacian period and culminating during the Magdalenian seems "explosive" from our distance of several thousand years. An extensive monograph by the distinguished painter and muralist Joaquín Vaquero Turcios (1995) provides detailed historical and artistic analysis of cave art, particularly wall paintings. He has identified 72 caves, distributed into eight regions in northern Spain and southern France (Fig. 13.7). Best known are Altamira and El Castillo in region 2 (Cantabria, Spain), Lascaux in region 5 (Dordogne-Gironde, France), and Chauvet in

**FIGURE 13.8**    A fire lamp. Over the stone, lumps of bone marrow (nearly exhausted) and the protruding wick (Vaquero Turcios, 1995, p. 25).

region 8 (Ardèche, France). All the caves seem to have experienced two distinct periods of artistic activity, the first around 40—34 kya and the second around 18—14 kya. Why these two periods of activity and why at relatively similar times in all caves is not known. One consideration is the overlap with the Würm glaciation, which reached its climax around 16 kya. Caves may have served as occasional refuge, as well as places for social, and perhaps religious, interaction.

One footnote that deserves attention is illumination. How did the artists, as well as visitors over many years in prehistoric times, manage to have light in the depth of caves, often hundreds of feet from the cave entrance, where no sunlight or any conceivable form of natural light could reach? Wax and various fats were the most common forms of lighting before modern times. But most fat or wax candles produce smoke that would have made it all but impossible to paint or to draw, and visitors would have obliterated with their smoke any paintings decorating the walls. Vaquero Turcios (1995) discovered that bone marrow, particularly from the calves of cattle and other large mammals, burns without producing smoke. Small hand-held stones covered with lumps of bone marrow and with an ignited wick could provide lighting for hours at a time (Fig. 13.8). The fire was started by rubbing stones or wooden sticks against each other. Vaquero Turcios (1995) showed how the artists changing their position around a particular spot could suitably illuminate it, and at times provide vantage points for identifying irregularities on the walls that could be used to represent different animal forms.

The Aurignacian cave paintings are mostly red ochre stencils of human hands and dot alignments with different configurations. It is difficult to ascertain *symbolism*; sometimes the hand stencils and dots are associated on the same wall with sketched and often incomplete animal designs, some of which may have been added later, during the Magdalenian period (Fig. 13.9).

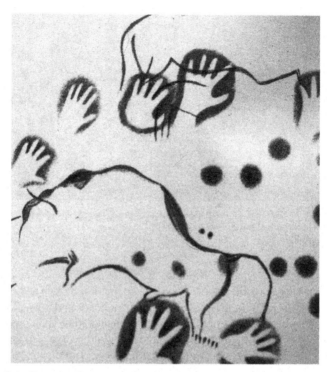

**FIGURE 13.9**  Fragment of the "panel of the hands," on the ceiling of a low cavern in El Castillo. The hand stencils start near the floor; the incomplete bisons would have been added later (Vaquero Turcios, 1995, p. 105).

**FIGURE 13.10**  Drawing in black and white of the ceiling of one of the main caverns in Altamira. The bisons, horses, deer, and boar are painted in vivid colors, with prevailing red ochre. Approximate sizing is provided by the added drawing of the artist himself (Vaquero Turcios, 1995, p. 140).

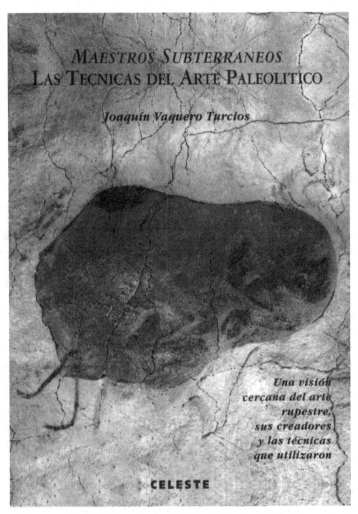

**FIGURE 13.11**    Cover of J. Vaquero Turcios's book (1995) about Paleolithic art, showing from Altamira a bison with the head turned. Translations: title, *Underground Masters. Techniques of Paleolithic Art*. Bottom right: "A close view of cave art, its creators and the techniques they used."

The Magdalenian paintings of the great "cave Sistine chapels," particularly those of Altamira (Figs 13.10 and 13.11), Chauvet, and Lascaux, are polychrome paintings of superb artistic value, even though their *symbolism* beyond their decorative value remains obscure. Some may have served conjuring, magic or even religious purposes.

Chapter 14

# Religion

## INTRODUCTION

Theologians and other religious authors have for centuries sought to demonstrate the existence of God by the *argument from design*, which asserts that organisms have been designed and only God could account for the *design* (Chapter 1). The most extensive formulation of the *argument from design* was advanced by William Paley in his *Natural Theology* (1802). Darwin's (1859) theory of evolution by natural selection disposed of Paley's arguments: the adaptations of organisms are outcomes of a natural process that causes the gradual accumulation of features beneficial to organisms and accounts for the *evolution* of new *species*.

There is "design" in the living world, but the *design* is not "intelligent," as expected from an engineer, but imperfect and worse: defects, dysfunctions, oddities, waste, and cruelty pervade the living world.

Science and religious faith need not be in contradiction. Science concerns processes that account for the natural world. Religion concerns the meaning and purpose of the world and of human life, the proper relation of humans to their Creator and to each other, and the moral values that inspire and govern people's lives.

## RELIGION

According to the great 20th-century Christian theologian Paul Tillich (1959), religion, in the largest and most basic sense of the word, is "ultimate concern." Tillich sees religion not as a special function of human spirit, but rather as the dimension of depth in all the human *creation* functions. Thus religion cannot exist separate from science, which is a creative significant human activity seeking to understand the meaning of the world and of human life, and of great consequence in technological developments and otherwise.

Teilhard de Chardin, a paleontologist and Christian theologian, asserted in *The Divine Milieu* (1960) that the separation of religion and science is a psychological absurdity. Seeking the congruence of science with religion is not only a possibility but a necessity. Teilhard believed that religion can be inspired by science. He proposed, as the fundamental religious question, how

Evolution, Explanation, Ethics, and Aesthetics. http://dx.doi.org/10.1016/B978-0-12-803693-8.00014-1

to "reconcile and provide mutual nourishment for the love of God and a healthy love of the world" (1960, p. 21). Theodosius Dobzhansky, one of the principal authors of the modern theory of *evolution*, contended in *The Biology of Ultimate Concern* (1967) that *evolution* can be a source of hope for humans and religion is, therefore, best formulated in an evolutionary context.

Daniel Dennett (1995), a distinguished philosopher and self-proclaimed atheist, consistent with the authors just cited, contends that religion has three purposes: to comfort us in our suffering and allay our fear of death, to explain things we cannot otherwise explain, and to encourage group cooperation in the face of trials and enemies. But he contrasts the claims of religion with the accomplishments of science. Religious answers are "skyhooks," which are "imaginary means of suspension on the sky," while science methods are "cranes," which "do the lifting work our imaginary skyhooks might do, and they do it in an honest, non-question-begging fashion" (Dennett, 1995, pp. 74–75).

This chapter does not seek to justify religion, or to explore in depth the diversity of religions and their distinct tenets. Rather, it explores the holdings of the three Abrahamic religions (Judaism, Christianity, and Islam) concerning the diversity and marvelous adaptations of organisms, particularly as stated by Christian authors. The prevailing theme is to show that religious beliefs and behaviors can be accounted for by means of a natural, scientific explanation, namely the theory of *evolution* by *natural selection*. Yet accepting this scientific explanation does not by itself imply that religion and faith must be rejected.

## CREATION

The Abrahamic religions, Judaism, Christianity, and Islam, proclaim the existence of a Supreme Being, a Creator, who accounts for the origin of the world and presides over everything that happens in it. Total world membership in the three Abrahamic religions amounts to over 4 billion people, more than half of the human *population*. Other major faiths with numerous members, accounting perhaps for about one-third of the human *population*, are the oriental religions. Among them Hinduism is the dominant religion in South Asia, particularly India and Nepal, with about 1 billion adherents worldwide. Hinduism is the synthesis of several religious traditions which do not include belief in a single God or Creator, although a majority of Hindus recognize Vishnu, Shiva, and Devi as different aspects of a Supreme Being or Brahman. Variations of Hinduism are Confucianism and Taoism, particularly in China. Shintoism is Japan's most distinctive religion, although about one-third of Japan's *population* is Buddhist. Buddhism is a nontheistic religion that includes a variety of beliefs and spiritual practices, mostly attributed to the teachings of Gautama Buddha, who lived around the fifth century BCE. There are in addition numerous tribal religions, particularly in Africa and the

America but also in Australia, Asia, and Indonesia, each with its own idio-syncratic myths, small membership, and limited geographic presence.

Most religions, other than Judaism, Christianity and Islam, do not profess faith in a single Supreme Being, or Creator, and are primarily concerned with moral behavior, rituals, and social customs and activities. The concept of a god, or gods, existed in Greek and Roman antiquity with the likes of Zeus, Athena, Poseidon, and many others; although they were not necessarily credited with the *creation* of the world, they had influence on human affairs, with each god's influence usually limited to particular concerns or human groups.

Other *creation* myths from antiquity are the Babylonian epic of Gilgamesh, which may date to the third millennium BCE and include a universal flood in which all humans perish by turning to clay, except those in a boat built by instruction of the god Ea. Also from Mesopotamia is the Enuma Elish, dated around 2000 BCE, which asserts that the world is not eternal but had been created on time. The Enuma Elish recognizes multiple gods. Even older are the first *creation* myths of Egypt, although these changed locally and over time. In one Egyptian account, in the beginning the sun god Atum spat out Shu, the god of air, and Tefnut, the goddess of moisture, who in turn gave birth to the earth god Geb and the sky god Nut, who mated and produced Osiris and his consort Isis. The Egyptian myths include multiple gods. However, the two largest divisions, Upper and Lower Egypt, each had a special god with a distinctive crown (Moore, 2002).[1]

## GOD'S WORLD

Religious concerns about the origin of the world and of living organisms are largely concentrated in the three Abrahamic religions. Christian authors over the centuries have explored God's actions over the world. The perceived conflicts between science and religious faith, in particular, gave rise to the metaphor of a war between religion and science which, according to the historian James R. Moore (1979), was actually introduced by Darwin's "bulldog," Thomas H. Huxley, and his followers "as part of their campaign to erect science as the new source of influence in modern society" (Bowler, 2007, p. 4). The iconic manifestation of the conflict is *History of the Conflict between Religion and Science*, published in 1875 by Huxley's American disciple J.W. Draper.

Christian authors have argued that the *order*, harmony, and *design* of the universe are incontrovertible evidence that it was created by an omniscient and omnipotent Creator. Notable Christian authors include Augustine (353−430 CE), who wrote in *The City of God* that the "world itself, by the perfect order of its changes and motions and by the great beauty of all things visible, proclaims ... that it has been created, and also that it could not have been made other than by a God ineffable and invisible in greatness, and ... in beauty." Thomas Aquinas (1224–74), considered by many to be the greatest Christian theologian, advances in his *Summa Theologiae* five ways to demonstrate by natural reason that God exists. The fifth way derives from the orderliness and designed purposefulness of the universe, which manifest that it has been created by a Supreme Intelligence: "Some intelligent being exists by which all natural things are directed to their end; and this being we call God."

This manner of seeking a natural demonstration of God's existence later became known as the "argument from design," which has two prongs. The first asserts that the universe evinces that it has been designed. The second prong affirms that only God could account for the complexity and perfection of the *design*. A forceful and elaborate formulation of the *argument from design* was *The Wisdom of God Manifested in the Works of Creation* (1691) by the English clergyman and naturalist John Ray (1627–1705). Ray regarded an incontrovertible evidence of God's wisdom that all components of the universe—the stars and the planets, as well as all organisms—are wisely contrived from the beginning and perfect in their operation. The "most convincing argument of the Existence of a Deity," writes Ray, "is the admirable Art and Wisdom that discovers itself in the Make of the Constitution, the Order and Disposition, the Ends and uses of all the parts and members of this stately fabric of Heaven and Earth."

The *design* argument was advanced, in greater or lesser detail, by a number of authors in the 17th and 18th centuries (see Chapter 1). John Ray's contemporary Henry More (1614–87) saw evidence of God's *design* in the succession of day and night, and of the seasons: "I say that the Phenomena of Day and Night, Winter and Summer, Spring-time and Harvest … are signs and tokens unto us that there is a God … things are so framed that they naturally imply a Principle of Wisdom and Counsel in the Author of them. And if there be such an Author of external Nature, there is a God." Robert Hooke (1635–1703), a physicist and eventual secretary of the Royal Society, formulated the watchmaker analogy: God had furnished each plant and animal "with all kinds of contrivances necessary for its own existence and propagation … as a Clock-maker might make a Set of Chimes to be a part of a Clock" (Hooke, 1665, p. 124). The clock *analogy*, among others such as temples, palaces, and ships, was used by Thomas Burnet (1635–1703) in his *Sacred Theory of the Earth* and would become common among natural theologians of the time. The Dutch philosopher and theologian Bernard Nieuwenfijdt (1654–1718) developed, at length, the *argument from design* in his three-volume treatise *The Religious Philosopher*, where in the Preface he introduces the watchmaker *analogy*. Voltaire (1694–1778), like other philosophers of the Enlightenment, accepted the *argument from design*. Voltaire asserted that in the same way as the existence of a watch proves the existence of a watchmaker, the *design* and purpose evident in nature prove that the universe was created by a Supreme Intelligence. The Irish archbishop of

Armagh, James Ussher, in his *Annals of the Old and New Testament* (*Annales veteris et novi testament*, 1650—1654), estimated that the Earth was created in 4004 BCE at midday on Sunday October 23.

Biblical skeptics included the French naturalist Comte de Buffon, head of the King's Botanical Gardens in Paris, who suggested in 1749 that the Earth might be 70,000 years old, 10 times older than calculated by Ussher. The editor of the *Encyclopedia*, the atheist Denis Diderot, thought there were not fixed *species*, but that nature was subject to constant flux producing organisms at random, without plan or purpose, rather than being divinely ordained creatures. The Scottish philosopher David Hume (1711—76), in his post-humously published *Dialogues on Natural Religion* written in the 1750s, criticized the *analogy* between human artifacts and organisms, denying that the element of *design* was involved in living *species*.

## NATURAL THEOLOGY

William Paley (1743—1805), one of the most influential English authors of his time, formulated at length in his *Natural Theology* (1802) the *argument from design*, based on the complex and precise *design* of organisms. Paley was an influential writer of works on Christian philosophy, *ethics*, and theology, such as *The Principles of Moral and Political Philosophy* (1785) and *A View of the Evidences of Christianity* (1794). With *Natural Theology*, Paley sought to update Ray's *Wisdom of God* of 1691. But Paley could now carry the argument much further than Ray, by taking advantage of one century of additional biological knowledge. Darwin, while he was an undergraduate at the University of Cambridge between 1827 and 1831, read Paley's *Natural Theology*, which was part of the university's canon up to nearly half a century after Paley's death. Darwin writes in his *Autobiography* of the "much delight" and profit he derived from reading Paley. "I was charmed and convinced [at that time] by the long line of argumentation."

Paley's keystone claim is that there "cannot be design without a designer; contrivance, without a contriver; order, without choice ... means suitable to an end, and executing their office in accomplishing that end, without the end ever having been contemplated." *Natural Theology* is a sustained argument for the existence of God based on the obvious *design* of humans and their organs, as well as the *design* of all sorts of organisms considered by themselves and in their relations to one another and to their environment. Paley's first analogical example in *Natural Theology* is the human eye. He points out that the eye and the telescope "are made upon the same principles; both being adjusted to the laws by which the transmission and refraction of rays of light are regulated." Specifically, there is a precise resemblance between the lenses of a telescope and "the humors of the eye" in their figure, their position, and the ability to converge the rays of light at a precise distance from the lens—on the retina, in the case of the eye.

Paley makes two remarkable observations, which enhance the complex and precise *design* of the eye. The first is that rays of light should be refracted by a more convex surface when transmitted through water than when passing out of air into the eye. Accordingly, "the eye of a fish, in that part of it called the crystalline lens, is much rounder than the eye of terrestrial animals. What plainer manifestation of design can there be than this difference? What could a mathematical instrument maker have done more to show his knowledge of [t] his principle …?" The second remarkable observation made by Paley in support of his argument is dioptric distortion: "Pencils of light, in passing through glass lenses, are separated into different colors, thereby tinging the object, especially the edges of it, as if it were viewed through a prism. To correct this inconvenience … a sagacious optician … [observed] that in the eye the evil was cured by combining lenses composed of different substances, that is, of substances which possessed different refracting powers." The telescope maker, accordingly, corrected the dioptric distortion "by imitating, in glasses made from different materials, the effects of the different humors through which the rays of light pass before they reach the bottom of the eye. Could this be in the eye without purpose, which suggested to the optician the only effectual means of attaining that purpose?" (Paley, 1802, p. 23).

Could the eye have come about without *design* or preconceived purpose, as a result of chance? Paley set the argument against chance in the very first paragraph of *Natural Theology*, arguing rhetorically by analogy: "In crossing a heath, suppose I pitched my foot against a *stone*, and were asked how the stone came to be there, I might possibly answer, that for any thing I knew to the contrary it had lain there forever … But suppose I had found a *watch* upon the ground, and it should be inquired how the watch happened to be in that place, I should hardly think of the answer which I had before given … Yet why should not this answer serve for the watch as well as for the stone; why is it not as admissible in the second case as in the first? For this reason, and for no other, namely, that when we come to inspect the watch, we perceive … that its several parts are framed and put together for a purpose … that if the different parts had been differently shaped from what they are, or placed after any other manner or in any other order than that in which they are placed, either no motion at all would have been carried on in the machine, or none which would have answered the use that is now served by it."

The strength of the argument against chance derives, Paley tells us, from what he names "relation," a notion akin to what some contemporary anti-evolutionist writers have named "irreducible complexity" (and some of them have given themselves credit for its discovery). "When several different parts contribute to one effect … the fitness of such parts … to one another for the purpose of producing, by their united action, the effect, is what I call *relation*; and wherever this is observed in the works of nature or of man, it appears to me to carry along with it decisive evidence of understanding, intention, art."

The outcomes of chance do not exhibit relation among the parts or, as we might say, organized complexity.

*Natural Theology* has chapters dedicated to the human frame, which displays a precise mechanical arrangement of bones, cartilage, and joints; to the circulation of the blood and the disposition of blood vessels; to the comparative anatomy of humans and animals; to the digestive tract, kidneys, urethra, and bladder; to the wings of birds and fins of fish; and much more. After detailing the precise organization and exquisite functionality of each biological entity, relationship, or process, Paley draws again and again the same conclusion: only an omniscient and omnipotent Deity could account for these marvels of mechanical perfection, purpose, and functionality, and for the enormous diversity of inventions that they entail.

Paley's *Natural Theology* fails, even in his time, when seeking an account of imperfections, defects, pain, and cruelty that would be consistent with his notion of the Creator. This is his general explanation for nature's imperfections: "Irregularities and imperfections are of little or no weight ... but they are to be taken in conjunction with the unexceptionable evidences which we possess of skill, power, and benevolence displayed in other instances" (Paley, 1802, p. 46). But if functional *design* manifests an intelligent designer, why should not deficiencies indicate that the designer is less than omniscient, less than omnipotent, or less than benevolent? We know that some deficiencies are not just imperfections but are outright dysfunctional, jeopardizing the very function the organ or part is supposed to serve. We now know, of course, that the explanation for dysfunction and imperfection is the process of *natural selection*, which can account for *design* and functionality but does not achieve any sort of perfection, nor is it omniscient or omnipotent. It is not only that organisms and their parts are less than perfect, but also that deficiencies and dysfunctions are pervasive, evidencing defective "design."

## THE BRIDGEWATER TREATISES

Francis Henry Egerton (1756—1829), the eighth Earl of Bridgewater, bequeathed in 1829 the sum of £8,000 sterling with instructions to the Royal Society to commission eight treatises that would promote *natural theology* by setting forth "The Power, Wisdom and Goodness of God as manifested in the Creation." Eight treatises were published in the 1830s, several of which artfully incorporate the best science of the time and had considerable influence on the public and among scientists. *The Hand, Its Mechanisms and Vital Endowments as Evincing Design* (1833) by Sir Charles Bell, a distinguished anatomist and surgeon, famous for his neurological discoveries, examines in considerable detail the wondrously useful *design* of the human hand, but also the perfection of *design* of the forelimb used for different purposes in different animals, serving in each case the particular needs and habits of its owner: the human arm for handling objects, the dog's leg for running, and the bird's wing for flying. He concludes that "Nothing less than the Power, which originally created, is equal to the effecting of those changes on animals, which are to adapt them to their conditions." William Buckland, professor of geology at the University of Oxford, notes in *Geology and Mineralogy* (1836) the world distribution of coal and mineral ores, and proceeds to point out that they had been deposited in remote parts where no humans had lived for centuries, yet obviously with the forethought of serving the larger human populations that would come much later. Another geologist, Hugh Miller, in *The Testimony of the Rocks* (1858) formulated what may be called the *argument from beauty*, which holds that it is not only the perfection of *design* but also the beauty of natural structures found in rock formations and in mountains and rivers that manifests the intervention of the Creator. One additional treatise, never completed, was authored by the notable mathematician and pioneer in the field of calculating machines, Charles Babbage (1791–1871). In the *Ninth Bridgewater Treatise: A Fragment*, published in 1838, he seeks to show how mathematics may be used to bolster religious belief.

In the 1990s a new version of the *design* argument was formulated in the United States, named *intelligent design* (ID), which refers to an unidentified Designer who accounts for the *order* and complexity of the universe, or in-tervenes from time to time in the universe so as to *design* organisms and their parts. The complexity of organisms, it is claimed, cannot be accounted for by natural processes. According to ID proponents, this intelligent designer could, but need not, be God. The intelligent designer could be an alien from outer space or some other creature, such as a "time-traveling cell biologist," with amazing powers to account for the universe's *design*. Explicit reference to God is avoided, so the "theory" of ID could be taught in public schools as an alternative to the theory of *evolution* without incurring conflict with the US Constitution, which forbids the endorsement of any religious beliefs in public institutions.

## ORGANISMS

Traditional Judaism, Christianity, and Islam account for the origin of living beings and their adaptations to life in their environments—wings, gills, hands, flowers—as they explain the origin of the universe as the handiwork of an omnipotent and omniscient God. Yet we can already see among early Christian authors gleams of evolutionary ideas; that is, acceptance of the possibility that some living *species* may have come about through natural processes. Gregory of Nyssa (335–394) and Augustine (354–430) maintained that not all of *creation* and all *species* of plants and animals were initially created by God; rather, some had evolved in historical times from God's creations.

According to Gregory of Nyssa, the world has come about in two suc-cessive stages. The first stage, the creative step, is instantaneous; the second stage, the formative step, is gradual and develops through time. According to Augustine, many plant and animal *species* were not directly created by God, but only indirectly, in their potentiality (their *rationes seminales*), so they would come about by natural processes later in the world. Gregory's and Augustine's motivation was not scientific but theological. For example, Augustine was concerned that it would have been impossible to hold representatives of all animal *species* in a single vessel, such as Noah's Ark; some *species* must have come into existence only after the Flood.

The notion that organisms may change by natural processes was not investigated as a biological subject by Christian theologians of the Middle Ages, but it was, usually incidentally, considered as a possibility by many, including Albertus Magnus (1200–80) and his student Thomas Aquinas (1224–74). Aquinas concluded, after consideration of relevant philosophical and theological arguments, that the popular belief that lower living creatures, such as maggots and flies, could develop from nonliving matter, such as decaying meat, was not incompatible with Christian faith or philosophy, but he left it to others (to scientists, in current parlance) to determine whether this actually happened.

The issue of whether living organisms could spontaneously arise from dead matter was not settled until four centuries later by the Italian Francesco Redi (1626–97), one of the first scientists to conduct biological experiments with proper controls. Redi set up flasks with various kinds of fresh meat; some were sealed, others covered with gauze so that air but not flies could enter, and others left uncovered. The meat putrefied in all flasks, but maggots appeared only in the uncovered flasks which flies had entered freely. Redi was a poet as well as a physician, chiefly known for his *Bacco in Toscana* (1685, *Baccus in Tuscany*).

The cause of putrefaction was discovered two centuries later by Darwin's younger contemporary, the French chemist Louis Pasteur (1822–95), one of the greatest scientists of all time. Pasteur demonstrated that fermentation and putrefaction were caused by minute organisms that could be destroyed by heat. Food decomposes when placed in contact with germs present in the air; the germs do not arise spontaneously within the food. We owe to Pasteur the process of pasteurization: the destruction by heat of microorganisms in milk, wine, and beer, which can thus be preserved if kept out of contact with microorganisms in the air. Pasteur also demonstrated that cholera and rabies are caused by microorganisms, and he invented vaccination, treatment with attenuated (or killed) infective agents that would stimulate the *immune system* of animals and humans, thus protecting them against infection.

The first broad theory of *evolution* was proposed by the French naturalist Jean-Baptiste de Monet, Chevalier de Lamarck (1744–1829). In his *Philosophie zoologique* (1809, *Zoological Philosophy*), Lamarck held the enlightened view, shared by the intellectuals of his age, that living organisms represent a progression, with humans as the highest form. Lamarck's theory of *evolution* asserts that organisms evolve through eons of time from lower to higher forms, a process still going on, always culminating in human beings. The remote ancestors of humans were worms and other inferior creatures, which gradually evolved into more and more advanced organisms, ultimately humans.

The "inheritance of acquired characteristics" is the theory most often associated with Lamarck's name. Yet this theory was actually a subsidiary construct of his theory of evolution: that *evolution* is a continuous process, and today's worms will yield humans as their remote descendants. It stated that as animals become adapted to their environments through their habits, modifications in their body plans occur by "use and disuse." Use of an organ or structure reinforces it; disuse leads to obliteration. The theory further asserted that the characteristics acquired by use or disuse would be inherited. This assumption was later called the inheritance of acquired characteristics (or Lamarckism). It was disproved in the 20th century.

Lamarck's evolutionary theory was metaphysical rather than scientific. He postulated that life possesses the innate property to improve over time, so progression from lower to higher organisms would continually occur, and always following the same path of transformation from lower organisms to increasingly higher and more complex organisms. A somewhat similar evolutionary theory was formulated one century later by another Frenchman, the philosopher Henri Bergson (1859−1940) in his *L'Evolution créatrice* (1907, *Creative Evolution*).

Erasmus Darwin (1731−1802), a physician and poet and the grandfather of Charles Darwin, proposed, in poetic rather than scientific language, a theory of the transmutation of life forms through eons of time (*Zoonomia, or the Laws of Organic Life*, 1794−1796). More significant for Charles Darwin was the influence of his older contemporary and friend, the eminent geologist Sir Charles Lyell (1797−1875). In his *Principles of Geology* (1830−1833), Lyell proposed that Earth's physical features were the outcome of major geological processes acting over immense periods of time, incomparably greater than the few thousand years since *Creation* that was commonly believed at the time.

## COPERNICUS AND DARWIN

Darwin occupies an exalted place in the history of Western thought, deservedly receiving credit for the theory of *evolution*. In *The Origin of Species* he laid out evidence demonstrating the *evolution* of organisms. However, Darwin accomplished something much more important for intellectual history than demonstrating *evolution*. His *Origin of Species* is, first and foremost, a sustained effort to solve the problem of how to account scientifically for the *design* of organisms, explaining their complexity, diversity, and marvelous contrivances as the result of natural processes.

The advances of physical science brought about by the Copernican Revolution had driven humankind's conception of the universe to a split-personality state of affairs. Scientific explanations derived from natural laws dominated the world of nonliving matter, on the Earth as well as in the heavens. However, supernatural powers and the unfathomable deeds of the Creator were accepted as explanations of the origin and configurations of living creatures. Authors such as William Paley argued that the complex *design* of organisms could not have come about by chance, or by the mechanical laws of physics, chemistry, and astronomy, but was rather accomplished by an omniscient and omnipotent Deity, just as the complexity of a watch, designed to tell the time, was accomplished by an intelligent watchmaker. It was Darwin's genius to resolve this conceptual schizophrenia. Darwin completed the Copernican Revolution by drawing out for biology the notion of nature as a lawful system of matter in motion that human reason can explain without recourse to supernatural agencies.

The conundrum faced by Darwin can hardly be overestimated. The strength of the *argument from design* to demonstrate the role of the Creator had been forcefully set forth by philosophers and theologians. Wherever there is function or *design*, we look for its author. It was Darwin's greatest accomplishment to show that the complex organization and functionality of living beings can be explained as the result of a natural process—*natural selection*—without any need to resort to a Creator or other external agent. The origin and adaptations of organisms in their profusion and wondrous variations were thus brought into the realm of science.

Organisms exhibit complex *design*, but it is not, in current language, "irreducible complexity" emerging all of a sudden in full bloom. Rather, according to Darwin's theory of *natural selection*, the *design* has arisen gradually and cumulatively, step by step, promoted by the reproductive success of individuals with incrementally more adaptive elaborations.

*Natural selection* accounts for the "design" of organisms, because adaptive variations tend to increase the probability of survival and reproduction of their carriers at the expense of maladaptive, or less adaptive, variations. The arguments of Paley against the incredible improbability of chance accounts of the adaptations of organisms are well taken as far as they go. But neither Paley nor any other author before Darwin was able to discern that there is a natural process (namely, natural selection) that is not random, but rather is oriented and able to generate *order* or "create." The traits that organisms acquire in their evolutionary histories are not fortuitous but determined by their functional utility to the organisms, "designed" as it were to serve their life needs.

## EVOLUTION AND THE BIBLE

To some Christians and other people of faith, the theory of *evolution* seems to be incompatible with their religious beliefs because it is inconsistent with the

Bible's narrative of *Creation*. The first chapters of the biblical Book of Genesis describe God's *creation* of the world, plants, animals, and human beings. A literal interpretation of Genesis seems incompatible with the gradual *evolution* of humans and other organisms by natural processes. Even independent of the biblical narrative, the Christian beliefs in the immortality of the soul and that humans are "created in the image of God" have appeared to many as contrary to the evolutionary origin of humans from nonhuman animals.

In 1874 Charles Hodge, an American Protestant theologian, published *What Is Darwinism?*, one of the most articulate assaults on evolutionary theory. Hodge perceived Darwin's theory as "the most thoroughly naturalistic that can be imagined and far more atheistic than that of his predecessor Lamarck." Echoing Paley, Hodge argued that the *design* of the human eye reveals that "it has been planned by the Creator, like the design of a watch evinces a watchmaker." He concluded that "the denial of design in nature is actually the denial of God."

Some Protestant theologians saw a solution to the apparent contradiction between *evolution* and *creation* in the argument that God operates through intermediate causes. The origin and motion of the planets could be explained by the law of gravity and other natural processes without denying God's *creation* and providence. Similarly, *evolution* could be seen as the natural process through which God brought living beings into existence and developed them according to his plan. Thus A. H. Strong, the president of Rochester Theological Seminary in New York state, wrote in his *Systematic Theology* (1885): "We grant the principle of evolution, but we regard it as only the method of divine intelligence." He explains that the brutish ancestry of human beings was not incompatible with their excelling status as creatures in the image of God. Strong drew an *analogy* with Christ's miraculous conversion of water into wine: "The wine in the miracle was not water because water had been used in the making of it, nor is man a brute because the brute has made some contributions to its creation." Arguments for and against Darwin's theory came from Roman Catholic theologians as well.

Gradually, well into the 20th century, *evolution* by *natural selection* came to be accepted by a majority of Christian writers. Pope Pius XII in his encyclical *Humani Generis* (1950, *Of the Human Race*) acknowledged that biological *evolution* was compatible with the Christian faith, although he argued that God's intervention was necessary for the *creation* of the human soul. Pope John Paul II, in an address to the Pontifical Academy of Sciences on October 22, 1996, deplored interpreting the Bible's texts as scientific statements rather than religious teachings. He added: "New scientific knowledge has led us to realize that the theory of evolution is no longer a mere hypothesis. It is indeed remarkable that this theory has been progressively accepted by researchers, following a series of discoveries in various fields of knowledge. The convergence, neither sought nor fabricated, of the results of work that was conducted independently is in itself a significant argument in favor of this theory."

Similar views have been expressed by other mainstream Christian denominations. The General Assembly of the United Presbyterian Church in 1982 adopted a resolution stating that "Biblical scholars and theological schools ... find that the scientific theory of evolution does not conflict with their interpretation of the origins of life found in Biblical literature" (National Academy of Sciences, 2008). The Lutheran World Federation in 1965 affirmed that "evolution's assumptions are as much around us as the air we breathe and no more escapable. At the same time theology's affirmations are being made as responsibly as ever. In this sense both science and religion are here to stay, and ... need to remain in a healthful tension of respect toward one another" (National Academy of Sciences, 2008).

Equally explicit statements have been made by Jewish authorities and leaders of other major religions. In 1984 the 95th Annual Convention of the Central Conference of American Rabbis adopted a resolution stating: "Whereas the principles and concepts of biological evolution are basic to understanding science ... we call upon science teachers and local school authorities in all states to demand quality textbooks that are based on modern, scientific knowledge and that exclude 'scientific' creationism" (National Academy of Sciences, 2008).

Christian denominations that hold a literal interpretation of the Bible have opposed these views. A succinct expression of this opposition is found in the statement of belief of the *Creation* Research Society, founded in 1963 as a "professional organization of trained scientists and interested laypersons who are firmly committed to scientific special creation": "The Bible is the Written Word of God, and because it is inspired throughout, all of its assertions are historically and scientifically true in the original autographs. To the student of nature this means that the account of origins in Genesis is a factual presentation of simple historical truths."

Many Bible scholars and theologians have long rejected a literal interpretation as untenable, however, because the Bible contains mutually incompatible statements. The very beginning of the Book of Genesis presents two different *creation* narratives. Extending through Chapter 1 and the first verses of Chapter 2 is the familiar six-day narrative, in which God creates human beings—both "male and female"—in his own image on the sixth day, after creating light, earth, firmament, fish, fowl, and cattle. In verse four of Chapter 2 a different narrative starts, in which God creates a male human, then plants a garden and creates the animals, and only then proceeds to take a rib from the man to make a woman.

Which one of the two narratives is correct and which is in error? Neither one contradicts the other if we understand the two narratives as conveying the same message: the world was created by God and humans are His creatures. But both narratives cannot be "historically and scientifically true," as postulated in the statement of belief of the Creation Research Society. Moreover, as William P. Brown pointed out in his scholarly but eminently readable

*The Seven Pillars of Creation. The Bible, Science and the Ecology of Wonder* (2010), the Bible encompasses five additional narratives of the world's *Creation*, which appear in Job 38–45, Psalm 104, Proverbs 8:27–31, Ecclesiastes 1:2–11, and Isaiah 40–55. These five *Creation* narratives are incompatible with the two narratives of Genesis and with each other, if literally interpreted.

There are numerous inconsistencies and contradictions in different parts of the Bible, for example in the description of the return from Egypt to the Promised Land by the chosen people of Israel, not to mention erroneous factual statements about the Sun circling around the Earth and the likes. Biblical scholars point out that the Bible should be held inerrant with respect to religious truth, not in matters that are of no significance to salvation. Augustine wrote in his *De Genesi ad litteram* (*Literal Commentary on Genesis*): "It is also frequently asked what our belief must be about the form and shape of heaven, according to Sacred Scripture ... Such subjects are of no profit for those who seek beatitude ... What concern is it of mine whether heaven is like a sphere and earth is enclosed by it and suspended in the middle of the universe, or whether heaven is like a disk and the Earth is above it and hovering to one side." He adds: "In the matter of the shape of heaven, the sacred writers did not wish to teach men facts that could be of no avail for their salvation." Augustine is saying that the Book of Genesis is not an elementary book of astronomy. The Bible is about religion, and it is not the purpose of the Bible's religious authors to settle questions about the shape of the universe that are of no relevance whatsoever to how to seek salvation.

In the same vein, Pope John Paul II said in 1981 that the Bible itself "speaks to us of the origins of the universe and its makeup, not in order to provide us with a scientific treatise but in order to state the correct relationships of man with God and with the universe. Sacred Scripture wishes simply to declare that the world was created by God, and in order to teach this truth, it expresses itself in the terms of the cosmology in use at the time of the writer."

## EVIL, PAIN, AND SIN

Christian scholars for centuries struggled with the problem of evil in the world. The Scottish philosopher David Hume (1711–76) outlined the problem succinctly with brutal directness: "Is he [God] willing to prevent evil, but not able? Then he is impotent. Is he able, but not willing? Then, he is malevolent. Is he both able and willing? Whence then evil?" If Hume's reasoning is valid, it would follow that God is not all-powerful or all-good. Christian theology accepts that evil exists, but denies the validity of Hume's argument.

Traditional theology distinguishes three kinds of evil: moral evil or sin, the evil originated by human beings; pain and suffering as experienced by human beings; physical evil, such as floods, tornados, earthquakes, and the imperfections of all creatures. Theology has a ready answer for the first two kinds. Sin is a consequence of free will; the flip side of sin is virtue, also a consequence of free will. Christian theologians expound that if humans are to enter into a genuinely personal relationship with their maker, they must first experience some degree of freedom and autonomy. The eternal reward of heaven calls for a virtuous life, as many Christians see it. Christian theology also provides a good accounting for human pain and suffering. To the extent that pain and suffering are caused by war, injustice, and other forms of human wrongdoing, they are also a consequence of free will; people choose to inflict harm on one another. On the flip side are good deeds by which people choose to alleviate human suffering.

What about earthquakes, storms, floods, droughts, and other physical catastrophes, as well as human birth defects and diseases? Enter modern science into the theologian's reasoning. Physical events are built into the structure of the world itself. Since the 17th century humans have known that the processes by which galaxies and stars come into existence, the planets are formed, the continents move, the weather and the seasons change, and floods and earthquakes occur are natural, not events specifically designed by God for punishing or rewarding humans. The extreme violence of supernova explosions and the chaotic frenzy at galactic centers are outcomes of the laws of physics, not the *design* of a fearsome deity.

Before Darwin, theologians encountered a seemingly insurmountable difficulty. If God is the designer of life, whence the lion's cruelty, the snake's poison, and the parasites that secure their existence only by destroying their hosts? *Evolution* came to the rescue. Jack Haught (1998), a Roman Catholic theologian, wrote of "Darwin's gift to theology." The Protestant theologian Arthur Peacocke referred to Darwin as the "disguised friend" by quoting the earlier theologian Aubrey Moore, who in 1891 wrote that "Darwinism appeared, and, under the guise of a foe, did the work of a friend" (Peacocke, 1998). Haught and Peacocke are acknowledging the irony that the theory of *evolution*, which at first seemed to remove the need for God in the world, now has convincingly removed the need to explain the world's imperfections as failed outcomes of God's *design*.

Indeed, a major burden was removed from the shoulders of believers when convincing evidence was advanced that the *design* of organisms, including humans, need not be attributed to the immediate agency of the Creator, but rather is an outcome of natural processes. If we claim that organisms and their parts have been specifically designed by God, we have to account for the incompetent *design* of the human jaw, the narrowness of the birth canal, and our poorly designed backbone, less than fittingly suited for walking upright. Imperfections and defects pervade the living world. Consider once again the human eye. The visual nerve fibers in the eye converge inside the eye cavity to form the optic nerve, which crosses the retina (to reach the brain) and thus creates a blind spot—a minor imperfection, but an imperfection of *design*, nevertheless. Squid and octopuses have compound eyes, similar in complexity and functionality to the human eye, but do not have this defect. Did the Designer have greater love for squid than for humans, and thus exhibit greater care in designing their eyes than ours? Or is it that God wanted humans to have this *design* defect as a punishment? It is not only that organisms and their parts are less than perfect, but also that deficiencies and dysfunctions are pervasive, evidencing incompetent rather than *intelligent design*. Consider the human jaw. We have too many teeth for the jaw's size, so wisdom teeth may need to be removed and orthodontists can make a decent living straightening the others. Would we want to blame God for this blunder? A human engineer would have done better.

*Evolution* gives a good account of this imperfection. Brain size increased over time in our ancestors; the remodeling of the *skull* to fit the larger brain entailed a reduction of the jaw, so the head of the newborn would not be too large to pass through the mother's birth canal. The birth canal of women is much too narrow for easy passage of the infant's head, so thousands upon thousands of babies and many mothers die during delivery. Surely we do not want to blame God for this dysfunctional *design* or the children's deaths? The theory of *evolution* makes it understandable, a consequence of the evolutionary enlargement of our brain. Females of other primates do not experience this difficulty. Theologians in the past struggled with the issue of dysfunction because they thought it had to be attributed to God's *design*. Science, much to the relief of theologians, provides an explanation that convincingly attributes defects, deformities, and dysfunctions to natural causes.

Consider the following. About 20% of all recognized human pregnancies end in spontaneous miscarriages during the first two months. This misfortune amounts at present to more than 20 million spontaneous abortions worldwide every year. Do we want to blame God for the deficiencies in the pregnancy process? Many people of faith would rather attribute this monumental mishap to the clumsy ways of the evolutionary process than to the incompetence or deviousness of an intelligent designer.

## EVOLUTION'S ACCOUNT

*Evolution* makes it possible to attribute these mishaps to natural processes rather than to the direct *creation* or specific *design* of the Creator. The response of some critics is that the process of *evolution* by *natural selection* does not discharge God's responsibility for the dysfunctions, cruelties, and sadism of the living world, because for people of faith God is the Creator of the universe and thus would be accountable for its consequences, direct or indirect, immediate or mediated. If God is omnipotent, the argument says, He could have created a world where such things as cruelty, parasitism, and human miscarriages would not occur.

One possible religious explanation goes along the following lines of reasoning. Consider, first, human beings, who perpetrate all sorts of misdeeds and sins, even perjury, adultery, and murder. People of faith believe that each human being is a *creation* of God, but this does not entail that God is responsible for human crimes and misdemeanors. Sin is a consequence of free will; the flip side of sin is virtue. The critics might say that this account does not excuse God, because God could have created humans without free will (whatever these "humans" may have been called and been like). But one could reasonably argue that "humans" without free will would be a very different kind of creature, a being much less interesting and creative than humans are. Robots are not a good replacement for humans; robots do not perform virtuous deeds.

This line of argumentation can be extended to the catastrophes and other events of the physical world and to the dysfunctions of organisms and the harms caused to them by other organisms and environmental mishaps. However, some authors do not find this extension fully satisfactory as an explanation to exonerate God from moral responsibility. The point made again is that the world was created by God, so God is ultimately responsible. God could have created a world without parasites or dysfunctionalities. But a world of life with *evolution* is much more exciting; it is a creative world where new *species* arise, complex ecosystems come about, and humans have evolved. These considerations may provide the beginning of an explanation for many people of faith, as well as for theologians.

The Anglican theologian Keith Ward (2008) stated the case even in stronger terms, arguing that the *creation* of a world without suffering and moral evil is not an option even for God: "Could [God] not actualize a world wherein suffering is not a possibility? He could not, if any world complex and diverse enough to include rational and moral agents must necessarily include the possibility of suffering ... A world with the sorts of success and happiness in it that we occasionally experience is a world that necessarily contains the possibility of failure and misery." A similar explanation was advanced by the physicist and theologian Robert J. Russell (2007), making the case why there should be natural (physical and biological) evil in the world, "including the pain, suffering, disease, death, and extinction that characterize the evolution of life."[2]

An additional point is that the physical or biological (other than human) events that cause harm are not moral evil actions, because they are not caused by moral agents but are the result of natural processes. If a terrorist blows up a bus full of schoolchildren, that is moral evil. If an earthquake kills several thousand people in China and destroys their homes and livelihood, there is no subject morally responsible, because the event was not committed by a moral agent but is the result of a natural process. If a mugger uses a vicious dog to brutalize a person, the mugger is morally responsible. But if a coyote attacks a person, there is no moral evil that needs to be accounted for. In the world of nature, physical and biological (again, excluding human deeds), no *morality* is involved. This claim, of course, may or may not satisfy everyone, but it deserves to be, and has been, explored by theologians and people of faith.

## THE NEW ATHEISTS

Some people of faith, such as the so-called Creationists and the proponents of *intelligent design*, assert that *evolution* is not compatible with religion because the theory of *evolution* makes assertions about the origin of humans and of the universe that contradict the Bible narrative and other religious beliefs. There are, on the other side, authors who assert that religion is incompatible with *evolution*, and indeed with science in general, because religious beliefs are outright false as well as toxic.

These authors include some who profess what has become known as "New Atheism," asserting that "the tenets of many religions are *hypotheses* that can, at least in principle, be examined by science and reason. If religious claims can't be substantiated with reliable evidence, the argument goes, they should like dubious scientific claims, be rejected" (Coyne, 2015, pp. xii–xiii; italics in the original). New Atheism includes distinguished scientists, philosophers, and other authors, such as Richard Dawkins, Sam Harris, Daniel Dennett, the late Christopher Hitchens, and, more recently, Jerry Coyne, quoted here (Pinker, 2015).

Richard Dawkins wrote that Darwinism makes it possible for someone to become "an intellectually fulfilled atheist": "The universe we observe has precisely the properties we should expect if there is, at bottom, no design, no purpose, no evil and no good, nothing but blind, pitiless indifference" (Dawkins, 1995, p. 133). Increasingly, in addition to being an outspoken proponent of *atheism*, Dawkins has become a severe critic of organized religion, arguing that the world religions are a positive danger to humanity (Dawkins, 2006). Daniel Dennett asserts that *natural selection*, "Darwin's idea," bears "an unmistakable likeness to universal acid: it eats through just about every traditional concept, and leaves in its wake a revolutionized worldview ... transformed in fundamental ways" (Dennett, 1995, p. 63). "Those evolutionists who see no conflict between evolution and their religious beliefs have been careful not to look as closely as we have been looking" (Dennett, 1995, p. 310). "[T]here is no Special Creation of language, and neither art nor religion has a literally divine inspiration" (Dennett, 1995, p. 144). The science historian William Provine not only affirms that there are no absolute principles of any sort, but draws the ultimate conclusion from a materialistic line of thinking that even free will is an illusion: "Modern science directly implies that there are no inherent moral or ethical laws, no absolute principles for human society ... free will as it is traditionally conceived—the freedom to make uncoerced and unpredictable choices among alternative courses of action—simply does not exist" (Provine, 1988). The subtitle of Christopher Hitchens's book *God Is Not Great* (2007) is *How Religion Poisons Everything*.

Other critics of religion include the Nobel Laureate physicist Steven Weinberg: "With or without religion, good people can behave well and bad people can do evil; but for good people to do evil—that takes religion" (Weinberg, 1999).

The New Atheists assert that valid knowledge about the natural world can be acquired through science and the scientific process of observation and experimentation. Thus religion is rejected because it does not pass the test of science. The evolutionary biologist Richard Lewontin adds an interesting twist: atheist scientists accept science and reject religion because of their uncompromising commitment to a materialistic philosophy. "It is not that the methods and institutions of science somehow compel us to accept a material explanation of the phenomenal world, but, on the contrary that we are forced by our a priori adherence to material causes to create an apparatus of investigation and a set of concepts that produce material explanations ... materialism is absolute, for we cannot allow a Divine foot in the door" (Lewontin, 1991).

The philosopher of science Michael Ruse disagrees: "Does this mean that science is just faith-based like religion? Not at all. The point about the assumptions of science is that they work. They justify themselves pragmatically ... Theories that correctly predict the unexpected do turn out to be powerful words of understanding" (Ruse, 2015, p. 243). Ruse, like other self-proclaimed atheists and agnostics, has been an articulate player, defending evolutionism against the attacks of Creationists and intelligent designers. Yet he has disagreed openly with Dawkins and Dennett. In the United States and other countries where a majority of the *population* hold religious beliefs, it is counterproductive to link the theory of *evolution* strongly to materialism. As Ruse has pointed out, there are, indeed, many people of faith who are evolutionists (see Ruse, 2001, 2008). The science historian Peter J. Bowler agrees. "As a historian who has spent decades studying the response to Darwin, and as an observer of modern debates in America and Europe, I too believe that the best defense of evolutionism is to show the complexity of the religious approach to science. There are many scientists who still have deeply held religious beliefs, and many religious thinkers who are happy to accept evolution" (Bowler, 2007, pp. 3—4).

Most famously, Stephen Jay Gould, who claimed to be an agnostic, asserted science and religion as nonoverlapping magisteria (NOMA): they deal with different subjects. Science is concerned with the facts of observed reality, whereas religion is concerned with *morality* and other human values. "Science covers the empirical realm: what is the universe made of (fact) and why does it work this way (theory). The magisterium of religion extends over questions of ultimate meaning and moral value. These two magisteria do not overlap, nor do they encompass all inquiry (consider, for example, the magisterium of art and the meaning of beauty)" (Gould, 1999, p. 6). "NOMA is no wimpish, wallpapering, superficial device, acting as a mere diplomatic fiction and smoke screen to make life more convenient by compromise in a world of divine and contradictory passions. NOMA is a proper and principled solution—based on sound philosophy—to an issue of great historical and emotional weight" (Gould, 1999, p. 92).

## EVOLUTION AND FAITH

*Evolution* and religious beliefs need not be in contradiction. Indeed, if science and religion are properly understood, they *cannot* be in contradiction because they concern different matters. Science and religion are like two different windows for looking at the world. The two windows look at the same world, but they show different aspects of it. Science concerns the processes that account for the natural world: how planets move, the composition of matter and the atmosphere, the origin and adaptations of organisms. Religion concerns the meaning and purpose of the world and of human life, the proper relation of people to the Creator and to each other, the moral values that inspire and govern people's lives. Apparent contradictions only emerge when either the science or the religious beliefs, or often both, trespass across boundaries and wrongfully encroach upon one another's subject matter.

The scope of science is the world of nature, the reality that is observed, directly or indirectly, by our senses. Science advances explanations concerning the natural world, explanations that are subject to the possibility of *corroboration* or rejection by observation and experiment. Outside that world, science has no authority, no statements to make, no business whatsoever taking one position or another. Science has nothing decisive to say about values, whether economic, aesthetic, or moral; nothing to say about the meaning of life or its purpose; nothing to say about religious beliefs (except in the case of beliefs that transcend the proper scope of religion and make assertions about the natural world that contradict scientific knowledge; such statements cannot be true).

Science is *methodologically* materialistic or, better, methodologically *naturalistic*. I prefer the second term because "materialism" often refers to a metaphysical conception of the world, a philosophy asserting that nothing exists beyond the world of matter, nothing exists beyond what our senses can experience. The question whether or not science is inherently materialistic depends on whether we are referring to the methods and scope of science, which remain within the world of nature, or to the metaphysical implications of materialistic philosophy asserting that nothing exists beyond the world of matter. Science does not imply metaphysical materialism.

Scientists and philosophers who assert that science excludes the validity of any knowledge outside science make a "categorical mistake," confusing the method and scope of science with its metaphysical implications. Methodological naturalism asserts the boundaries of scientific knowledge, not its universality. Science transcends cultural, political, and religious differences because it has no assertions to make about these subjects (except, again, to the extent that scientific knowledge is negated). That science is not constrained by cultural or religious differences is one of its great virtues. Science does not transcend those differences by denying them or taking one position rather than another. It transcends cultural, political, and religious differences because these matters are none of its business (Scott, 2009).

Science is a way of knowing, but it is not the only way. Knowledge also derives from other sources. Common experience, imaginative literature, art, and history provide valid knowledge about the world; and so do revelation and religion for people of faith. The significance of the world and human life, as well as matters concerning moral or religious values, transcends science. Yet these matters are important; for most people, including scientists, they are at least as important as scientific knowledge *per se*.

The proper relationship between science and religion can be, for people of faith, mutually motivating and inspiring. Science may inspire religious beliefs and religious behavior, as we respond with awe to the immensity of the universe, the glorious diversity and wondrous adaptations of organisms, and the marvels of the human brain and human mind. Religion promotes reverence for the *Creation*, for humankind as well as for the world of life and the environment. Religion often is, for scientists and others, a motivating force and source of inspiration for investigating the marvelous world and solving the puzzles with which it confronts us.

## Endnotes

1. Two very informative books explaining in depth, as well as elegantly, the world's *Creation* stories are Karen Armstrong, *A History of God. The 4,000-Year Quest of Judaism, Christianity and Islam* (1993) and Jonathan Kirsch, *God against the Gods. The History of the War between Monotheism and Polytheism* (2004). See also Karen Armstrong's *The Case for God* (2009). Kirsch argues that monotheism was not first due to the Israelites; rather, "the fact is that an eccentric young pharaoh of Ancient Egypt was apparently inspired to worship a single god even before … [the idea] was embraced by the ancient Israelites. The first recorded experiment in monotheism took place in ancient Egypt in the 14th century B.C.E. under the reign of a pharaoh called Akhenaton, and it is likely that the Israelites borrowed the idea from the Egyptians" (Kirsch, 2004, p. 4). However, "Akhenaton failed to win the hearts and minds of the ordinary men and women of ancient Egypt … Promptly upon his death, the monotheistic revolution that he imposed from above on ancient Egypt was wholly undone" (Kirsch, 2004, p. 27). Kirsch's *God against the Gods* is an extensive and well-documented history of the Roman Empire's transition from pagan worship to the official acceptance of Christianity, associated with Constantine the Great (274—337 CE), who became Augustus Emperor in 312 CE, and the publication of the Edict of Milan in 313 CE. Constantine was not baptized until 337, on his deathbed. It also covers the restoration of paganism and persecution of Christians by Julian the Apostate, nephew of Constantine, who became Augustus after the death of Constantine's two sons, but whose reign lasted only 18 months before his death in battle on June 26, 363; and the

restoration of Christian tolerance by Julian's successor, Jovian, who "reigned as a Christian emperor ... [and] issued an edict of toleration ... promising that no one would be persecuted for the practice of any religion, pagan or Christian" (Kirsch, 2004, p. 271).

2. See also Robert J. Russell, *Cosmology, Evolution, and Resurrection Hope* (2006) and J. David Pleins, *The Evolving God. Charles Darwin on the Naturalness of Religion* (2013).

# Chapter 15

# Law and the Courts

## INTRODUCTION

In September 1943, Joan Berry gave birth to a baby daughter, Carol Ann. The father, she claimed in court, was Charlie Chaplin, something ardently denied by the great movie star. The court decided against Chaplin, who was ordered to pay for Carol Ann's support. Charlie Chaplin's blood group was O, Joan Berry's was A. Chaplin could not have been the biological father of Carol Ann, who had blood group B. The verdict was nevertheless affirmed by a California appellate court, although with the vigorous dissent of Justice McComb, who argued that "modern science [has] brought new aids ... [that] have revised the judicial guessing game of the past into an institution approaching accuracy in portraying the truth as to the actual fact ... If the courts do not utilize these unimpeachable methods for acquiring accurate knowledge of pertinent facts they will neglect the employment of available, potent agencies which serve to avoid miscarriages of justice" (Huber, 1991, pp. 148–149).

More and more recent cases of "junk science in the courtroom" (Huber, 1991) can readily be cited. For example, in 1963 it was found that Jack Murdock had cancer in the left testicle. Some time earlier he had been violently thrust against a car's seatbelt by a rear-end collision. A medical doctor testified that a "possibility" existed that the trauma produced by pressure from the seatbelt upon impact "might" have caused inflammation of the testicles, which "conceivably" might have triggered the cancer. A Georgia court of appeals agreed. In 1964, Jerome Baker's chest was bruised against the steering wheel when his car was hit from behind. Two months later, he died of lung cancer. The Pennsylvania Supreme Court upheld a jury verdict that attributed the cancer and death to the collision.

The Ohio Supreme Court has an even loftier record. It upheld compensation for a worker who developed cancer of the right lung and claimed it was caused when a piece of equipment hit him on the left side of his chest. The claim was supported by an "expert" who proposed a "contre coup" theory, according to which trauma causes cancer by crossing the sides of the body (Huber, 1991, p. 44).

Evolution, Explanation, Ethics, and Aesthetics. http://dx.doi.org/10.1016/B978-0-12-803693-8.00015-3
Copyright © 2016 Elsevier Inc. All rights reserved.

## SCIENCE AND THE COURTS

Human knowledge takes many forms, from art to religion, philosophy to athletic skills. But scientific knowledge is unique in the systematic understanding it provides of the world within and around us and the wondrous technologies derived from it. Although most people appreciate that scientific knowledge is special, they know science only by its fruits. They do not understand how science works through the formulation and testing of hypotheses, nor do they understand the institutional mechanisms science has developed for sharing and evaluating scientific results.

Small wonder, then, that legal and public policy decisions based on scientific knowledge present a persistent problem. This problem is apparent in US courtrooms, where scientific evidence is becoming increasingly important and controversial.

Judges and lawyers generally exhibit little enthusiasm for scientific matters when they appear in the courts of law. Scientific knowledge is highly specialized and often esoteric, which makes it paradoxical that nonscientist judges and jurors are the ones to decide disputes about science when it is introduced in the courts. Recognizing this paradox, the legal profession mostly agrees on the need for special measures to ensure the reliability of scientific evidence. Agreement breaks down, however, over how rigorously courts should screen this evidence and what tests for admissibility judges should apply.

No coherent conceptual framework emerged for courts to use authoritatively in determining whether or not information offered into evidence constitutes valid scientific knowledge until the US Supreme Court decision *Daubert v. Merrell Dow Pharmaceuticals* of June 28, 1993. Though validity was often recognized as the primary concern, there was very little guidance about what validity means or how it is established. Instead, most courts avoided coming to grips with science by embracing various surrogate tests based on factors like general acceptance, peer review, or error rate, which resulted in legal verdicts at odds with scientific reality. Courts can and must do better, but improvement depends on judges and lawyers learning how scientists themselves evaluate scientific claims. Though the details of science may be remote from common experience, nonscientists can understand the fundamental characteristics that separate valid science from pale imitations. Courts can also avail themselves of the institutional mechanisms science has developed to screen and disseminate scientific information.

There are, of course, the Federal Rules of Evidence, but most important is the US Supreme Court decision in the *Daubert* case. *Daubert* clearly establishes that trial judges must evaluate expert scientific testimony "at the outset," and their analysis should focus on whether the testimony of the expert witnesses constitutes "scientific knowledge that ... will assist the trier of fact to understand or determine a fact in issue." Justice Harold Blackmun's opinion requires federal judges to undertake "a preliminary assessment of whether the reasoning or methodology underlying the testimony is scientifically valid and of whether that reasoning or methodology properly can be applied to the facts in issue," but it provides only scant guidance on what factors should be considered and what procedures should be followed in conducting this review. The admonition to concentrate on the fundamental question of whether testimony derives from valid scientific knowledge points in the right direction and should bring legal decisions more in line with the realities of science. To realize this promise, however, courts must develop an analytical framework based on the criteria and quality-control mechanisms that scientists themselves rely on when they assess each other's work (Black et al., 1994).

Properly applied, the *Daubert* test should mean a deeper and more detailed preliminary review of scientific claims than most courts usually undertake.

## THE *FRYE* DECISION: SEVEN DECADES WITHOUT APPROPRIATE SCIENCE

Expert witnesses in common law were first used in England in the 14th century as court-appointed witnesses. In the United States, by the 19th century, experts were usually hired by the disputing parties. However, the courts reined in the experts, limiting their role to conveying the consensual view of their profession. This practice became sanctioned in *Frye v. United States* by a federal appellate court that ruled in 1923 that only testimony founded on theories, methods, and procedures generally accepted by other experts in the same field could be presented in court. This "general acceptance" rule, known as the *Frye* test, was one of the two principles followed by the US courts in deciding the admissibility of expert testimony as evidence. The other principle, known as the "helpfulness standard," comes from the Federal Rules of Evidence, first codified in 1975. Rule 702 states that expert testimony is allowed when "scientific, technical, or other specialized knowledge will assist the trier of fact to understand the evidence or to determine a fact."

Although commentary and complaints about expert witnesses predate the 1923 decision in *Frye*, recognition of the peculiar problems of scientific evidence originated with that seminal case. The *Frye* rule conditioned the admissibility of scientific evidence on whether the principle or discovery from which the evidence derived is "sufficiently established to have gained general acceptance in the particular field in which it belongs." For almost 30 years *Frye* remained largely unnoted, but as courts began to apply the general acceptance rule, criticism mounted. The foremost critic was C. T. McCormick, whose 1954 treatise on evidence declared general scientific acceptance an improper criterion for admissibility (McCormick, 1954).

*Daubert* ended the longstanding debate about *Frye* (at least in federal courts) by discarding general acceptance as the sole test of admissibility. The *Daubert* decision also affirms, however, that trial judges must screen scientific evidence before admitting it, which rekindles the question of how a court should determine scientific validity. Building on *Daubert* without repeating past mistakes thus requires an understanding of which approaches have worked and which have not.

Neither *Frye* nor any of the various relevance/reliability alternatives produced consistent and rational decisions, primarily because they all avoid the real question: does the evidence at issue derive from valid scientific knowledge? In most cases, the legal "tests" for admissibility became labels for results rather than rules for analysis. *Frye*'s only salutary effect was to institutionalize hesitation to embrace scientific claims too quickly, which is hardly a substitute for thoughtful consideration of scientific merit. Some courts would undertake a more thorough assessment of scientific validity, but this approach reflects their willingness to grapple with scientific details, not the rule they applied.

The pre-*Daubert* standards for reviewing scientific evidence were all essentially surrogates for evaluating whether scientific claims are really scientific. Though most of them derive from some element of scientific practice, none by itself provided an adequate basis for determining validity. Simply combining several such elements into an undifferentiated list advanced the analysis only slightly.

## *DAUBERT*: A NEW FOCUS ON SCIENTIFIC KNOWLEDGE

The plaintiffs in *Daubert v. Merrell Dow Pharmaceuticals* claimed that the antinausea drug Bendectin taken by thousands of pregnant women caused birth defects. The oral arguments were presented to a Supreme Court overflowing with public onlookers on March 30, 1993. Justice Harold Blackmun delivered the opinion of the court three months later on June 28, 1993. The *Daubert* case attracted 22 amicus briefs from over 100 organizations and individuals. The widespread interest in *Daubert* demonstrates the crucial role science plays in modern litigation (Black et al., 1994).

The parties and the amici in *Daubert* focused on two competing policy concerns: first, that an overly lax evidentiary standard might impede the judicial process by cluttering many trials with confusing and prejudicial nonsense cloaked as science; and second, that an overly strict standard could hamstring the courts' efforts to reach decisions as fully informed as possible by the latest scientific discoveries and developments. In two paragraphs at the end of *Daubert*, the court addressed both these concerns. The first paragraph seems to support relaxation of limits on admissibility, but the second seems to affirm the judge's role as gatekeeper.

The court directed the first paragraph to concerns that "abandonment of 'general acceptance' as the exclusive requirement for admission will result in a 'free-for-all' in which befuddled juries are confounded by absurd and irrational pseudoscientific assertions." Justice Blackmun responded to such concerns by expressing faith in juries and the adversary system. "Vigorous cross-examination, presentation of contrary evidence, and careful instruction on the burden of proof are the traditional and appropriate means of attacking shaky but admissible evidence." The opinion also pointed out that admissibility is not the only hurdle that evidence must clear; if scientific testimony proves insufficient, a court may direct judgment or grant summary judgment.

In the next paragraph, the court responded to the concern that screening by judges might "sanction a stifling and repressive scientific orthodoxy and ... be inimical to the search for truth." Justice Blackmun pointed out that whereas scientific conclusions "are subject to perpetual revision," the law "must resolve disputes finally and quickly." Science can entertain a wide range of hypotheses, he noted, but "[c]onjectures that are probably wrong are of little use ... in the project of reaching a quick, final, and binding legal judgment." In other words, courts must rely on validated hypotheses, not mere speculations.

*Daubert* recognizes that having judges apply this standard in their gatekeeping role may exclude questionable evidence that later proves to be correct. "That, nevertheless, is the balance that is struck by Rules of Evidence designed not for the exhaustive search for cosmic understanding but for the particularized resolute of legal disputes." Thus the court acknowledges that trial judges will often exclude evidence even though exclusion might limit the search for the truth.

## DETERMINING SCIENTIFIC VALIDITY

The *Daubert* court avoided providing a definitive checklist or test for evaluating scientific validity, but did make four general observations. First, it recognized that a theory or technique constitutes valid scientific knowledge only if it is testable and has in fact been tested. "Scientific methodology today is based on generating hypotheses and testing them to see if they can be falsified; indeed, this methodology is what distinguishes science from other fields of human inquiry."

Second, the court acknowledged the important role of peer review and publication in the *development* of theories and techniques on which testimony or evidence is based. Though peer review (or lack thereof) will not always be dispositive on the issue of admissibility, "submission to the scrutiny of the scientific community is a component of 'good science,' in part because it increases the likelihood that substantive flaws in methodology will be detected."

Third, the court set forth special considerations applicable to scientific techniques, including the known or potential error rate and the standards controlling a technique's operation.

Finally, the court noted that "general acceptance," the *Frye* criterion, could still have a bearing on the inquiry into the validity of scientific evidence.

"Widespread acceptance can be an important factor in ruling particular evidence admissible." Furthermore, the court recognized that the converse is also true. Indeed, a technique that is well known but not widely recognized "may properly be viewed with skepticism."

Some commentators have read these observations as essentially constituting a new four-factor test: *falsifiability*, peer review, error rate and standards, and general acceptance. After briefly discussing the four "factors," Justice Blackmun emphasized the need to focus on the "overarching subject [of] the scientific validity—and thus the evidentiary relevance and reliability—of the principles that underlie a proposed submission." Heeding this admonition requires far more than the kind of label and checklist approaches that had worked so poorly in the past. If lawyers and judges want to apply the new *Daubert* test rationally, they have to learn what distinguishes science from other forms of knowledge—what it is that makes science scientific.

## SCIENTIFIC CRITERIA FOR SCIENTIFIC CLAIMS

Scientific knowledge is unique and special in the way it affects modern life. Computers, vaccines, *gene* therapies, and countless other technological achievements derived from science attest to its validity.

Scientific knowledge is also unique in the way it emerges by way of consensus and agreement among scientists, and in the way new knowledge builds upon past accomplishments rather than starting anew with each practitioner or each generation. Modern scientists do not challenge the existence of atoms or of a universe with myriad stars, and they do not deny that deoxyribonucleic acid (*DNA*) determines heredity. Scientists, who build upon matters resolved in the past to formulate new questions for resolution, differ from philosophers, who interminably debate the questions they seek to answer without ever achieving resolution. Nor does one find among scientists anything like the radically disparate and irreconcilable views held by different religions or the ever-changing means of artistic expression.

The legal profession has long recognized the uniqueness of the knowledge that science can provide, but has struggled to understand this knowledge and the way it is produced. The law also has had trouble drawing boundaries to delineate what falls within scientific knowledge and what does not.

Lawyers and judges will accomplish the task of understanding science best if they look for the same traits that are important to scientists and rely on the same process of review that scientists use. Though the details of science may be complex, the characteristics of valid scientific knowledge and the kind of reasoning that produces it are not difficult to grasp. Moreover, the cumulative nature of scientific knowledge and the need for constant communication among scientists have given rise to identifiable institutional mechanisms like peer review, which in many ways lie at the heart of the scientific endeavor. By using these scientific criteria and relying on the review mechanisms of science, the law can take full advantage of scientific expertise while avoiding the pitfalls of false claims posed in scientific garb (see Chapter 9).

Among the characteristics that distinguish scientific knowledge from other forms of knowledge, the single most salient characteristic is *falsifiability* (Ayala, 1977). For scientists, a new idea or explanation is not valid unless there is the possibility that empirical testing can prove it false and until it has withstood thoughtful efforts at *falsification*. Because people typically think science works by verifying the "truth" of hypotheses rather than by attempting to falsify them, the requirement that scientific hypotheses survive attempts at *falsification* may seem surprising, counterintuitive, and perhaps trivial. The distinction between verification and *falsification* is, however, crucial to an understanding of scientific reasoning.

The logical nature of universal statements creates an asymmetry between *falsifiability* and *verifiability*. A universal statement can be shown to be false if it is found inconsistent with even one singular statement about a particular event or occurrence. But the reverse is not true; a universal statement can never be proven true by virtue of the truth of particular statements, no matter how numerous.

A simple example is used in Chapter 9 to illustrate why *falsification* is so pivotal to science. Consider the hypothesis that apples are made of iron, from which we derive the prediction that they should fall to the ground when cut off a tree. Apples do in fact fall when cut off, but it would be wrong to conclude that we have verified they are made of iron. In fact, if we conduct another experiment, based on the prediction that iron apples will sink in water, our hypothesis will quickly be falsified when the apples float. *Falsification* is possible, but verification is not. Thus no hypothesis can ever be proven absolutely true. But a hypothesis may become well corroborated if it survives a variety of appropriate tests that fail to falsify it.

The proper form of logical inference for conditional statements is what logicians call the *modus tollens* (*modus* = mode; *tollens* = taking away, rejecting). The *modus tollens* is a logically conclusive form of inference. If the two premises of an argument are true, but the conclusion derived from a particular hypothesis is false, it follows that the hypothesis is false.

Hypothesis testing is not a simple pass-or-fail proposition. In practice, what constitutes adequate *corroboration* depends on what decision will follow from accepting or rejecting a new idea. Provisional acceptance as the basis for further research is one thing; a decision to license a new drug *therapy* is quite different. In either case, a scientist would weigh not so much the number of different tests attempted as their severity (discussed later).

It follows from the requirement of *falsifiability* that any scientific hypothesis must divide particular statements of fact into two subclasses: "potential falsifiers"—statements with which the hypothesis is inconsistent— and "permitted statements"—statements that do not contradict the hypothesis. A hypothesis is scientific only if it has at least some potential falsifiers because it can make empirically meaningful assertions about only such statements. Testing a hypothesis either establishes or fails to establish falsehood; it never establishes absolute truth. "Not for nothing do we call the laws of nature 'laws': the more they prohibit the more they say" (Popper, 1959, p. 41). The empirical or informational content of a hypothesis is measured by the *class* of its potential falsifiers (Chapter 9).

Because the truth of scientific hypotheses can never be established conclusively, they can only be accepted contingently. The degree of confidence that scientists have in a hypothesis that has survived *falsification* depends on the variety and severity of the tests performed. Variety is required because performing the same test again and again provides little or no new information. But if the tests are varied, each will provide an incremental piece of *corroboration* for the hypothesis. Severity refers to the likelihood that the outcome will be incompatible with the hypothesis if this is false. The more precise the predictions derived from a hypothesis, the more severe a test will be. The best tests are the most severe because surviving those tests provides the most valuable information about a hypothesis. A so-called "critical" test is an experiment for which competing hypotheses predict alternative, mutually exclusive outcomes. A critical test will corroborate one hypothesis and falsify the others.

An example of a critical test is the experiment of Matthew Meselson and Franklin Stahl (1958) described in Chapter 9, testing the double-helix model of *DNA* proposed by James Watson and Francis Crick (1953). This model predicts that the replication of *DNA* is "semiconservative," (each daughter *DNA* molecule will consist of one parental strand and a newly synthesized strand). Two other possible models of *DNA* replication were the "conserved" model (the parental *DNA* molecule is fully conserved and the daughter molecule consists wholly of newly synthesized *DNA*), and the "dispersive" model (the parental molecule becomes degraded into its component fragments, nucleotides, which are then used, together with additional nucleotides, in the synthesis of the two strands of the daughter *DNA* molecule). Meselson and Stahl's experiment confirmed the semiconservative model. Scientists sometimes refer to hypotheses or models as "facts" or "laws" after they have become extensively corroborated.

## SCIENCE IN THE COURTS

Lawyers and judges need to understand that the scientific landscape encompasses many hypotheses that lie between the poles of speculative conjecture and established scientific fact. There is no particular level of certainty at which a proposition becomes "scientific" for legal purposes. Deciding if a hypothesis is corroborated enough for legal purposes depends on a variety of factors, including the kind of case. In civil litigation, for example, requiring *corroboration* sufficient to establish that a hypothesis is more likely correct than incorrect would seem to be in line with the plaintiff's ultimate burden of proof. Whether testimony that accords with scientific practice but falls below this more-likely-than-not degree of *corroboration* should be held inadmissible or insufficient is unclear in the wake of *Daubert*.

Lawyers and judges must also recognize that different scientific hypotheses may emphasize different aspects of science. In some cases a hypothesis may so elegantly and precisely clarify a previously murky area that it is quickly accepted with only minimal observational or experimental *corroboration*. Newton's theory of gravity and his laws of mechanical motion, for example, crystallized the scientific understanding of a host of seemingly disparate phenomena. The *development* of quantum mechanics during the first half of the 20th century provided similar clarification about the behavior of matter at an atomic and subatomic level.

Most scientific discoveries, however, are far less universal than Newton's laws or quantum mechanics in what they explain, and most are based far more on observational or experimental *corroboration* than on explanatory power. The relationship between cigarette smoking and lung cancer is a classic example. Medical science has yet to identify fully the chemical or chemicals in tobacco that have this effect and has not developed a detailed and precise explanation of how tobacco smoke causes cancer. Nonetheless, extensive epidemiologic results have established quite conclusively that cigarette smoke causes lung cancer.

## EXAMPLES OF GOOD SCIENCE: GREGOR MENDEL

Countless examples from the history of science demonstrate how the model of science sketched out here and in Chapter 9 actually works in practice. Most educated people know such famous instances as the experiments of Galileo Galilei and Sir Isaac Newton demonstrating the laws of motion, Blaise Pascal's measurements of atmospheric pressure, William Harvey's demonstration of the circulation of the blood, Antoine Lavoisier's rejection of the *phlogiston* theory and demonstration of the existence of oxygen, and Louis Pasteur's experiments on fermentation and putrefaction showing that they are caused by living organisms. Other examples abound, such as the Meselson and Stahl experiment cited earlier.

In every case one sees the two episodes—creative and critical—that characterize scientific knowledge. In every case science advanced through daring hypotheses and experiments cleverly designed to falsify them. One distinguished example is the discovery of the laws of biological heredity by Gregor Mendel. This major discovery in the history of biological inheritance contrasts with the abuses of scientific methodology and the interference of politics in the "theory" of biological inheritance proposed by Trofim Lysenko and sanctioned by Stalin. Gregor Mendel's discovery of the laws of heredity and his formulation of a theory of genetics provide a classic illustration of science at its best. At the other extreme, we find Trofim Lysenko's effort to bend scientific reality to fit communist ideology (see Chapter 9).

## GOOD AND BAD SCIENCE: ROBERT KOCH AND TUBERCULOSIS

Errors in science do not usually result from such aberrations as the Lysenko affair. Far more commonly, errors occur because well-qualified and well-intentioned scientists sometimes violate the proper procedures of scientific investigation. The process of empirical testing may be bent or ignored even when scientists act with the best of motives. The case of Robert Koch (1843–1910) and his tuberculosis vaccine described in Chapter 9 provides a particularly instructive example. Koch discovered the tubercle *bacillus*, the agent that causes tuberculosis, a discovery he announced on March 24, 1882 and for which he would later (in 1905) receive the Nobel Prize. A few years after the discovery of the tuberculosis bacillum, he announced that he had discovered a substance that could both protect against tuberculosis and cure it. Koch had anticipated, on the basis of limited evidence, that injecting people with dead bacilli would result in a local reaction that might protect them if they were infected with live organisms. Koch became persuaded that his hypothesis regarding diagnosis and cure would prove correct, so he proceeded to announce his vaccine before conducting appropriate tests. The *British Medical Journal*, which had earlier celebrated the original announcement, published devastating criticisms of Koch for having recommended a remedy (which proved to be ineffective) without adequate testing (Brock, 1999).

Koch was not a charlatan like the Russian Lysenko, but wishful thinking about a discovery can be just as destructive as forcing science to fit political ends. The story of the failed tuberculosis vaccine shows why science emphasizes replication, review, and the free exchange of ideas and information.

Though empirical testing usually serves very well to sort out correct from incorrect hypotheses and would have prevented Koch's worthless vaccine, the process is subject to error. The sources of error fall roughly into two categories: implicit problems and performance problems. By implicit problems, one means that a scientist may reach the wrong conclusion because the prediction being tested is not really a logical consequence of the hypothesis or because of erroneous assumptions in an experiment's *design*. Performance problems refer to mistakes that intrude into the actual conduct of an experiment and may result from negligence or even fraud.

## INSTITUTIONAL MECHANISMS

The practice of science is largely self-policing and depends on the free exchange of accurately reported information. To minimize the occurrence and effect of errors, scientists have developed mechanisms, such as peer review (see Chapter 9, final section on "Social Mechanisms: Peer Review and

Publication"). Scientists generously invest their time and effort in the peer-review process precisely because of the need to eliminate erroneous hypotheses and procedures as quickly as possible. An experiment may take several months and require the investment of tens of thousands of dollars in materials, labor, and equipment. Scientists report the materials, conditions, and procedures they have used in full detail, so others will not misallocate their time and resources as a result of misunderstanding. In the standard format of a scientific paper, a detailed section on materials and methods usually appears between the introduction of the problem and the presentation of results.

The peer-review system represents an effort both to police scientific claims and to ensure their widest possible dissemination. The pressure on scientists to publish derives not only from narrow concerns about recognition and career advancement, but also from the desire of all scientists to learn of new developments that may guide their own work. Because submitting a paper for peer review is the best way to spread word of a new discovery or idea and establish priority for it, the process serves to get new information out fast as well as to control its quality. The comments of peer reviewers contribute to the advancement of science by helping proponents of new hypotheses improve their research and interpretations.

Peer review takes place in a variety of contexts. Informal review can occur when scientists discuss their work with one another at the laboratory bench, at seminars, and at scientific meetings. Formal peer review is generally an integral part of the scientific publication process and the process by which funds are allocated for the conduct of research. Any claim that would significantly add to or change the body of scientific knowledge must be regarded skeptically if it has not been subject to some form of peer scrutiny, preferably by submission to a reputable journal.

Going through the review process may slow down the publication of results, but the delay and the large investment of time that reviewers and journal editors dedicate to peer review are justified by the need to weed out erroneous work. As with any other human activity, this process is subject to error and prejudice, but it is the most accessible and often most dependable element of the process of invention, validation, and refinement by which scientific knowledge advances.

Publication in a peer-reviewed journal does not by itself guarantee the validity of the published results, nor is there reason for outright rejection of unpublished work or work not published in a reputable journal. But one should treat with great suspicion a scientifically significant proposition that has not been submitted for publication and has not successfully undergone review.

## CONCERN FOR REJECTING IDEAS THAT ARE RIGHT

Lawyers opposed to limiting the admissibility of questionable science often raise the specter of entrenched scientists rejecting valid new ideas. Peer review in particular is seen as stultifying *progress*. The *National Law Journal*'s op-ed page once ran a piece captioned "Peer Panels Can Stomp out Truth" (Speiser, 1993). Despite this perception by some lawyers, the proliferation of scientific journals makes it doubtful that a new idea of any merit would not be accepted somewhere.

The prestigious journal *Nature* once reluctantly published a report seemingly supportive of the possibility that infinitesimally low dilutions of reactants might have some effect (Davenas et al., 1998). Such a finding would support homeopathic medicine, but follow-up studies indicated that the original report was based on shoddy work and perhaps self-deception (Maddox et al., 1988; Pool, 1988). This incident shows why the hesitation of scientists to accept dramatically new ideas without adequate review represents reasonable caution rather than conservative intransigence.

The examples of stultifying conservatism cited by the critics of scientific standards are generally ancient and inapposite. The example most often given by those who oppose scientific standards in legal proceedings is Galileo, who in 1616 was forced by the Catholic Church to recant his view that the Earth revolved around the Sun. His problem, however, was not with other scientists, and his treatment argues not for abandoning scientific standards but for separating the scientific process from both church and state control.

Also off the mark is Speiser's (1993) example of the rejection of Louis Pasteur's germ theory of disease by the French medical establishment. Pasteur's work exemplifies the methods of science. His discoveries and ideas were widely published during his career and would easily pass both modern peer review and the kind of judicial screening procedure we advocate. He was hardly a lone outcast railing against the establishment; he *was* the establishment. He was appointed university professor at age 26 and dean of the science faculty at 31, and elected to the French Academy of Sciences at 39. A large research organization, the Pasteur Institute, was named after him, and he directed it until his death at age 72.

The misinterpretation of the histories of Galileo and Pasteur notwithstanding, there are instances in which the scientific community has delayed for years the acceptance of theories that ultimately proved valid. Mendel's discovery of the laws of inheritance arguably falls into this *category.* Two more recent examples are the discovery by Oswald Avery and his colleagues that *DNA* is the hereditary substance (rather than *protein*, as was generally believed at the time) and the theory of continental *drift* first proposed by Alfred Wegener. They illustrate not scientific intransigence, but understandable reluctance to abandon well-established ideas without adequate justification.

## *DAUBERT'S* AFTERMATH

The US Supreme Court's *Daubert* decision empowered the federal and other courts to ascertain the scientific expertise and other relevant attributes of witnesses proposed by the plaintiff or defendant. Most scientists would endorse a decision that would enable the courts to avoid aberrations, such as those mentioned at the beginning of this chapter. The legal profession also mostly agrees with the need for special measures to ensure the reliability of scientific evidence presented at trials (Black et al., 1994). Although specific scientific knowledge may be remote from common experience, most people are able to understand the fundamental features that separate good science from the aberrations asserted by ignorant or unscrupulous self-proclaimed experts. *Daubert* clearly establishes that trial judges must evaluate expert scientific testimony "at the outset" and that their analysis should focus on whether the testimony constitutes "scientific knowledge that … will assist the trier of fact to understand or determine a fact in issue" (Black et al., 1994, p. 721). Juries as well as judges can evaluate scientific claims intelligently, but one would agree with *Daubert*'s preference for preliminary judicial screening because judges are in a better position than juries to evaluate the kind of information that bears on the resolution of disputed cases where expert scientific knowledge is relevant.

More than two decades after *Daubert* the record of the courts, federal and state, is mixed. The issue whether or not the *Daubert* standards are accepted is significant because "As many as 86% of all cases involve expert evidence and, consequently, this evidence is of central importance to modern civil and criminal litigation. With certain civil cases of action— such as product liability or medical malpractice—the percentage of cases utilizing expert evidence can be even higher" (Jurs and DeVito, 2013, p. 681). Nevertheless, it is often difficult for trial judges, who must ultimately decide, to determine which expert evidence is admissible and which is not. The scientific issues at stake may be complex to understand for persons who are not trained in science. Yet the decision to admit or exclude expert testimony impacts the outcome of civil as well as criminal cases. In civil trials an injured plaintiff may be denied compensation in spite of noxious harm or a defendant may be penalized for harm that was not caused. In criminal cases expert testimony may be decisive, depriving a defendant of freedom or even life; or, alternatively, may free or reduce the penalty of an egregious criminal.

In the United States, federal and state courts until 1923 used common laws and common sense to determine whether expert evidence would be admissible, mostly as the basis of the qualifications of the expert and whether the evidence seemed valid and relevant to the case. In 1923 the DC Circuit Court explored the standards of admissibility used in various courts and formulated in the *Frye* case the "general acceptance" rule: expert testimony could not be considered at trial unless it had "gained general acceptance in the particular field to which it belongs." The issue for the courts over the seven following decades was how to determine "general acceptance" and what level of proof would be required. It was in the context of this debate that the US Supreme Court decided in 1993 the *Daubert v. Merrell Dow Pharmaceuticals, Inc.* case, determining that the courts should ascertain the qualifications of proposed expert witnesses and follow four particular rules to evaluate evidence.

Since 1993, the success of *Daubert* has been differently evaluated by different authors. According to Edward K. Cheng and Albert H. Yoon (2005, p. 472–473), "the enduring legacy of the *Daubert* decision is now relatively clear. In federal courts, where the decision is legally binding, *Daubert* has become a potent weapon of tort reform by causing judges to scrutinize evidence more closely." Tort reform efforts often focus on medical malpractice, product liability, and toxic torts—all cases in which scientific evidence is likely to play a decisive or at least highly influential role. However, according to the same authors, only in about half of the states have the state courts formally adopted the *Daubert* standard. "In those states, one might expect results similar to those observed in the federal context ... [H]owever, many states have expressly rejected *Daubert* and chosen to retain the *Frye* standard" (Cheng and Yoon, 2005, p. 471). These authors conclude that debates about the practical merits and drawbacks of *Daubert* versus *Frye* may be largely superfluous: "In addition, our findings lend support to those scholars advocating for the uniform adoption of *Daubert* by the states" (Cheng and Yoon, 2005, p. 511).

Jurs and DeVito (2013) analyzed the most extensive survey of state court cases up to the present: "a database of approximately 4 million cases ... during the period from 1990 to 2000" (p. 680). "The results of this analysis demonstrated that ... civil defendants believe the *Daubert* standard to be a stricter one" (p. 707). They point out that after the *Daubert* case, defendants, seeking stricter rules for the admission of evidence, increasingly removed their cases from state to federal courts. Moreover, whenever a "state adopts *Daubert* and in so doing returns the state and federal court to the same admissibility standard, the removal rate then drops in response ... [supporting] the conclusion that defendants perceive *Daubert* as an advantageous, stricter standard ... [T]he answer is clear: *Daubert* is the stricter standard of admissibility" (p. 706).

Chapter 16

# Intelligent Design

## INTRODUCTION

In his *Natural Theology* of 1802, the English theologian William Paley advanced the "argument from design." The living world, he argues, provides compelling evidence of being designed by an omniscient and omnipotent Creator. Paley's first example is the human eye, which he compares with a telescope: they are both made upon the same principles and bear a complete resemblance to one another in their configuration, position of the lenses, and effectiveness in bringing each pencil of light to a point at the right distance from the lens. Could these attributes be in the eye without purpose, he asks? "There cannot be *design* without designer; contrivance, without a contriver."

The *argument from design* has two parts. In the familiar form, such as that formulated by Paley and other Christian authors, the argument asserts first that organisms evince to have been designed, and second that only God could account for the *design*. This argument was advanced in a variety of forms in classical Greece and early Christianity. In the 13th century, it was proposed by Thomas Aquinas as one of five arguments to demonstrate the existence of God. It was elaborated by numerous authors over ensuing centuries, but was mercilessly criticized by David Hume. Its most extensive formulation is in Paley's *Natural Theology*. The eye—as well as all sorts of organs, organisms, and their interactions—manifests to be the outcome of *design* and not of chance, thus it shows itself to have been created by God.

Darwin's (1859) theory of *evolution* by *natural selection* disposed of Paley's arguments: the adaptations of organisms are outcomes not of chance but of natural processes that, over time, cause the gradual accumulation of features beneficial to organisms whenever these features increase the organisms' chances of surviving and reproducing.

In the 1990s the *design* argument was revived in the United States by several authors. The *flagellum* used by bacteria for swimming and the *immune system* of mammals, as well as some improbability calculations, were advanced as evidence of "intelligent design" (ID), on the grounds that natural processes could not account for the phenomena to be explained. Scientists have refuted these arguments with extensive evidence. ID is not a scientifically

Evolution, Explanation, Ethics, and Aesthetics. http://dx.doi.org/10.1016/B978-0-12-803693-8.00016-5

acceptable proposal, because it cannot be empirically tested, nor has it produced any scientific results.

There is "design" in the living world: eyes are designed for seeing, wings for flying, and kidneys for regulating the composition of the blood. The *design* of organisms comes about not by ID, but by the interaction of *mutation* and *natural selection* in a process that is creative through the interaction of chance and necessity.

## THE DESIGN ARGUMENT

The *argument from design*, as stated, is a two-tined. The first prong, as formulated for example by Paley (1743—1805), asserts that organisms in their wholes, their parts, and in their relations to one another and to the environment appear to have been designed to serve certain functions and fulfill certain ways of life. The second prong of the argument affirms that only God, an omnipotent and omniscient Creator, could account for the diversity, perfection, and functionality of living organisms and their parts.

The *argument from design* has been repeatedly advanced through history, formulated in different versions, with variable scope for each of the two prongs. The first prong comes, importantly, in at least two flavors. One version refers to the order and harmony of the universe as a whole, as for example in St. Augustine (354—430), "The world itself, by the perfect order of its changes and motions and by the great beauty of all things visible" (Augustine, 1998, pp. 452—453) or St. Thomas Aquinas (1225—1274), "It is impossible for contrary and discordant things to fall into one harmonious *order* except under some guidance, assigning to each and all parts a tendency to a fixed end. But in the world we see things of different natures falling into harmonious *order*" (Aquinas, 1905, p. 12). The second version of the first prong refers to the living world, the intricate organized complexity of organisms, as formulated by, among others, Paley (1802) and the modern proponents of ID.

The second prong of the *argument from design* has been formulated in at least three important versions. One formulation of the Designer appears in classical Greece, including Plato (1997a), who postulates the existence of a Demiurge, a creator of the universe's order, who is a universal and impersonal ordering principle rather than the personalized Judeo-Christian God. Plato's Demiurge is an orderer of the world who accounts for its rationality but not necessarily its *creation*. A second version of the Designer is the familiar Judeo-Christian God, as formulated by Paley and other Christian philosophers and theologians (such as Aquinas, 1905, 1964; see Swinburne, 1994): a "person," the Creator and steward of the universe, who creates a world from nothing and is omniscient, omnipotent, omnibenevolent, and provident for humans.

David Hume, in his posthumously published *Dialogues Concerning Natural Religion* (1779 [2006]), summarizes the *argument from design*, for the purpose of criticizing it: "Look round the world: Contemplate the whole and every part of it: You will find it to be nothing but one great machine, subdivided into an infinite number of lesser machines, which again admit of subdivisions, to a degree beyond what human senses and faculties can trace and explain. All these various machines, and even their most minute parts, are adjusted to each other with an accuracy, which ravishes into admiration all men, who have ever contemplated them. The curious adapting of means to ends, throughout all nature, resembles exactly, though it much exceeds, the productions of human contrivance; of human *design*, thought, wisdom, and intelligence. Since therefore the effects resemble each other, we are led to infer, by all the rules of *analogy*, that the causes also resemble; and that the Author of Nature is somewhat similar to the mind of man; though possessed of much larger faculties, proportioned to the grandeur of the work, which he has executed. By this argument a posteriori, and by this argument alone do we prove at once the existence of a Deity, and his similarity to human mind and intelligence" (Hume, 1779 [2006], p. 22).

Proponents of ID have in recent years formulated a third version of the second prong of the argument: an unidentified Designer who may account for the order and complexity of the universe, or may simply intervene from time to time in the universe so as to *design* organisms and their parts, because the complexity of organisms cannot be accounted for by natural processes. According to ID proponents, this intelligent designer could, but need not, be God. The intelligent designer could be an alien from outer space or some other creature, such as a "time-traveling cell biologist" with amazing powers to account for the universe's *design* (Dembski, 1995; Behe, 1996; Johnson, 2002). Explicit reference to God is avoided, so the "theory" of ID can be taught in US public schools as an alternative to the theory of *evolution* without incurring conflict with the US Constitution, which forbids the endorsement of any religious beliefs in public institutions.

## THE DESIGN ARGUMENT IN ANTIQUITY

Anaxagoras of Clazomenae (c.500–428 BCE) was among the early Presocratic Greek philosophers who formulated versions of the *argument from design*. Anaxagoras primarily concerned himself with astronomical and meteorological questions, but also addressed biological doctrines. He postulated a Mind that accounts for order in the world: "All things that were to be... those that were and those that are now and those that shall be... Mind arranged them all, including this rotation in which are now rotating the stars, the sun and moon" (Kirk et al., 1983, p. 363).

Other Presocratic Greek philosophers who saw the presence of a Mind or "ordering principle" in the harmony of the cosmos and the Earth include Diogenes of Apollonia, a near contemporary of Anaxagoras: "Men and the other living creatures live by means of air through breathing it. And this is for them both soul [ie, life principle] and intelligence... And it seems to me that that which has intelligence is what men call air, and that all men are steered by this and that it has power over all things. For this very thing seems to me to be a god and to have reached everywhere and to dispose all things and to be in everything" (Kirk et al., 1983, p. 442).

The *argument from design* is attributed to Socrates (470–399 BCE) by Xenophon. In the *Phaedo*, Plato (427–347 BCE) puts the argument in Socrates' mouth: "I have heard someone reading, as he said, from a book by Anaxagoras, and saying that it is Mind that directs and is cause of everything and placed each thing severally as it was best that it should be" (Plato, 1997a, 97c, p. 84). In the *Timaeus*, Plato attributes creative powers to this Mind, which does not create by making something out of nothing, but accounts for rational *order* in the world and the configuration of organisms: "Prior to the coming to be of time, the universe had already been made to resemble in various respects the model in whose likeness the god was making it, but the resemblance still fell short in that it didn't yet contain all the living things that were to have come to be within it. This he [Mind] went on to perform" (Plato, 1997b, 39e, p. 1243).

Among the ancient Romans, it was particularly Marcus Tullius Cicero (106–43 BCE), the great statesman and orator, who argued that the purposeful complexity of the living world, such as we see in the eye, could not come about by chance or without guidance. The *design* argument was, however, dismissed by Lucrecius (99–55 BCE): "This world was made naturally, and without *Design*, and the Seeds of Things of their own accord jostling together by Variety of Motions, rashly sometimes, in vain often, and to no purpose" (Lucrecius, 1743, p. 185).

## CHRISTIAN AND ISLAM AUTHORS

The *argument from design* was advanced in the early centuries of the Christian era on the basis of the overall harmony and perfection of the universe. Augustine (1998, pp. 452–453), as quoted earlier, affirms that the "world itself, by the perfect order of its changes and motions and by the great beauty of all things visible, proclaims... that it has been created, and also that it could not have been made other than by a God ineffable and invisible in greatness, and... in beauty." According to St. Thomas Aquinas, the *argument from design* had also been proposed by St. John of Damascus (675–749 CE).

Aquinas formulated the *argument from design*, as pointed out earlier, as a way to demonstrate the existence of God. Aquinas distinguished between truths, such as the Incarnation and the Trinity, which can be known only by divine revelation, and truths accessible by human reason, which include God's existence. In his *Summa Theologiae* he advances five ways to demonstrate by natural reason that God exists. The fifth way derives from the orderliness and designed purposefulness of the universe, which manifest that it has been created by a Supreme Intelligence: "Some intelligent being exists by which all natural things are directed to their end; and this being we call God" (Aquinas, 1964; I, 2, 3).[1]

The *argument from design* was formulated in the Middle Ages by not only Christian theologians but also Islamic and Jewish writers. For example, al-Kasim ibn Ibrahim (785−860) uses the concept of a painter, among other analogies, to argue for the existence of a Creator: "whoever sees a painting knows certainly that there is someone who painted it... This is also [for] this wonderful *creation* that is seen; whoever considers and thinks, knows certainly that it has a creator" (Abrahamov, 1986, p. 261). The Spanish Jewish philosopher Moses Maimonides (1135−1204) argued that the universe gave evidence of *design* and rejected the possibility that it would have come about as the result of natural laws: "all things in the Universe are the result of *design*... How, then, can any reasonable person imagine that the position, magnitude, and number of the stars, or the various courses of their spheres, are purposeless or the result of chance?... it is extremely improbable that these things should be the necessary result of natural laws, and not that of *design*" (Maimonides, 1956, pp. 184, 188).

The most forceful and elaborate formulation of the *argument from design*, before Paley's, was *The Wisdom of God Manifested in the Works of Creation* (1691) by the English clergyman and naturalist John Ray (1627−1705). Ray regarded as incontrovertible evidence of God's wisdom that all components of the universe—the stars and the planets, as well as all organisms—are so wisely contrived from the beginning and perfect in their operation. The "most convincing argument of the Existence of a Deity," writes Ray, "is the admirable Art and Wisdom that discovers itself in the Make of the Constitution, the *Order* and Disposition, the Ends and uses of all the parts and members of this stately fabric of Heaven and Earth" (Ray, 1691, p. 33).

The *design* argument was advanced, in greater or lesser detail, by a number of authors in the 17th and 18th centuries (see eg, Arp, 1999; Klocker, 1968). John Ray's contemporary Henry More (1614−1687) saw evidence of God's *design* in the succession of day and night and of the seasons: "I say that the Phenomena of Day and Night, Winter and Summer, Spring-time and Harvest... are signs and tokens unto us that there is a God... things are so framed that they naturally imply a Principle of Wisdom and Counsel in the Author of them. And if there be such an Author of external Nature, there is a God" (More, 1662, p. 38). Robert Hooke (1635−1703), a physicist and eventual Secretary of the Royal Society, formulated the watchmaker *analogy*: God had furnished each plant and animal "with all kinds of contrivances necessary for its own existence and propagation... as a Clock-maker might make a Set of Chimes to be a part of a Clock" (Hooke, 1665, p. 124). The clock *analogy*, among others, such as temples, palaces, and ships, was also used by Thomas Burnet (1635−1703) in his *Sacred Theory of the Earth* (1651[1965]), and would become common among British natural theologians of the time.

In Europe the Dutch philosopher and theologian Bernard Nieuwentijdt (1654−1718) developed the *argument from design* at length in his three-volume treatise *The Religious Philosopher,* where in the Preface he introduces the watchmaker *analogy* (Nieuwenfijdt, 1718[2007]). Voltaire (1694−1778), like other philosophers of the Enlightenment, accepted the *argument from design.* Voltaire asserted that in the same way as the existence of a watch proves the existence of a watchmaker, the *design* and purpose evident in nature prove that the universe was created by a Supreme Intelligence (Voltaire, 1967, pp. 262−270).

## HUME'S CRITIQUE

The great Scottish philosopher David Hume (1711−1776) wrote a multi-pronged attack against the *design* argument in his posthumously published *Dialogues Concerning Natural Religion* (1779), the dialogues in question being with an imaginary interlocutor, Cleanthes. First, Hume (1779[2006], p. 23) denies the first prong of the *design* argument, namely that the universe gives evidence of *design,* just as a house does: "If we see a house, Cleanthes, we conclude, with the greatest certainty, that it had an architect or builder, because this is precisely that *species* of effect, which we have experienced to proceed from that *species* of cause. But surely you will not affirm, that the universe bears such a resemblance to a house."

Indeed, the world is more like a vegetable or an animal, rather than a machine, writes Hume (1779[2006], p. 62), in which case we may have infinite regress, one world generated by a previous one, which in turn would have been generated by another previous world, and so on: "The world plainly resembles more an animal or a vegetable, that it does a watch or a knitting-loom. Its cause, therefore, it is more probable, resembles the cause of the former." An additional argument against the world giving evidence of *design* is that the world may contain, in itself, an ordering principle or power: "*order,* arrangement, or the adjustment of final causes, is not, of itself, any proof of *design*; but only so far as it has been experienced to proceed from that principle. For aught we can know a priori, matter may contain the source or spring of order originally within itself, as well as mind does" (Hume, 1779 [2006], p. 25).

Hume further points out that the inference from *design* to a designer would not be warranted in this case because it is based on the observation of a single instantiation—namely, the existence of the world—and lacks the repetition of observations that warrants valid inference. The inference to a designer is therefore based only on an *analogy,* and a weak *analogy* at that: "When two *species* of objects have always been observed to be conjoined together, I can *infer,* by custom, the existence of one wherever I *see* the existence of the other; and this I call an argument from experience. But how this argument can have place where the objects, as in the present case, are single, individual, without parallel or specific resemblance, may be difficult to explain... That a stone will fall, that a fire will burn, that the earth has solidity, we have observed a thousand and a thousand times; and when any new instance of this nature is presented, we draw without hesitation the accustomed inference. The exact similarity of the cases gives us a perfect assurance of a similar event; and a stronger evidence is never desired nor sought after. But wherever you depart, in the least, from the similarity of the cases, you diminish proportionately evidence; and may at last bring it to a very weak *analogy*, which is confessedly liable to error and uncertainty" (Hume, 1779[2006], pp. 30, 23).

The *design* argument, says Hume (1779 [2006], pp. 37−38) is anthropomorphic, and as such it diminishes a proper conception of God, reducing it to the human level: the *design* argument, "may it not render us presumptuous, by making us imagine we comprehend the Deity and have some adequate idea of his nature and attributes?... by representing the Deity as so intelligible and comprehensible, and so similar to a human mind, we are guilty of the grossest and most narrow partiality, and make ourselves the model of the whole universe" (also see Arp, 1998).

## NATURAL THEOLOGY

William Paley, one of the most influential English authors of his time, argued forcefully in his *Natural Theology* (1802) that the complex and precise *design* of organisms and their parts could be accounted for only as the deed of an intelligent and omnipotent "Designer." The *design* of organisms, he argued, was incontrovertible evidence of the existence of the Creator.

Paley was an influential writer of works on Christian philosophy, *ethics*, and theology, such as *The Principles of Moral and Political Philosophy* (1785) and *A View of the Evidences of Christianity* (1794). With *Natural Theology*, Paley sought to update Ray's *Wisdom of God Manifested in the Works of the Creation* (1691). But Paley could carry the argument much further than Ray by taking advantage of a century of additional biological knowledge. Paley's (1802) keystone claim is that there "cannot be *design* without a designer; contrivance, without a contriver; *order*, without choice;... means suitable to an end, and executing their office in accomplishing that end, without the end ever having been contemplated" (pp. 15−16).

*Natural Theology* is a sustained argument for the existence of God based on the obvious *design* of humans and their organs, as well as the *design* of all sorts of organisms considered by themselves and in their relations to one another and to their environment. The argument has two parts: first, that organisms give evidence of being designed; second, that only an omnipotent God could account for the perfection, multitude, and diversity of the *design*.

Paley's first analogical example in *Natural Theology* is the human eye. Early in Chapter 3, he points out that the eye and the telescope "are made upon the same principles; both being adjusted to the laws by which the transmission and refraction of rays of light are regulated" (p. 20). Specifically, there is a precise resemblance between the lenses of a telescope and "the humors of the eye" in their figure, their position, and the ability to converge the rays of light at a precise distance from the lens—on the retina, in the case of the eye.

Paley makes two remarkable observations regarding the complex and precise *design* of the eye. The first is that rays of light should be refracted by a more convex surface when transmitted through water than when passing out of air into the eye. Accordingly, "the eye of a fish, in that part of it called the crystalline lens, is much rounder than the eye of terrestrial animals. What plainer manifestation of *design* can there be than this difference? What could a mathematical instrument maker have done more to show his knowledge of [t] his principle...?" (p. 20).

The second remarkable observation made by Paley in support of his argument is dioptric distortion: "Pencils of light, in passing through glass lenses, are separated into different colors, thereby tinging the object, especially the edges of it, as if it were viewed through a prism. To correct this inconvenience has been long a desideratum in the art. At last it came into the mind of a sagacious optician, to inquire how this matter was managed in the eye, in which there was exactly the same difficulty to contend with as in the telescope. His observation taught him that in the eye the evil was cured by combining lenses composed of different substances, that is, of substances which possessed different refracting powers" (pp. 22–23). The telescope maker, accordingly, corrected the dioptric distortion "by imitating, in glasses made from different materials, the effects of the different humors through which the rays of light pass before they reach the bottom of the eye. Could this be in the eye without purpose, which suggested to the optician the only effectual means of attaining that purpose?" (p. 23).

Paley summarizes his argument by asserting the complex functional anatomy of the eye. The eye consists, "first, of a series of transparent lenses—very different, by the by, even in their substance, from the opaque materials of which the rest of the body is, in general at least, composed." Second, the eye has the retina, which, as Paley points out, is the only membrane in the body that is black, spread out behind the lenses so as to receive the image formed by pencils of light transmitted through them, and "placed at the precise geometrical distance at which, and at which alone, a distinct image could be formed, namely, at the concourse of the refracted rays." Third, he writes, the eye possesses "a large nerve communicating between this membrane [the retina] and the brain; without which, the action of light upon the membrane, however modified by the organ, would be lost to the purposes of sensation" (p. 48).

*Natural Theology* has chapters dedicated to the human frame, which displays a precise mechanical arrangement of bones, cartilage, and joints; the circulation of the blood and the disposition of blood vessels; the comparative anatomy of humans and animals; the digestive tract, kidneys, urethra, and bladder; the wings of birds and the fins of fish; and much more. For 352 pages, *Natural Theology* conveys Paley's expertise: extensive and accurate biological knowledge, as detailed and precise as was available in the year 1802. After detailing the precise organization and exquisite functionality of each biological entity, relationship, or process, Paley draws again and again the same conclusion: only an omniscient and omnipotent Deity could account for these marvels of mechanical perfection, purpose, and functionality, and for the enormous diversity of inventions that they entail.

The strength of the argument against chance derives, Paley tells us, from what he names "relation," a notion akin to what Michael Behe (1996), a modern proponent of ID, has named "irreducible complexity." This is how Paley (1802, pp. 175–176) formulates the argument for *irreducible complexity*: "When several different parts contribute to one effect, or, which is the same thing, when an effect is produced by the joint action of different instruments, the *fitness* of such parts or instruments to one another for the purpose of producing, by their united action, the effect, is what I call *relation*; and wherever this is observed in the works of nature or of man, it appears to me to carry along with it decisive evidence of understanding, intention, art."

The outcomes of chance do not exhibit relation among the parts or, as we might say, do not display organized complexity. Paley writes: "The feet of the mole are made for digging; the neck, nose, eyes, ears, and skin, are peculiarly adapted to an under-ground life... this is what I call relation" (p. 183).

Throughout *Natural Theology*, Paley displays extensive and profound biological knowledge. He discusses the fish's air bladder, the viper's fang, the heron's claw, the camel's stomach, the woodpecker's tongue, the elephant's proboscis, the bat's wing hook, the spider's web, insects' compound eyes and metamorphosis, the glowworm, univalve and bivalve mollusks, seed dispersal, and on and on, with as much accuracy and detail as known to the best biologists of his time. The complex disposition and purposeful function of the organisms' components reveal, in each case, an intelligent designer; and the diversity, richness, and pervasiveness of the designs show that only the omnipotent Creator could be this designer.

In 1829, nearly three decades after the publication of *Natural Theology*, Francis Henry Egerton (1756–1829), the eighth Earl of Bridgewater, bequeathed the sum of £8,000 to the Royal Society with instructions that it commission eight treatises that would promote *natural theology* by setting forth "The Power, Wisdom and Goodness of God as manifested in the *Creation*." Eight treatises were published in the 1830s, several of which artfully incorporate the best science of the time and had considerable influence on the public and among scientists. One additional treatise, never completed, was authored by the notable mathematician and pioneer in the field of calculating machines, Charles Babbage: *The Ninth Bridgewater Treatise: A Fragment* (1838), where he seeks to show how mathematics may be used to bolster religious belief. He advances the unexpected proposition that it "is more probable that any law, at the knowledge of which we have arrived by observation, shall be the subject to one of those violations, which, according to Hume's definition, constitutes a miracle, than that it should not be so subjected" (Jones and Cohen, 1963, pp. 69−70).

## A POLITICAL MOVE

The publication in 1859 of Darwin's *On the Origin of Species* very much disposed of *natural theology* as an attempt to prove the existence of God based on the *argument from design*. Nevertheless, in the 1990s several authors, notably biochemist Michael Behe (1996), theorist William Dembski (1995, 2002), and law professor Phillip Johnson (1993, 2002), revived the *argument from design* (see Ayala, 2006, 2007, 2010; Dembski and Ruse, 2004; Pennock, 2001; Scott, 2005; Young and Edis, 2004). Often, however, these modern proponents of ID sought to hide their real agenda—and thus avoided explicit reference to God, so the "theory" of *intelligent design* could be taught in public schools as a "scientific" rather than a religious alternative to the theory of *evolution*. These modern proponents at times claim that the intelligent designer need not be God, but could, as suggested earlier, be a space alien or some other intelligent superpower unknown to us (Scott, 2009). The folly of this pretense is apparent to anyone who takes the time to consider the issue seriously. It is nothing but a vulgar charade. Federal Judge John E. Jones III in his 2005 *Dover* decision wrote: "although proponents of IDM [the ID movement] occasionally suggest that the designer could be a space alien or a time-traveling cell biologist, no serious alternative to God as a designer has been proposed by members of the IDM." Further, Professor Behe's "testimony at trial indicated that ID is only a scientific, as opposed to a religious, project for him; however, considerable evidence was introduced to refute this claim... ID's religious nature is evident because it involves a supernatural designer... expert witness ID proponents confirmed that the existence of a supernatural designer is a hallmark of ID" (*Kitzmiller v. Dover Area School District*, 2005, pp. 25, 28−29; slip opinion, p. 18).

Proponents of ID call for an intelligent designer to explain the supposed *irreducible complexity* in organisms. An *irreducibly complex system* is defined as an entity "composed of several well-matched, interacting parts that contribute to the basic function, wherein the removal of any one of the parts causes the system to effectively cease functioning" (Behe, 1996, p. 39). The claim is that irreducibly complex systems cannot be the outcome of *evolution*, which proceeds by small steps, slowly accumulating over thousands or millions of generations the components of complex systems. According to Behe (1996, p. 72): "An irreducibly complex system cannot be produced directly... by slight, successive modifications of a precursor system, because any precursor to an irreducible complex system that is missing a part is by definition nonfunctional... Since *natural selection* can only choose systems that are already working, then if a biological system cannot be produced gradually it would have to arise as an integrated unit, in one fell swoop, for *natural selection* to have anything to act on." In other words, unless all parts of the eye come simultaneously into existence, the eye cannot function; it does not benefit a precursor organism to have just a retina, or a lens, if the other parts are lacking. The human eye, according to this argument, could not have evolved one small step at a time, in the piecemeal manner by which *natural selection* works.

Evolutionists have pointed out again and again, with supporting evidence, that organs and other components of living beings are not irreducibly complex—they do not come about suddenly, or in one fell swoop (see, for example, Ayala, 2007, 2010; Brauer and Brumbaugh, 2001; Miller, 1999, 2004; Perakh, 2004a; Pennock, 2002). Evolutionists have shown that organs, such as the human eye are not irreducible at all; rather, less complex versions of the same systems have existed in the past, and some can be found in today's organisms as well.

## MOLLUSKS' EYES

Eyes evolved gradually and achieved very different configurations in different organisms, all serving the function of vision. The simplest "organ" of vision occurs in some single-celled organisms living in water that have enzymes or spots sensitive to light, which help them move toward the surface of their pond, where they feed on the algae growing there. Some multicellular animals exhibit light-sensitive spots on their epidermis. Further steps—deposition of pigment around the spot, configuration of cells into a cuplike shape, thickening of the epidermis leading to the *development* of a cornea, *development* of muscles to move the eyes and nerves to transmit optical signals to the brain—gradually led to the highly developed eyes of vertebrates and cephalopods (octopuses and squid) and the compound eyes of insects.

A record of the major stages in the *evolution* of a complex eye has survived in living mollusks (clams, snails, and squid) (Fig. 16.1). The eye of octopuses and squid is as complex as the human eye, with cornea, iris, refractive lens, retina, vitreous internal substance, optic nerve, and muscle. Limpets (*Patella*) have about the simplest imaginable eye: just an eye spot consisting of a few pigmented cells with nerve fibers attached to them. Several intermediate stages are found in other living mollusks. One step in complexity above the limpet eye is found in slit-shell mollusks (*Pleurotomaria*), which have a cup eye: one layer of pigmented cells curved like a cup with a wide opening through which light enters, with each pigmented cell in the back of the cup attached to a nerve fiber. More complex is the pinhole-lens eye found in *Nautilus*, a marine snail. The layer of pigmented cells is considerably more extensive than in slit-shell mollusks: the pigmented cells are covered toward the front with epithelium (skin) cells that are nearly closed except for a small opening

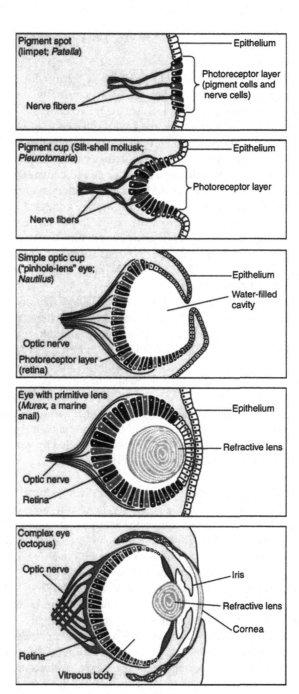

**FIGURE 16.1** Eyes in living mollusks. The octopus eye (bottom) is quite complex, with components similar to those of the human eye, such as cornea, iris, refractive lens, and retina. Other mollusks have simpler eyes. The simplest eye is found in limpets (top), consisting of only a few pigmented cells, slightly modified from typical epithelial (skin) cells. Slit-shell mollusks (second from top) have a slightly more advanced organ, consisting of some pigmented cells shaped as a cup. Further elaborations and increasing complexity are found in the eyes of *Nautilus* and *Murex*, but not yet as complex as the eyes of the squid and octopus. *Adapted from "Evolution, The Theory of," courtesy of Encyclopedia Britannica, Inc.*

("pinhole") for the passage of light, creating a cavity filled with water. *Murex*, another marine snail, has an eye with a primitive refractive lens covered with epithelium cells (serving as a primitive cornea); the pigmented cells extend through the back of the eye cavity (thus serving as a retina) and the nerve fibers are collected into an optic nerve that goes to the brain. The most advanced mollusk eye is found in the octopus and the squid, an eye that, as mentioned, is just as complex and effective as the human eye. Moreover, the octopus's and squid's eyes lack the human blind spot, an imperfection due to the fact that the nerve fibers of the human eye are collected inside the eye cavity, so the optic nerve has to cross the retina on its way to the brain; the nerve fibers and optic nerve of the octopus eye are outside the eye cavity and travel to the brain without crossing the retina (Ayala, 2007; Blake and Truscianko, 1990; Cronly-Dillon, 1991; DeDuve, 1996; Horridge, 1987; Salvini-Plawen and Mayr, 1977).

The gradual process of *natural selection* adapting organs to functions occurs in a variety of ways, reflecting the haphazard characteristics of the evolutionary process due to *mutation*, past history, and the vagaries of environments (Ayala, 1982; Ayala and Valentine, 1979; Berra, 1990; Mayr, 2001). In some cases, the changes of an organ amount to a shift of function, as in the *evolution* of the forelimbs of vertebrates, which first evolved in amphibians as adapted for walking, but which are now used by birds for flying, by whales for swimming, and by humans for handling objects. Other cases, like the *evolution* of eyes, exemplify gradual advancement of the same function—seeing, in the case of eyes. In all cases, however, the process is impelled by *natural selection* favoring through time individuals exhibiting functional advantages over others of the same *species*.

Examples of functional shifts are many and diverse. Some transitions at first seem unlikely because of the difficulty in identifying which possible functions may have been served during the intermediate stages. These cases are eventually resolved with further research and, often, by the discovery of intermediate *fossil* forms or living organisms with intermediate stages of *development*, as in the case of mollusks' eyes. It is similarly the case with bacterial *flagellum*, a favorite example of Behe (1996, 2007). In different *species* of bacteria there are different kinds of flagella, some simpler than that described by Behe, others just different, even very different, as in the archaea, a very large bacteria-like group of organisms. Moreover, mobility in many bacteria is accomplished without flagella at all. Still more, biochemists have shown that some *flagellum* components may have evolved from secretory systems, which are very similar to the *flagellum* but lack some of its components (see, for example, Liu and Ochman, 2007; Pallen and Matzke, 2006). The argument for the *irreducible complexity* of the *flagellum* is formulated, like other ID arguments, as an *argument from ignorance*: because one author does not know how a complex organ may have come about, it must be the case that it is irreducibly complex. This argument from ignorance dissolves as scientific knowledge advances, or when preexisting scientific knowledge is taken into account (see next section).

## NO EVIDENCE, NO SCIENCE

The proposition of ID as a scientific alternative to *evolution* brings to mind Gertrude Stein's quip about Oakland, her native city: "There is no there there." The call for an intelligent designer is predicated by ID proponents on the existence of *irreducible complexity* in organisms. *Irreducible complexity*, the claim goes, cannot come about by the stepwise manner of *natural selection*, the main process by which the theory of *evolution* accounts for *adaptation*. But the ID "explanation" is not a scientific hypothesis that can be tested by observation and experiment. Indeed, ID does not advance any explanation, but amounts only to a negative claim: that the relevant evolutionary explanations are not satisfactory.

ID's lack of scientific cogency is unwittingly displayed by its proponents when confronted with the dysfunctions, imperfections, cruelty, and even sadism of the living world. ID proponents affirm that these failings are not a valid argument against organisms being the creations of an intelligent designer. According to Behe (1996, p. 223), the "argument from imperfection overlooks the possibility that the designer might have multiple motives, with engineering excellence oftentimes relegated to a secondary role... the reasons that a designer would or would not do anything are virtually impossible to know unless the designer tells you specifically what those reasons are."

It is clear, then, that ID is not a scientific hypothesis because this claim provides it with an empirically impenetrable shield against any predictions, since we know not how "intelligent" or "perfect" a *design* should be, because the designer has not told us about it. ID, therefore, cannot be empirically tested (Ayala, 2008, 2010; National Academy of Sciences and Institute of Medicine, 2008; Pennock, 2002; Sober, 2007; Sonleitner, 2006). Science tests its hypotheses by observing whether or not predictions derived from them hold true in the observable world. A hypothesis that cannot be tested empirically—that is, by observation or experiment—is not scientific. ID as an explanation for the adaptations of organisms could be (natural) theology, as Paley would have it; but, whatever it is, it is not a scientific hypothesis.

There is a fundamental fallacy in the ID logic. The claim is that if *evolution* fails to explain some biological phenomenon, ID must be the correct explanation. This is a misunderstanding of the scientific process. If one explanation fails, it does not necessarily follow that some other proposed explanation is correct. Explanations must stand on their own evidence, not on the failure of their alternatives. Scientific explanations or hypotheses are creations of the mind, conjectures, imaginative exploits about the makeup and operation of the natural world. It is the imaginative preconception of what might be true in a particular case that guides observations and experiments designed to test whether a hypothesis is correct. The degree of acceptance of a hypothesis is related to the severity of the tests that it has passed (see Ayala, 1994; Giere et al., 2005; Gott and Duggan, 2003; Hung, 1996).

It is not sufficient for a theory to be accepted because some alternative theory has failed. Oxygen was not discovered simply because it was shown that *phlogiston* does not exist.[2] Nor is the periodic table of chemical elements accepted just because chemical substances react and yield a variety of components. Similarly, Darwin's theory of *evolution* by *natural selection* became generally accepted by scientists not because other evolutionary theories, such as those of Lamarck, Bergson, or Darwin's grandfather Erasmus, had failed the tests of science, but because it has sustained innumerable tests and has been fertile in yielding new knowledge (see, for example, Ghiselin, 1969; Quammen, 2007). Theodosius Dobzhansky (1967, p. 13) one of the greatest evolutionists of the 20th century, wrote: "There are people… to whom the gaps in our understanding of nature are pleasing for a different reason. These people hope that the gaps will be permanent, and that what is unexplained will also remain inexplicable. By a curious twist of reasoning, what is unexplained is then assumed to be the realm of divine activity. The historical odds are all against the 'God of the gaps' being able to retain these shelters in perpetuity. There is nothing, however, that can satisfy the type of mind which refuses to accept this testimony of historical experience."

Science is a complex enterprise that essentially consists of two interdependent episodes: one imaginative or *creative*, the other *critical*. To have an idea, advance a hypothesis, or suggest what might be true is a creative exercise, but scientific conjectures or hypotheses must also be subject to critical examination and empirical testing. Scientific knowledge may be characterized as a process of invention or discovery followed by validation or confirmation. One episode concerns the formulation of new ideas, sometimes referred to as the *acquisition of knowledge*; the other concerns the validation of these new ideas, or the *justification of knowledge*.

New ideas in science are advanced in the form of hypotheses. Hypotheses are mental constructs, imaginative exploits, that provide guidance as to what is worth observing and encourage the scientist to seek observations that would corroborate or falsify the hypothesis. The tests to which scientific ideas are subjected include contrasting hypotheses with the world of experience in a manner that must leave open the possibility that anyone might reject any particular hypothesis if it leads to wrong predictions about the world of experience. The possibility of empirical *falsification* of a hypothesis involves ascertaining whether or not precise predictions derived as logical consequences from the hypothesis agree with the state of affairs found in the empirical world. A hypothesis that cannot be subjected to the possibility of rejection by observation and experiment cannot be regarded as scientific. The possibility of empirical *falsification* of hypotheses has been called by the philosopher Karl Popper (1959) the "criterion of demarcation" that sets scientific knowledge apart from other forms of knowledge. ID offers no propositions to test, but makes only the lame claim that the theory of *evolution* by *natural selection* cannot account for the complexity of organisms (see Chapter 9).

## BACTERIAL FLAGELLUM, BLOOD CLOTTING, IMMUNE SYSTEM

The traditional favorite instantiation of *irreducible complexity* is the eye of humans and other vertebrates. I explained earlier that the complexity of the eye is not irreducible but, rather, may come about by steps, starting with the simplest of eyes—just one or a few pigmented cells sensitive to light. The evidence of how the complexity of the eye may arise by gradual steps is there for all to see in living mollusks. Michael Behe, a biochemist, the only proponent of ID who qualifies as a *bona fide* biologist, has argued other cases of supposed *irreducible complexity*: the bacterial *flagellum*, the blood-clotting process, and the *immune system*. That these biological systems are not irreducible has been shown by numerous authors, including Barbara Forrest and Paul Gross (2004), Renyi Liu and Howard Ochman (2007), Kenneth Miller (2004), Ian Musgrave (2004), Mark J. Pallen and Nicholas J. Matzke (2006), Robert Pennock (2002), Mark Perakh (2004a,b), David Ussery (2004), Matt Young and Taner Edis (2004), and others. The bacterial *flagellum* is, according to Behe, irreducibly complex because it consists of several parts, so if any part is missing the *flagellum* will not function. It could not, therefore, says Behe, have evolved gradually, one part at a time, because the function belongs to the whole; the separate parts cannot function by themselves. "Because the bacterial *flagellum* is necessarily composed of at least three parts—a paddle, a rotor, and a motor—it is irreducibly complex" (Behe, 1996, p. 72). This inference is, of course, erroneous.

The argument that the different components of the *flagellum* must have come about "in one fell swoop"—because the parts cannot function separately and thus could not have evolved independently—is reminiscent of Paley's argument about the eye. Of what possible use would the iris, cornea, lens, retina, or optic nerve be without the others? Yet we know that component elements of the octopus eye can evolve gradually, cumulatively, and that simple eyes, as they exist in a limpet, in slit-shell mollusks, and in marine snails, are functional.

The components of the bacterial *flagellum* have come about gradually in *evolution*, by genetic evolutionary changes that have been reconstructed in technical publications, such as the articles by Pallen and Matzke (2006) and Liu and Ochman (2007). The components of the *flagellum* are encoded by *gene* clusters that may in some *species* include upwards of 50 genes. The number of genes, and the genes themselves, greatly vary among different groups of bacteria. Liu and Ochman (2007) identified all the flagellar proteins in 41 *species* from 11 quite diverse groups of bacteria: 24 of the genes encoding the flagellar proteins were already present in the remote common ancestor of all the bacterial *species* studied. The other genes have come about by *duplication* and *evolution* of preexisting genes. Moreover, many of the core 24 ancestral genes are also derived from a few preexisting ones by successive *gene* duplications that gradually increased their number. The sequence similarity among all the flagellar genes in the 41 bacterial *species* allowed Liu and Ochman to reconstruct the successive steps of addition and modification by which modern bacterial flagella have arisen.

An injured person bleeds for a short time until a clot forms; this soon hardens and the bleeding stops. Blood clotting is a very complex process with many interacting components. According to Behe (1996, p. 97), *"no one on earth has the vaguest idea how the coagulation cascade came to be."* As I have written elsewhere (Ayala, 2007, p. 151), this is a remarkable statement, particularly because of the numerous scientific papers about the *evolution* of the various components of the blood-clotting mechanism in vertebrates, including "The *Evolution* of *Vertebrate* Blood Coagulation: A Case of Yin and Yang" by the eminent biochemist Russell F. Doolittle, published in 1993. Several of the authors cited earlier have provided accounts in layman's terms.

Behe (1996) made the outlandish claim that there is "no publication in the scientific literature—in prestigious journals, specialty journals, or books—that describes how *molecular evolution* of any real, complex, biochemical system either did occur or even might have occurred," and, in particular, "the scientific literature has no answers to the origin of the *immune system*" (pp. 185, 138). In fact, examples of complex biochemical structures or systems that have arisen from simpler components are very, very numerous. One concerns the phylogenetic enigma of snail hemoglobin. It turns out that the complex hemoglobin of the planorbid snail *Biomphalaria glabrata* has evolved from pulmonate myoglobin by a simple evolutionary mechanism that creates a high molecular mass respiratory *protein* from 78 similar globin domains (see Lieb et al., 2006). An example that involves the *evolution* of a particular *protein* from a single-cell ancestor to animal descendants appears in the same journal issue (Segawa et al., 2006). I suspect that if I were to examine relevant scientific journals, I would find each month one or more examples showing biochemical structures that have evolved from simpler ones.

As a further response to Behe, it may suffice to quote Judge John Jones III in the *Kitzmiller v. Dover* decision: "Professor Behe was questioned concerning his 1996 claim that science would never find an evolutionary explanation for the *immune system*. He was presented with 58 peer-reviewed publications, nine books, and several immunology textbook chapters about the *evolution* of the *immune system*; however, he simply insisted that this was still not sufficient evidence of *evolution*, and that it was not 'good enough.'" Judge Jones concludes: "We therefore find that Professor Behe's claim for *irreducible complexity* has been refuted in peer-reviewed research papers and has been rejected by the scientific community at large" (*Kitzmiller v. Dover*, 2005, p. 18).

## GAMBLING WITHOUT NATURAL SELECTION

The mathematically trained William Dembski (1995, 2002) is another ID proponent who has used supposedly scientific arguments to demonstrate the *irreducible complexity* of organisms. According to Dembski, organisms exhibit "complex specified information," which is information that has a very low prior probability and, therefore, high information content. Dembski argues that *mutation* and *natural selection* are incapable of generating such highly improbable states of affairs. Consider the 30 proteins that make up the bacterial *flagellum*. Assuming that each *protein* has about 300 amino acids, he calculates that the probability of one such *protein* is $20^{-300}$. After some refinements, he calculates that the probability of origination for the *flagellum* is $10^{-1170}$ (one divided by 1 followed by 1170 zeroes; to get some idea of the magnitude of this number, consider that the number of atoms in the universe is estimated to be $10^{77}$, or one followed by 77 zeroes). Dembski concludes that, even if one takes into account that life has existed on Earth for 3.5 billion years, the assembly of a functioning *flagellum* is impossibly improbable.

This numerological exercise is, however, irrelevant because Dembski does not take into account the role of *natural selection* and makes a number of erroneous assumptions, as pointed out in several of the works quoted earlier and others (see the papers in Dembski and Ruse, 2004; Pennock, 2001). Dembski, in his book *No Free Lunch* (2002), devotes many pages to the optimization of the "no free lunch" theorems of Wolpert and Macready (1997). Dembski's misunderstanding of the theorems and their implications has been pointed out, for example by Mark Perakh (2004a,b), and ridiculed by one of the theorems' authors, Wolpert, in his "William Dembski's Treatment of No Free Lunch Theorems is Written in Jell-O" (2003). The fundamental fallacies of Dembski's improbability calculations were exposed in an insightful critique by the distinguished evolutionist Joe Felsenstein (2007, p. 24), who showed that there "can be no theorem saying that adaptive information is conserved and cannot be increased by *natural selection*... Specified information, including complex specified information, can be generated by *natural selection*."

Dembski's numerological exercises suffer from several fatal flaws, including the assumption of a specified final outcome. This point was amusingly made by developmental biologist Scott F. Gilbert and the Swarthmore College *Evolution* and Developmental Seminar (2007): "Some of *Intelligent Design*'s most powerful arguments depend on a simple fallacy: the assumption of an end point... [ID proponents claim that] it is impossible to evolve a particular *protein* because it has 100 amino acids and the chance of this occurring randomly is one in $20^{100}$... But such supporters of ID don't know a billionth of how impossible it is! Let's say that your mother ovulated 500 eggs during her life and that your father produced $2 \times 10^{12}$ sperm. The chances of *you* being born, then, are one in $10^{15}$... [and] the chances of your grandparents giving rise to you is one in $10^{45}$. Another reason not to argue with the *Intelligent Design* people, then, is that, by their own logic, they cannot exist" (pp. 44–45).[3]

It is worthwhile pointing out, once again, the vacuity of the claims made by ID proponents. Their arguments consist of the tiresome repetition that something in the world of life cannot have come by natural processes because of its high improbability, or because no satisfactory scientific account exists of that something. Of course, Dembski and the other ID proponents exist. The playful argument about their nonexistence is based on the high improbability of mutations and other chance events, while ignoring *natural selection*, as they do, and assuming a predetermined end point, as the ID proponents do as well.

ID proponents have anticipated that, in time, they would make scientific discoveries and even "breakthroughs." The science writer Gordy Slack (2007) tells how he first encountered in January 1998 Phillip Johnson, professor of law at the University of California, Berkeley, and often considered as the founding father of the ID movement. Slack and Johnson had lunch at the university's Faculty Club. In response to Slack's (2007, p. vii) criticisms, Johnson advanced the following promise: "Give us five or 10 years, and you'll see scientific breakthroughs biologists hadn't dreamed of before ID." Ten years have passed since Johnson's promise was made. No scientific "breakthroughs" have been delivered, nor, indeed, any sort of scientific knowledge. Once again, there is no *there* there.

## COUNTERPOINT

The vacuity of ID can be contrasted with the substantial and fully corroborated evolutionary explanations by making three relevant points, if only very briefly because they are developed in earlier chapters: *evolution* is a fact; *natural selection* explains *design*; and the dysfunctions and cruelty of the living world are incompatible with an explicit *design* by an omnipotent and benevolent Creator.

Most biologists agree that the evolutionary origin of organisms is today a scientific conclusion beyond reasonable doubt, established with the kind of certainty attributable to such scientific concepts as the roundness of the Earth, the motions of the planets, and the molecular composition of matter. This degree of certainty beyond reasonable doubt is what is implied when biologists say that *evolution* is a "fact"; the evolutionary origin of organisms is accepted by the immense majority of biologists.

How is this factual claim compatible with the accepted view that science relies on observation, replication, and experimentation, since nobody has observed the *evolution* of *species*, much less replicated it by experiment? What scientists observe are not the concepts or general conclusions of theories, but their consequences. Copernicus's heliocentric theory affirms that the Earth revolves around the Sun. Nobody has observed this phenomenon, but we accept it because of numerous confirmations of its predicted consequences. We accept that matter is made of atoms, even though nobody has seen them, because of corroborating observations and experiments in physics and chemistry.

Darwin in *Origin of Species* gathered convincing evidence of revolution based on the current knowledge from paleontology, *biogeography*, comparative anatomy, and embryology. The evidence coming from these and other biological disciplines has enormously increased since Darwin's time. Most notable is the evidence from molecular biology, a discipline that came into existence only after the discovery of the structure of *DNA* by James Watson and Francis Crick in 1953. Molecular biology provides overwhelming evidence for the fact of *evolution*: that organisms are related by common descent with modification. *Molecular evolution*, moreover, provides detailed and redundant information about the evolutionary history of organisms to the extent that it is now possible to assert that gaps of knowledge in this field no longer need to exist. The virtually unlimited evolutionary information encoded in the *DNA* sequence of living organisms allows evolutionists to reconstruct the evolutionary relationships leading to present-day organisms with as much detail and precision as desired (see Chapter 4).

*Natural selection* is an incremental process that accounts for the "design" of organisms because adaptive variations increase the probability of survival and reproduction of their carriers at the expense of maladaptive, or less adaptive, variations (see Chapters 6 and 7). Adaptive variations arise as a consequence of genetic changes due to new mutations or new combinations of preexisting genes and genotypes. Numerous variations are known in all sorts of organisms that involve only one or a few genes. Adaptations that involve complex structures, functions, or behaviors involve numerous genes. Many common mammals, but not marsupials, have a placenta. Marsupials include the familiar kangaroo and other mammals native primarily to Australia and South America. Dogs, cats, mice, donkeys, and primates are placental. The placenta makes it possible to extend the time the developing embryo is kept inside the mother, and thus the newborn is better prepared for independent survival. However, the placenta requires complex adaptations, such as the suppression of harmful immune interactions between mother and embryo, delivery of suitable nutrients and oxygen to the embryo, and the disposal of embryonic wastes. The mammalian placenta evolved more than 100 million years ago and proved a successful *adaptation*, contributing to the explosive diversification of placental mammals in the Old World and North America. The placenta also has evolved in some fish groups, such as *Poeciliopsis*. Some *Poeciliopsis species* hatch eggs. The females supply the yolk in the egg, which furnishes nutrients to the developing embryo (as in chickens).

Other *Poeciliopsis species*, however, have evolved a placenta through which the mother provides nutrients to the developing embryo. Molecular biology has made possible the reconstruction of the evolutionary history of *Poeciliopsis species*. A surprising result is that the placenta evolved independently three times in this fish group. The required complex adaptations accumulated in each case in less than 750,000 years (Reznick et al., 2002; see Avise, 2006, 2010).

ID leads to conclusions about the nature of the designer quite different from those of omniscience, omnipotence, and omnibenevolence that people of faith attribute to the Creator. Organisms and their parts are less than perfect. Moreover, deficiencies and dysfunctions are pervasive, evidencing defective *design*. Humans have too many teeth for their jaw size, so wisdom teeth need to be removed. Would we want to blame God for such defective design? A human engineer could have done better. *Evolution* gives a good account of this imperfection (see Chapter 14). Brain size increased over time in our ancestors, and the remodeling of the *skull* to fit the larger brain entailed a reduction of the jaw. *Evolution* responds to the organisms' needs through *natural selection*, not by optimal *design* but by tinkering, as it were, by slowly modifying existing structures. Consider now the birth canal of women, much too narrow for easy passage of the infant's head, so that thousands upon thousands of babies die during delivery. Do we want to blame God for this defective *design* or the children's deaths. Science makes it understandable, a consequence of the evolutionary enlargement of our brain. Females of other mammals do not experience this difficulty. Theologians in the past struggled with the issue of dysfunction because they thought it had to be attributed to God's *design*. Science, much to the relief of many theologians, provides an explanation that convincingly attributes defects, deformities, and dysfunctions to natural causes.

As asserted in Chapter 6, examples of deficiencies, dysfunctions and "oddities" in all sorts of organisms can be endlessly multiplied. The opportunistic character of natural selection accounts also for the "cruelty" of predators and the "pitilessness" of parasites, like the agents of malaria, leprosy and tuberculosis, which can only reproduce at the expense of their human or other animal hosts.

## CREATIONISM

The word "creationism" has many meanings. In its broadest and traditional sense *creationism* is a religious belief: the idea that a supernatural power, God, created the universe as well as everything that exists in the universe, including humans. In a narrower sense it has come to mean the doctrine that the universe and all that is in it were created by God, essentially in their present form, a few thousand years ago. It is a doctrine that was largely formulated in reaction to Darwin's theory of *evolution*, based on a literal interpretation of the Bible.

*Creationism*, in this sense, denies the discoveries of astronomy concerning the *evolution* of the universe and the discoveries of biology concerning the *evolution* of humans.

In modern times biblical fundamentalists, although a minority of Christians in the United States, have periodically gained considerable public and political influence. Opposition to the teaching of *evolution* can largely be traced to two movements with 19th-century roots, Seventh-Day Adventism and Pentecostalism. Consistent with their emphasis on the seventh-day Sabbath as a memorial of the biblical *Creation*, Seventh-Day Adventists insist on the recent *creation* of life and the universality of the Flood, which they believe deposited the fossil-bearing rocks. This distinctively Adventist interpretation of Genesis became the hard core of "*creation* science" in the late 20th century and was incorporated into the "balanced treatment" laws of Arkansas and Louisiana (discussed later). Many Pentecostals, who generally endorse a literal interpretation of the Bible, have also adopted and endorsed the tenets of *creation* science, including the recent origin of Earth and a geology interpreted in terms of the Flood. They differ from Seventh-Day Adventists and other adherents of *creation* science, however, in their tolerance of diverse views and the limited import they attribute to the *evolution—creation* controversy.

During the 1920s, biblical fundamentalists helped influence more than 20 state legislatures to debate antievolution legislation, and four states— Arkansas, Mississippi, Oklahoma, and Tennessee—prohibited the teaching of *evolution* in their public schools. A spokesman for the antievolutionists was William Jennings Bryan, three times the unsuccessful Democratic candidate for the US presidency, who said in 1922, "We will drive Darwinism from our schools." In 1925, Bryan took part in the prosecution of John T. Scopes, a high-school teacher in Dayton, Tennessee, who admitted violating the state's law forbidding the teaching of *evolution*.

In 1968, the US Supreme Court declared unconstitutional any law banning the teaching of *evolution* in public schools (*Epperson v. Arkansas* 393 US.97, 1968). Thereafter, Christian fundamentalists introduced legislation in a number of state legislatures ordering that the teaching of "*evolution* science" be balanced by allocating equal time to "*creation* science." *Creation* science, it was asserted, propounds that all kinds of organisms abruptly came into existence when God created the universe, that the world is only a few thousand years old, and that the biblical Flood was an actual event only survived by one pair of each animal *species*. The legislatures of Arkansas in 1981 and Louisiana in 1982 passed statutes requiring the balanced treatment of *evolution* science and *creation* science in their schools, but opponents successfully challenged the statutes as violations of the constitutionally mandated separation of church and state.

The Arkansas statute was declared unconstitutional in federal court in 1982 after a public trial in Little Rock (*McLean v. Arkansas Board of Education*, 1982).[3] The Louisiana law was appealed all the way to the US Supreme Court, which in 1987 ruled Louisiana's *Creationism* Act unconstitutional because, by advancing the religious belief that a supernatural being created humankind, which is embraced by the phrase "*creation* science," the act impermissibly endorses religion (*Edwards v. Aguilar*, 1987).

To some people the theory of *evolution* seems incompatible with religious beliefs, particularly those of Christians, because it is inconsistent with the Bible's narrative of *Creation*. The first chapters of the biblical Book of Genesis describe God's *creation* of the world, plants, animals, and human beings. A literal interpretation of Genesis seems incompatible with the gradual *evolution* of humans and other organisms by natural processes.

Many Bible scholars and theologians have long rejected a literal interpretation as untenable, however, because the Bible contains mutually incompatible statements. The very beginning of the Book of Genesis presents two different *Creation* narratives. In Chapter 1, is the familiar 6-day account in which God creates human beings, both male and female, after creating light, earth, fish, fowl, and cattle is illustrated. A different version appears in Chapter 2: "And the Lord God formed man of the dust of the ground... And the Lord God planted a garden eastward of Eden; and there he put the man whom he had formed... And out of the ground the Lord God formed every beast of the field and every fowl of the air... And the Lord God caused a deep sleep to fall upon Adam, and he slept; and he took one of his ribs... And the rib, which the Lord God had taken from man, made he a woman, and brought her unto the man."

In this second narrative, Adam is created first, before the Garden of Eden and before plants and animals. Only afterward does God create the first woman, out of Adam's rib. Which of the two narratives is correct? Neither one would contradict the other if we understand them as conveying the same message: that the world was created by God and that humans are His creatures. But both narratives cannot be "historically and scientifically true," as postulated by the *Creation* Research Society.

There are five additional *Creation* narratives in the Bible: Job 38—41; Psalm 104; Proverbs 8:22—31; Ecclesiastes 1:2—11, 12:1—3; and Isaiah 40—55. These five narratives, if taken literally, are incompatible with each other as well as with the two narratives of Genesis, Chapters 1 and 2 (Brown, 2010).

There are numerous inconsistencies and contradictions throughout the Bible—for example in the description of the return from Egypt to the Promised Land by the chosen people of Israel—not to mention erroneous factual statements about the Sun circling around the Earth and the like. Biblical scholars point out that the Bible is infallible with respect to religious truth, not in matters that are of no significance to salvation. Augustine, one of the greatest Christian theologians, wrote in his *De Genesi ad litteram* (*Literal Commentary on Genesis*): "It is also frequently asked what our belief must be about the form and shape of heaven, according to Sacred Scripture... Such subjects are of no profit for those who seek beatitude... What concern is it of mine whether heaven is like a sphere and Earth is enclosed by it and suspended in the middle of the universe, or whether heaven is like a disc and the Earth is above it and hovering to one side." He adds: "In the matter of the shape of heaven, the sacred writers did not wish to teach men facts that could be of no avail for their salvation." Augustine is saying that the Book of Genesis is not an elementary book of astronomy. Noting that in the Genesis narrative of *Creation* God creates light on the first day but does not create the Sun until the fourth day, he concludes that "light" and "days" in Genesis make no literal sense.

Pope John Paul II said in 1981 that the Bible "speaks to us of the origins of the universe and its makeup, not in *order* to provide us with a scientific treatise but in *order* to state the correct relationships of man with God and with the universe. Sacred Scripture wishes simply to declare that the world was created by God, and in *order* to teach this truth, it expresses itself in the terms of the cosmology in use at the time of the writer."

As stated in Chapter 14, scientific knowledge and religious belief need not be in contradiction. If they are correctly assessed, they *cannot* be in contradiction, because they concern nonoverlapping realms of knowledge. It is only when assertions are made beyond their legitimate boundaries that science and religious belief appear to be antithetical.

Specifically, does the theory of *evolution* exclude religious belief? Is it not true that science is fundamentally materialistic and thus excludes any spiritual values? The answer to both questions is "no." The scope of science is the world of nature, the reality that is observed, directly or indirectly, by our senses. Science advances explanations concerning the natural world, explanations that are subject to the possibility of *corroboration* or rejection by observation and experiment. Outside that world, it has no authority, no statements to make, and no business whatsoever taking one position or another. Science has nothing decisive to say about values, whether economic, aesthetic, or moral; nothing to say about the meaning of life or its purpose; and nothing to say about religious beliefs (except in the case of beliefs that transcend the proper scope of religion and make assertions that contradict scientific knowledge; such statements cannot be true).

It is possible to believe that God created the world while also accepting that the planets, mountains, plants, and animals came about, after the initial *Creation*, by natural processes. As the National Academy of Sciences asserts in the document *Teaching Evolution and the Nature of Science*: "Within the Judeo-Christian religions, many people believe that God works through the process of *evolution*. That is, God has created both a world that is ever-changing and a mechanism through which creatures can adapt to environmental change over time."

In theological parlance, God may act through secondary causes. Similarly, at the level of the individual, a person can believe he is God's creature without denying that he developed from a single cell in his mother's womb. For the believer, the providence of God impacts personal life and world events through natural causes. The point, once again, is that scientific conclusions and religious beliefs concern different sorts of issues and belong to different realms of knowledge; they do not need to stand in contradiction.

## Endnotes

1. A bilingual edition of Aquinas's works in English and Latin, in 60 volumes, has been published by Blackfriars & McGraw-Hill. *Existence and Nature of God* I, 2–11, is vol. 2 (1964).
2. In the 17th century and beyond most chemists accepted that every combustible substance, such as wood or coal, was in part composed of *phlogiston*. Burning was thought to be caused by the liberation of *phlogiston*, with the remaining substance left as ash. Between 1770 and 1790 Antoine Lavoisier, the founder of modern chemistry, studied the gain or loss of weight when various substances were burned, and demonstrated that *phlogiston* does not exist; eventually, he demonstrated that oxygen, which had been recently discovered by Joseph Priestley, was always involved in combustion and he went on to propose a general theory of oxidation.
3. This point was made, also ironically, by Conway Zirkle in the symposium of April 25–26, 1966 at the Istar Institute on "Mathematical Challenges to the Neo-Darwinian Interpretation of *Evolution*." See Chapter 6, section on "Mathematical Challenges."

Chapter 17

# Human Evolution, Genetic Engineering, and Cloning

## INTRODUCTION

In mankind, there are two kinds of heredity: biological and cultural. Cultural inheritance makes possible for humans what no other organism can accomplish: the cumulative transmission of experience from generation to generation. In turn, cultural inheritance leads to *cultural evolution*, the prevailing mode of human *adaptation*. For the past few millennia, humans have been adapting the environments to their genes more often than adapting their genes to the environments. Nevertheless, *natural selection* persists in modern humans, both as differential mortality and as differential fertility, although its intensity has decreased in the past with respect to certain features and may decrease further in the future. More than 2000 human diseases and abnormalities have a genetic causation. Healthcare and the increasing feasibility of genetic *therapy* will, though slowly, augment the future incidence of hereditary ailments. Germ-line *gene therapy* could halt this increase, but at present it is not technically feasible. The proposal to enhance the human genetic endowment by genetic *cloning* of eminent individuals is not warranted. Genomes can be cloned; individuals cannot. *Therapeutic cloning* will in the future bring enhanced possibilities for organ transplantation, nerve cell and tissue healing, and other health benefits. Mitochondrial replacement *therapy* is a recent technique, still under *development*, that could "cure" various severe diseases transmitted to the progeny through the mother's *mitochondria*.

Chimpanzees are the closest relatives of *Homo sapiens*, our *species*. There is a precise correspondence bone by bone between the skeletons of a chimpanzee and a human. Humans bear young, like apes and other mammals. Humans have organs and limbs similar to birds, reptiles, and amphibians; these similarities reflect the common evolutionary origin of vertebrates. But it does not take much reflection to notice the distinct uniqueness of our *species*. Conspicuous anatomical differences between humans and apes include bipedal gait and enlarged brain. Much more conspicuous than the anatomical differences are the distinct behaviors and institutions. Humans have symbolic language, elaborate social and political institutions, codes of law, literature and

Evolution, Explanation, Ethics, and Aesthetics. http://dx.doi.org/10.1016/B978-0-12-803693-8.00017-7
**339**

art, *ethics* and religion; humans build roads and cities, travel by motorcars, ships, and airplanes, and communicate by means of telephones, computers, and televisions. These distinctive human behaviors and institutions may be collectively designated as human culture. When considering the potential health benefits of genetic engineering, we must keep in mind the cultural dimension of human life, which transcends its biological dimension. As we seek to enhance humankind's biological endowment, we must seek to preserve, and indeed enhance, humans' *cultural evolution*.

## HUMAN ORIGINS

The *hominin* lineage diverged from the chimpanzee lineage 6—7 million years ago (Mya) and evolved exclusively in the African continent until the *emergence* of *Homo erectus*, somewhat before 1.8 Mya. Shortly after its *emergence* in tropical or subtropical Africa, *H. erectus* spread to other continents. *Fossil* remains of *H. erectus* (*sensu lato*) are known from Africa, Indonesia (Java), China, the Middle East, and Europe. *H. erectus* fossils from Java have been dated 1.81 ± 0.04 and 1.66 ± 0.04 Mya, and from Georgia 1.6—1.8 Mya (Lordkipanidze et al., 2013). Anatomically distinctive *H. erectus* fossils have been found in Spain, deposited more than 780,000 years ago, the oldest in southern Europe (Cela-Conde and Ayala, 2007, 2016).

The transition from *H. erectus* to *H. sapiens* occurred around 400,000 years ago, although this date is not well determined owing to uncertainty as to whether some fossils are *erectus* or "archaic" forms of *sapiens*. *H. erectus* persisted for some time in Asia—until 250,000 years ago in China and perhaps until 100,000 years ago in Java—and thus was contemporary with early members of its descendant *species, H. sapiens*. *Fossil* remains of Neanderthal hominids (*Homo neanderthalensis*), with brains as large as those of *H. sapiens*, appeared in Europe earlier than 200,000 years ago and persisted until 30,000—40,000 years ago (Higham et al., 2014; Pääbo, 2014).

There is controversy about the origin of modern humans. Some anthropologists argue that the transition from *H. erectus* to archaic *H. sapiens* and later to anatomically modern humans occurred consonantly in various parts of the Old World. Proponents of this "multiregional model" emphasize *fossil* evidence showing regional continuity in the transition from *H. erectus* to archaic and then modern *H. sapiens*. Most anthropologists argue instead that modern humans first arose in Africa somewhat prior to 100,000 years ago and from there spread throughout the world, eventually replacing elsewhere the preexisting populations of *H. erectus, H. neanderthalensis*, and archaic *H. sapiens*. The African origin of modern humans is supported by a wealth of recent genetic evidence, and is therefore favored by many evolutionists (Cela-Conde and Ayala, 2007, 2016; Pääbo, 2014).

The colonization of Eurasia by modern humans was thought until recently to have occurred around 65,000 years ago. Recent evidence suggests, however, that there may have been at least two independent early colonizations of Eurasia by African immigrants. Archaeological and a defendant (*DNA*) data indicates a very ancient first exit from Africa as early as 100,000 years ago, and another exit around 65,000 years ago (Mellars, 2006a,b). A recent discovery in the Fuyan cave in Daoxian (southern China) of 47 human teeth dated somewhat more than 80,000 (but less than 120,000) years old (Liu et al., 2015) supports the notion of an early first colonization.

Mitochondrial and nuclear sequencing of *DNA* from Eurasian human fossils have recently brought some unanticipated results. One is that the Neanderthals are genetically more different from (particularly southern) African than from European and Asian individuals (Green et al., 2010). This state of affairs might be accounted for if modern humans had evolved in Africa for several thousand years before even the first Eurasian colonization, and thus become regionally differentiated. The Eurasian colonizers would have come from northern African populations. But there are also some relevant discoveries, including that between 2% and 4% of the *genome* of non-African or north Saharan Africans is of Neanderthal origin, indicating that some *hybridization* between Neanderthals and humans occurred after the Eurasian colonization (Green et al., 2010). There is also the discovery in the cave of Denisova in the Altai mountains (Russia) of a partial finger bone, dated 48,000—30,000 years old (Krause et al., 2010), belonging to a new *species*, closely related to modern humans and Neanderthals but different from both. Modern human populations, aboriginals from Australia and New Guinea, share about 3% of their *genome* with the Denisovan *fossil*, indicating that some *hybridization* occurred between Denisovan humans and *H. sapiens* (Reich et al., 2011; Rasmussen et al., 2011; Skoglund and Jakobsson, 2011).

## CULTURAL EVOLUTION

Humans live in groups that are socially organized, and so do other primates. But other *primate* societies do not approach the complexity of human social organization. A distinctive human social trait is culture, which may be understood as the set of nonstrictly biological human activities and creations. Culture includes social and political institutions, ways of doing things, religious and ethical traditions, language, common sense, scientific knowledge, art and literature, technology, and in general all the creations of the human mind. The advent of culture has brought with it *cultural evolution* superimposed on the organic mode, that has become the *dominant* mode of human *evolution*. *Cultural evolution* has come about because of cultural inheritance, a distinctively human mode of achieving *adaptation* to the environment (Cela-Conde and Ayala, 2007; Boyd and Richerson, 1985, 2005).

As expounded in Chapter 5, there are in humankind two kinds of heredity—the biological and the cultural. Biological inheritance in humans is very much like that in any other sexually reproducing organisms based on the transmission of genetic information encoded in *DNA* from one generation to the next by means of the sex cells. Cultural inheritance, on the other hand, is based on transmission of information by a teaching—learning process, and is independent of biological parentage. Culture is transmitted by instruction and learning, by example and imitation, through books, newspapers, radio, television, and motion pictures, electronically, through works of art, and by any other means of communication. Individuals acquire their "culture" not only from parents but also from other relatives and neighbors, and from the whole human environment. Acquired cultural traits may be beneficial but also toxic; for example, racial prejudice or religious bigotry.

Biological heredity is Mendelian, or vertical; it is transmitted from parents to their children, and only inherited traits can be transmitted to the progeny. (New mutations are insignificant in the present context.) Cultural heredity is Lamarckian: acquired characters can be transmitted to the progeny. But cultural heredity goes beyond Lamarckian heredity because it is horizontal and oblique, not only vertical. Traits can be acquired from and transmitted to other members of the same generation, whether or not they are relatives, and also from and to all other individuals with whom a person has contact, whether they are from the same or any previous or ensuing generation.

Animals can learn from experience, but they do not transmit their experiences and "discoveries" (at least not to any large extent) to the following generations. Animals have individual memory, but they do not have a "social memory." Humans, on the other hand, have developed a culture because they can transmit cumulatively their experiences through the generations.

Cultural inheritance makes possible *cultural evolution*, a new mode of *adaptation* to the environment that is not available to nonhuman organisms (Chapter 5). Organisms adapt to the environment by means of *natural selection*, by changing over generations their genetic constitution to suit the demands of the environment. But humans, and humans alone, can also adapt by changing the environment to suit the needs of their genes. At least since the origins of *H. sapiens*, and to some limited extent at least since our *Homo habilis* ancestors made their earliest tools, humans have been adapting their environments to their genes so as to meet their biological needs.

To extend its geographical *habitat*, or survive in a changing environment, a *population* of organisms must become adapted, through many generations of *natural selection*, to new climatic conditions, different sources of food, different competitors, and so on. The discovery of fire and the use of shelter and clothing allowed *H. sapiens* to spread from the African tropical and subtropical regions where it originated to the whole Earth. Humans have conquered the air without evolving wings, but much more effectively by building flying machines. People travel the rivers and the seas by means of boats, without having evolved gills or fins. Our *species, H. sapiens*, evolved in Africa, but it has become the most widespread and abundant *species* of mammal on Earth because of the *evolution* of culture, a superorganic form of *adaptation* (Chapter 5).

Cultural *adaptation* has prevailed in humankind over biological *adaptation* because it is a more effective mode of *adaptation*; it is more rapid and it can be directed. A favorable genetic *mutation* newly arisen in an individual can be transmitted to a sizable part of the human *species* only through innumerable generations. However, a new scientific discovery or technical achievement can be transmitted to the whole of humankind, potentially at least, in less than one generation. Witness the rapid spread of personal computers, iPhones, and the internet. Moreover, whenever a need arises, culture can directly pursue the appropriate changes to meet the challenge, while biological *adaptation* depends on the accidental availability of favorable mutations (Cela-Conde and Ayala, 2007; Boyd and Richerson, 1985, 2005).

## BIOLOGICAL EVOLUTION IN MODERN HUMANS

There is no scientific basis to the claim sometimes made that the biological *evolution* of humankind has stopped, or nearly so, in technologically advanced countries. It is asserted that the *progress* of medicine, hygiene, and nutrition has largely eliminated death before middle age; that is, most people live beyond reproductive age, after which death is inconsequential for *natural selection*. That humankind continues to evolve biologically can be shown because the necessary and sufficient conditions for biological *evolution* persist. These conditions are genetic variability and differential reproduction. There is a wealth of genetic variation in humankind. With the trivial exception of identical twins developed from a single fertilized egg, no two people who live now, lived in the past, or will live in the future are likely to be genetically identical. Much of this variation is relevant to *natural selection* (Ayala, 1986, 2007; Richerson and Boyd, 2005).

*Natural selection* is simply differential reproduction of alternative genetic variants. *Natural selection* will occur in humankind if the carriers of some genotypes are likely to leave more descendants than the carriers of other genotypes. *Natural selection* consists of two main components, differential mortality and differential fertility; both persist in modern humankind, although the intensity of *selection* due to postnatal mortality has been somewhat attenuated.

Death may occur between conception and birth (prenatal) or after birth (postnatal). The proportion of prenatal deaths is not well known. Death during the early weeks of embryonic *development* may go totally undetected. But it is known that no less than 20% of all ascertained human conceptions end in spontaneous abortion during the first 2 months of pregnancy. Such deaths are often due to deleterious genetic constitutions, and thus they have a selective effect in the *population*. The intensity of this form of *selection* has not changed substantially in modern humankind, although it has been slightly reduced with respect to a few genes, such as those involved in Rh blood group incompatibility.

Postnatal mortality has been considerably reduced in recent times in technologically advanced countries. For example, in the United States somewhat less than 50% of those born in 1840 survived to age 45, while the average life expectancy for people born in the United States in 1960 is 78 years (Table 17.1; Ayala, 2015). In some regions of the world postnatal mortality remains quite high, although it has generally decreased in recent decades. Mortality before the end of reproductive age, particularly where it has been considerably reduced, is largely associated with genetic defects, and thus it has a favorable selective effect in human populations. Several thousand genetic variants are known that cause diseases and malformations in humans; such variants are kept at low frequencies due to *natural selection*.

It might seem at first that *selection* due to differential fertility has been considerably reduced in industrial countries as a consequence of the reduction in the average number of children per *family* that has taken place. However, this is not so. The intensity of fertility *selection* depends not on the mean number of children per *family*, but on the *variance* in the number of children per *family*. It is clear why this should be so. Assume that all people of reproductive age marry and all have exactly the same number of children. In this case, there would not be fertility *selection* whether couples all had very few or all had very many children. Assume, on the other hand, that the mean number of children per *family* is low, but some families have no children at all or very few while others have many. In this case, there would be considerable opportunity for *selection*—the genotypes of parents producing many children would increase in frequency at the expense of those having few or none. Studies of human populations have shown that the opportunity for *natural*

**TABLE 17.1** Percentage of Americans Born Between 1840 and 1960 Surviving to Age 15 and to Age 45

| Birth | Surviving to Age 15 | | Surviving to Age 45 | |
|---|---|---|---|---|
| | Men | Women | Men | Women |
| 1840 | 62.8 | 66.4 | 48.2 | 49.4 |
| 1880 | 71.5 | 73.1 | 58.3 | 61.1 |
| 1920 | 87.6 | 89.9 | 79.8 | 85.8 |
| 1960 | 99.0 | 99.2 | 94.1 | 96.1 |

After Ayala, F.J., 1986. Whither mankind? The choice between a genetic twilight and a moral twilight. American Zoologist 26, 895—905 and Arias, E., 2010. United States Life Tables, 2006. National Vital Statistical Reports 58 (21), 1—40.

*selection* often increases as the mean number of children decreases. An extensive study published years ago showed that the index of opportunity for *selection* due to fertility was four times larger among US women born in the 20th century, with an average of fewer than three children per woman, than among women in the Gold Coast of Africa or in rural Quebec, who had three times more children on average (Table 17.2; Ayala, 2015). There is no evidence that *natural selection* due to fertility has decreased in modern human populations.

## FUTURE NATURAL SELECTION

*Natural selection* may decrease in intensity in the future, but it will not disappear altogether. So long as there is genetic variation and the carriers of some genotypes are more likely to reproduce than others, *natural selection* will continue operating in human populations. Cultural changes, such as the *development* of agriculture, migration from the country to the cities, environmental pollution, and many others, create new selective pressures. The pressures of city life are partly responsible for the high incidence of mental disorders in certain human societies. The point to bear in mind is that human environments are changing faster than ever, owing precisely to the accelerating rate of cultural change; and environmental changes create new selective pressures, thus fueling biological *evolution*.

**TABLE 17.2** Mean Number of Children per Family and Index of Opportunity for Fertility Selection $I_f$, in Various Human Populations

| Population | Mean Number of Children | $I_f$ |
|---|---|---|
| Rural Quebec, Canada | 9.9 | 0.20 |
| Gold Coast, Africa | 6.5 | 0.23 |
| New South Wales, Australia (1898–1902) | 6.2 | 0.42 |
| United States, women born in 1839 | 5.5 | 0.23 |
| United States, women born in 1871–1875 | 3.5 | 0.71 |
| United States, women born in 1928 | 2.8 | 0.45 |
| United States, women born in 1909 | 2.1 | 0.88 |
| United States, Navajo Indians | 2.1 | 1.57 |

$I_f$ is calculated as the variance divided by the square of the mean number of children. The opportunity for selection often increases as the mean number of children decreases (Ayala, 1986).
After Crow, J.F., 1958. Some possibilities for measuring selection intensities in man. Human Biology 30 (1), 1–13.

*Natural selection* is the process of differential reproduction of alternative genetic variants. In terms of single genes, variation occurs when two or more alleles are present in the *population* at a given *gene locus*. How much genetic variation exists in the current human population? The answer is "quite a lot," as will be shown, but *natural selection* will take place only if the alleles of a particular *gene* have different effects on *fitness*—that is, if alternative alleles differentially impact the probability of survival and reproduction.

The two genomes that we inherit from each parent are estimated to differ at about one or two nucleotides per 1000. The human *genome* consists of somewhat more than 3 billion nucleotides (Lynch, 2007). Thus about 3—6 million nucleotides are different between the two genomes of each human individual, which is a lot of genetic *polymorphism*. Moreover, the process of *mutation* introduces new variation in any *population* every generation. The rate of *mutation* in the human *genome* is estimated to be about $10^{-8}$, one *nucleotide mutation* for every 100 million nucleotides, or about 30 new mutations per *genome* per generation. Thus every human has about 60 new mutations (30 in each *genome*) that were not present in the parents. If we consider the total human *population*, that is 60 mutations per person multiplied by 7 billion people, which is about 420 billion new mutations per generation added to the preexisting 3—6 million polymorphic nucleotides per individual.

That is a lot of mutations, even if many are redundant. Moreover, and most important, we must remember that the polymorphisms that count for *natural selection* are those that impact the probability of survival and reproduction of their carriers. Otherwise, the variant nucleotides may increase or decrease in frequency by chance, a process that evolutionists call "*genetic drift*," but will not be impacted by *natural selection* (Ayala, 1982a; Cela-Conde and Ayala, 2007, 2016; Lynch, 2007).

## GENETIC DISORDERS

More than 2000 human diseases and abnormalities that have a genetic cause have been identified in the human *population*. Genetic disorders may be *dominant, recessive*, multifactorial, or chromosomal. *Dominant* disorders are caused by the presence of a single copy of the defective *allele*, so the disorder is expressed in heterozygous individuals—those having one normal and one defective *allele*. In *recessive* disorders the defective *allele* must be present in both alleles, so it must be inherited from each parent to be expressed. Multifactorial disorders are caused by interaction among several *gene* loci, and chromosomal disorders are due to the presence or absence of a full *chromosome* or a fragment of a *chromosome* (Harrison, 1993; Mascie-Taylor, 1993).

Examples of *dominant* disorders are some forms of retinoblastoma and other kinds of blindness, achondroplastic dwarfism, and Marfan syndrome (which is thought to have affected President Lincoln). Examples of *recessive* disorders are cystic fibrosis, Tay-Sachs disease, and sickle-cell anemia (caused by an *allele* that in heterozygous conditions protects against malaria). Examples of multifactorial diseases are spina bifida and cleft palate. Among the most common chromosomal disorders are Down's syndrome, caused by the presence of an extra *chromosome* 21, and various syndromes due to the absence of one *sex chromosome* or the presence of an extra one, beyond the normal condition of XX for women and XY for men. Examples are Turner's syndrome (XO) and Klinefelter's syndrome (XXY) (Nesse and Williams, 1994).

The incidence of genetic disorders expressed in the living human *population* is estimated to be no less than 2.56%, affecting about 180 million people. *Natural selection* reduces the incidence of the genes causing disease—more effectively in the case of *dominant* disorders, where all carriers of the *gene* will express the disease, than for *recessive* disorders, which are expressed only in homozygous individuals. Consider, for example, phenylketonuria (PKU), a lethal disease if untreated, due to homozygosis for a *recessive gene*, which has an incidence of 1 in 10,000 newborns or 0.01%. PKU is due to an inability to metabolize the amino acid phenylalanine, with devastating mental and physical consequences. A very elaborate diet free of phenylalanine allows the patient to survive and reproduce, if started early in life. As stated earlier, the frequency of the PKU *allele* is about 1%, so in heterozygous conditions it is present in about 140 million people, but only the 0.01% of people who are homozygous, or about 1.4 million people, express the disease and are subject to *natural selection*. The reduction of genetic disorders due to *natural selection* is balanced with their increase due to the incidence of new mutations.

Let us consider another example. Hereditary retinoblastoma is a disease attributed to a *dominant mutation* of the *gene* RB1; it is actually due to a deletion in *chromosome* 13. The unfortunate child with this condition develops during infancy a tumorous growth that, without treatment, starts in one eye and often extends to the other eye and then to the brain, causing death before puberty. Surgical treatment now makes it possible to save the life of the child if the condition is detected sufficiently early, although often one or both eyes may be lost. The treated person can live a more or less normal life, marry, and procreate. However, because the genetic determination is *dominant* (a *gene* deletion), half of the progeny will, on average, be born with the same genetic condition and will have to be treated. Before modern medicine, every *mutation* for retinoblastoma arising in the human *population* was eliminated from the *population* in the same generation owing to the death of its carrier. With surgical treatment, the *mutant* condition can be preserved and new mutations arising each generation are added to those arisen in the past (Bainbridge et al., 2008; Cideciyan et al., 2009; Gene Therapy Clinical Trials Worldwide Database, 2014).

The proportion of individuals affected by any one serious hereditary infirmity is relatively small, but there are more than 2000 known serious physical infirmities determined by genes. When all these hereditary ailments are considered together, the proportion of persons born who will suffer from a serious handicap during their lifetimes owing to their heredity is more than 2% of the total *population*, as pointed out earlier (Mascie-Taylor, 1993; Nesse and Williams, 1994; Gene Therapy Clinical Trials Worldwide Database, 2014; Sheridan, 2011).

The problem becomes more serious when mental defects are taken into consideration. More than 2% of people in the world are affected by schizophrenia or a related condition known as schizoid disease, ailments which may be in some cases determined by a single *mutant gene*. Another 3% or so of the *population* suffer from mild mental retardation (IQ below 70). More than 100 million people in the world suffer from mental impairments due in good part to the genetic endowment they inherited from their parents.

## THE FUTURE

*Natural selection* also acts on a multitude of genes that do not cause disease. Genes affect skin pigmentation, hair color and configuration, height, muscle strength and body shape, and many other anatomical polymorphisms that are apparent, as well as many that are not externally obvious, such as variations in the blood group, the *immune system*, and the heart, liver, kidney, pancreas, and other organs. It is not always known how *natural selection* impacts these traits, but surely it does, and does so differently in different parts of the world and at different times as a consequence of the *development* of new vaccines, drugs, and medical treatments, and also as a result of changes in lifestyle, such as the reduction in the number of smokers or the increase in the rate of obesity in a particular country.

Where is human *evolution* going? Biological *evolution* is directed by *natural selection*, which is not a benevolent force guiding *evolution* toward sure success. *Natural selection* brings about genetic changes that often appear purposeful because they are dictated by the requirements of the environment. The end result may, nevertheless, be extinction—more than 99.9% of all *species* which ever existed have become extinct. *Natural selection* has no purpose; humans alone have purposes, and they alone may introduce them into their *evolution*. No *species* before humankind could select its evolutionary destiny; humankind possesses techniques to do so, and more effective techniques for directed genetic change are becoming available. Because we are self-aware, we cannot refrain from asking what lies ahead, and because we are ethical beings we must choose between alternative courses of action, some of which may appear as good, others as bad.

The argument has been advanced that the biological endowment of humankind is rapidly deteriorating owing precisely to the improving conditions of life and the increasing power of modern medicine. The detailed arguments that support this contention involve some mathematical exercises, but their essence can be simply presented. Genetic changes (that is, point or *chromosome* mutations) arise spontaneously in humans as well as in other living *species*. The great majority of newly arising mutations are either neutral or harmful to their carriers; only a very small fraction is likely to be beneficial. In a human *population* under so-called "natural" conditions—that is, without the intervention of modern medicine and technology—the newly arising harmful mutations are eliminated from the *population* more or less rapidly depending upon how harmful they are. The more harmful the effect of a *mutation*, the more rapidly it will be eliminated from the *population* by the process of *natural selection*. However, owing to medical intervention and, more recently, because of the possibility of genetic *therapy*, the elimination of some harmful mutations from the *population* is no longer taking place as rapidly and effectively as it did in the past.

## GENETIC THERAPY

Molecular biology has opened up for modern medicine a new way to cure diseases, namely genetic *therapy*, direct intervention in the genetic makeup of an individual. *Gene therapy* can be somatic or germ line. Germ-line genetic *therapy* seeks to correct a genetic defect not only in the organs or tissues impacted but also in the germ line, so the person treated would not transmit the genetic impairment to the descendants. To date, no germ-line *therapy* interventions except mitochondrial replacement *therapy* (MRT, discussed later) have seriously been contemplated by scientists, physicians, or pharmaceutical companies. Convincing medical and ethical arguments against germ-line *gene* modification, at least for the time being, have been raised by a distinguished group of scholars, including Nobel Laureates (Baltimore et al., 2015).

The possibility of *gene therapy* was anticipated in 1972 (Friedmann and Roblin, 1972). The possible objectives are to correct the *DNA* of a defective *gene* or to insert a new *gene* that would allow the proper function of the *gene* or *DNA* to take place. In the case of a harmful *gene*, the objective would be to disrupt or neutralize the *gene* that is not functioning properly. The eminent biologist E. O. Wilson (2014) stated, somewhat hyperbolically, that the issue of how much to use genetic engineering to direct our own *evolution* is "the greatest moral dilemma since God stayed the hand of Abraham" (Wilson, 2014).

The first successful interventions of *gene therapy* concerned patients suffering from severe combined immunodeficiency (SCID). The first was performed in a 4-year-old girl at the National Institutes of Health in 1990 (Blaese et al., 1995), soon followed by successful trials in other countries (Abbott, 1992). Treatments were halted temporarily from 2000 to 2002 in Paris, when two of about 12 treated children developed a leukemia-like condition which was attributed to the *gene therapy* treatment. Since 2004, successful clinical trials for SCID have been performed in the United States,

United Kingdom, France, Italy, and Germany (Cavazzana-Calvo et al., 2004; Fischer et al., 2010).

*Gene therapy* treatments are still considered experimental. Successful clinical trials have been performed in patients suffering from adrenoleukodystrophy, Parkinson's disease, chronic lymphocytic leukemia, acute lymphocytic leukemia, multiple myeloma, and hemophilia (Cartier and Aubourg, 2009; LeWitt et al., 2011). Initially the prevailing *gene therapy* methods involved recombinant viruses, but nonviral methods (transfection molecules) have become increasingly successful. Since 2013, US pharmaceutical companies have invested over $600 million in *gene therapy* (Herper, 2014). Yet in addition to the huge economic costs, technical hurdles remain. Frequent negative effects include immune response against an extraneous object introduced into human tissues, leukemia, tumors, and other disorders provoked by vector viruses. Moreover, the genetic *therapy* corrections are often short-lived, which calls for multiple rounds of treatment, thereby increasing costs and other handicaps. In addition, many of the most common genetic disorders are multifactorial, and thus beyond current *gene therapy* treatment. Examples are diabetes, high blood pressure, heart disease, arthritis, and Alzheimer's disease, which at the present state of knowledge and technology are not suitable for *gene therapy*.

The most significant recent technological advance concerning *gene therapy* is the *development* of CRISPR ("clustered regularly interspersed short palindrome repeats") and associated proteins, known as Cas, particularly Cas9, a nuclease or *enzyme* specialized for cutting *DNA*. CRISPR are segments of bacterial *DNA* encompassing short repetitions of base sequences (range 24−48 base pairs) with each repetition followed by short segments of spacer *DNA*. With the CRISPR−Cas9 system, animal and other genomes, including human, can be cut at specific desired sites, thus making possible *gene* editing, activating or silencing specific genes in human cells. At present CRISPR−Cas9 related systems are seen by expert scientists, as well as ethicists, as being in an experimental stage, although with enormously promising possibilities (Larson et al., 2013; Pennisi, 2013; Doudna and Charpentier, 2014; Baltimore et al., 2015; Ledford, 2015; Morange, 2015; Zimmerman, 2015; see also Shalem et al., 2014; van der Oost et al., 2014; Doudna, 2015; Gagnon and Corey, 2015; Genome Editing, 2015; Lamphier et al., 2015; Travis, 2015; van Erp et al., 2015; Slaymaker et al., 2016).

## H. J. MULLER'S FORECAST

If a genetic defect is corrected in the affected cells, tissues, or organs but not in the germ line, the ova or sperm produced by the individual will transmit the defect to the progeny. A deleterious *gene* that might have been reduced in frequency or eliminated from the *population*, owing to the death or reduced fertility of the carriers, will now persist in the *population* and be added to its load of hereditary diseases. A consequence of genetic *therapy* is that the more hereditary diseases and defects are cured today, the more of them will be there to be cured in succeeding generations. This consequence follows not only from *gene therapy* but also from typical medical treatments.

The Nobel Laureate geneticist H. J. Muller eloquently voiced this concern about the cure, whether through genetic *therapy* or traditional medical treatment, of genetic ailments. "The more sick people we now cure and allow them to reproduce, the more there will be to cure in the future." The fate toward which humankind is drifting is painted by Muller in somber colors. "The amount of genetically caused impairment suffered by the average individual ... must by that time have grown ... people's time and energy ... would be devoted chiefly to the effort to live carefully, to spare and to prop up their own feebleness, to soothe their inner disharmonies and, in general, to doctor themselves as effectively as possible. For everyone would be an invalid, with his own special familial twists ..." (Muller, 1950; Fig. 17.1).

It must be pointed out that the *population* genetic consequences of curing hereditary diseases are not as immediate ("a few centuries hence") as Muller anticipated. Consider, as a first example, the *recessive* hereditary condition of PKU. The estimated frequency of the *gene* is $q = 0.01$; the expected number of humans born with PKU is $q^2 = 0.0001$, one for every 10,000 births. If all PKU individuals are cured all over the world and all of them leave as many descendants, on average, as other humans, the frequency of the PKU *allele* will double after $1/q = 1/0.01 = 100$ generations. If we assume 25 years per generation, we conclude that after 2500 years the frequency of the PKU *allele* will be $q = 0.02$, and $q^2 = 0.0004$, so four persons in every 10,000, rather than only one, will be born with PKU. It needs to be noted that not now, nor in the foreseeable future, is it anticipated that all newborn PKU patients in the world will be cured.

In the case of *dominant* lethal diseases, the incidence is determined by the *mutation* frequency of the normal to the disease *allele*, which is typically of the order of $m = 10^{-6}-10^{-8}$, or between one in a million and one in 100 million. Assuming the highest rate of $m = 10^{-6}$ for a particular disease, the incidence of the disease after 100 generations, if all newborn patients are cured, would become one for every 10,000 births. It would seem likely that much earlier than 2500 years hence humans are likely to find ways of correcting hereditary ailments in the germ line, thereby stopping their transmission.

It must be pointed out that although the proportion of individuals affected by any one serious hereditary infirmity is relatively small, there are many such hereditary ailments, which on aggregate makes the problem very serious. It becomes even more serious when mental defects are taken into consideration. As pointed out earlier, more than 100 million people in the world suffer from mental impairments due in good part to the genetic endowment they inherited from their parents.

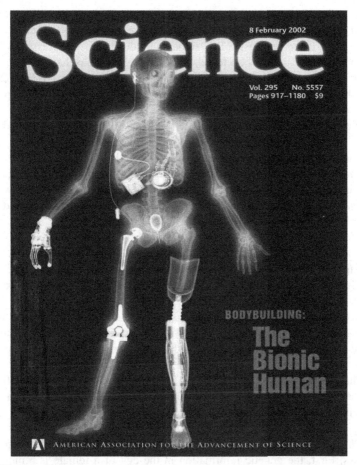

**FIGURE 17.1**    The bionic human, on the cover of *Science* (Cover Image, 2002): an image that could represent how H. J. Muller anticipates the human condition, a few centuries hence, showing the accumulation of physical handicaps as a consequence of the medical cure of hereditary diseases.

## CLONING

Human *cloning* may refer to "*therapeutic cloning*," particularly the *cloning* of embryonic or stem cells to obtain organs for transplantation or for treating injured nerve cells and other health purposes. Human *cloning* more typically refers to "reproductive cloning," the use of somatic cell nuclear transfer (SCNT) to obtain eggs that could develop into adult individuals. A distinctive genetic variant of genetic *therapy* that involves nuclear transplant is MRT, mentioned earlier.

Human *cloning* has occasionally been suggested as a way to improve the genetic endowment of humankind by *cloning* individuals of great achievement, for example in sports, music, the arts, science, literature, politics, and the like, or of acknowledged virtue. These suggestions seemingly have never been taken seriously. But some individuals have expressed a wish, however unrealistic, to be cloned, and some physicians have on occasion advertised that they were ready to carry out the *cloning* (Staff Writer, 2001). The obstacles and drawbacks are many and insuperable, at least at the present state of knowledge.

Biologists use the term *"cloning"* with variable meanings, although all uses imply obtaining more or less precise copies of a biological entity. Three common uses refer to *cloning* genes, *cloning* cells, and *cloning* individuals. *Cloning* an individual, particularly in the case of a multicellular organism, such as a plant or an animal, is not strictly possible. The genes of an individual, the *"genome,"* can be cloned, but the individual as such cannot be cloned, as is made clear later.

*Cloning* genes or, more generally, *cloning DNA* segments is routinely done in genetics and pharmaceutical laboratories around the world (Lynch, 2007; Jobling et al., 2004). Technologies for *cloning* cells in the laboratory are seven decades old, and are used for reproducing a particular type of cell, for example a skin or a liver cell, to investigate its characteristics.

Individual human *cloning* occurs naturally in the case of identical twins, when two individuals develop from a single fertilized egg. These twins are called "identical" precisely because they are genetically identical to each other.

The sheep Dolly, cloned in July 1996, was the first *mammal* artificially cloned using an adult cell as the source of the *genotype*. Frogs and other amphibians had been obtained by artificial *cloning* as much as 50 years earlier (King, 1996).

*Cloning* an animal by somatic cell nuclear transfer (SCNT) proceeds as follows. First, the genetic information in the egg of a female is removed or neutralized. Somatic (ie, body) cells are taken from the individual selected to be cloned, and the cell *nucleus* (where the genetic information is stored) of one cell is transferred with a micropipette (or otherwise) into the host oocyte. The egg, so "fertilized", is stimulated to start embryonic *development* (McLaren, 2000).

Can a human individual be cloned? The correct answer is, strictly speaking, "no." What is cloned are the genes, not the individual; the *genotype*, not the *phenotype*, as is further elaborated later. But the technical obstacles are immense even for *cloning* a mammal's *genotype*.

Ian Wilmut, the British scientist who directed the sheep *cloning* project, succeeded with Dolly only after 270 trials. The rate of success for *cloning* mammals has notably increased over the years without ever reaching 100%. The animals cloned to date include mice, rats, goats, sheep, cows, pigs, horses, and other mammals. The great majority of pregnancies end in spontaneous abortion (Jabr, 2013). Moreover, as Wilmut noted, in many cases the death of the fetus occurs close to term, which would have devastating eco-nomic, health, and emotional consequences in the case of humans (Highfield, 2007).

In mammals, in general, the animals produced by *cloning* suffer from serious health handicaps; among others, gross obesity, early death, distorted limbs, dysfunctional immune systems and organs, including liver and kidneys, and other mishaps. Even Dolly had to be euthanized early in 2003, after only 6 years of life, because her health was rapidly decaying, including progressive lung disease and arthritis (Highfield, 2007; Shields et al., 1999).

The low rate of *cloning* success may improve in the future. It may be that the organ and other failures of those that are born will be corrected by tech-nical advances. But human *cloning* would still face ethical objections from a majority of concerned people, as well as opposition from diverse religions (Caplan and Arp, 2013; Levick, 2013). Moreover, there remains the limiting consideration asserted earlier: it might be possible to clone a person's genes, but the individual cannot be cloned. The character, personality, and the fea-tures other than anatomical and physiological that make up the individual are not precisely determined by the *genotype*.

## THE GENOTYPE AND THE INDIVIDUAL

The genetic makeup of an individual is its "*genotype*." The "*phenotype*" refers to what the individual is, which includes not only the individual's external appearance or anatomy but also its physiology and behavioral predispositions and attributes, encompassing intellectual abilities, moral values, aesthetic preferences, religious values, and, in general, all other behavioral character-istics or features acquired by experience, imitation, learning, or in any other way throughout the individual's life, from conception to death. The *phenotype* results from complex networks of interactions between the genes and the person's history, including the human and physical environment.

A person's environmental influences begin, importantly, in the mother's womb and continue after birth, through childhood, adolescence, and the whole life. Behavioral experiences are associated with *family*, friends, schooling, social and political life, readings, aesthetic and religious experiences, and every event in the person's life, whether conscious or not. The *genotype* of a person has an unlimited number, virtually infinite, of possibilities to be real-ized; these have been called the *genotype*'s "norm of reaction," only one of which will be realized in a particular individual (Ayala and Kiger, 1984). If an adult person is cloned, the disparate life circumstances experienced by the cloned individual would surely result in a very different individual, even if anatomically the clone resembled the *genome*'s donor at a similar age. As Oliver Sacks, the eminent neurologist and author, wrote in the *New York Times* a few months before his anticipated imminent death on August 30, 2015: "When people die, they cannot be replaced ... for it is the fate—the genetic and neural fate—of every human being to be a unique individual, to find his own path, to live his own life, to die his own death" (Sacks, 2015).

An illustration of environmental effects on the *phenotype*, and of interactions between the *genotype* and the environment, is shown in Fig. 17.2. Three plants of the cinquefoil, *Potentilla glandulosa*, were collected in California—one on the coast at about 100 feet above sea level (Stanford), the second at about 4600 feet (Mather), and the third in the Alpine zone of the Sierra Nevada at about 10,000 feet above sea level (Timberline). From each plant, three cuttings were obtained in each of several replicated experiments, which were planted in three experimental gardens at different altitudes, the same gardens from which the plants were collected. The division of one plant ensured that all three cuttings planted at different altitudes had the same *genotype*—that is, they were genetic clones from one another. (*P. glandulosa,*

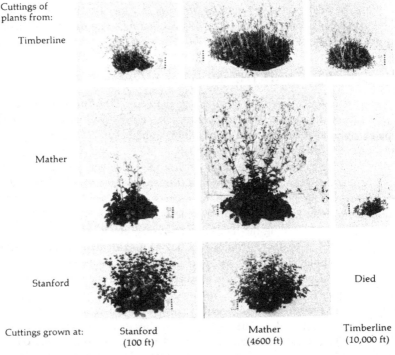

**FIGURE 17.2**  Interacting effects of the genotype and the environment on the phenotype of the cinquefoil, *Pontentilla glandulosa*. Cuttings of plants collected at different altitudes were planted in three different experimental gardens. Plants in the same row are genetically identical because they have been grown from cuttings of a single plant; plants in the same column are genetically different but have been grown in the same experimental garden (Ayala, 1982a). *After Claussen, J., Keck, D.D., Hiesey, W.M., 1940. Experimental Studies on the Nature of Species. I. The Effect of Varied Environment on Western North American Plants. Carnegie Institution of Washington Publication 520, Washington, DC.*

like many other plants, can be reproduced by "cuttings," which are genetically identical.)

Comparison of the plants in any row shows how a given *genotype* gives rise to different phenotypes in different environments. Genetically identical plants (for example, those in the bottom row) may prosper or not, or even die, depending on the environmental conditions. Plants from different altitudes are known to be genetically different. Hence comparison of the plants in any column shows that in a given environment different genotypes result in different phenotypes. An important inference derived from this experiment is that there is no single *genotype* that is "best" in all environments.

The interaction between the *genotype* and the environment is similarly significant, or even more so, in the case of animals. In one experiment, two strains of rats were selected over many generations; one strain for brightness at finding their way through a maze, and the other for "dullness" (Fig. 17.3). *Selection* was done in the bright strain by using the brightest rats of each generation to breed the following generation, and in the dull strain by breeding the dullest rats every generation. After many generations of *selection*, the descendant bright rats made only about 120 errors running through the maze, whereas dull rats averaged 165 errors. That is a 40% difference. However, the differences between the strains disappeared when rats of both strains were

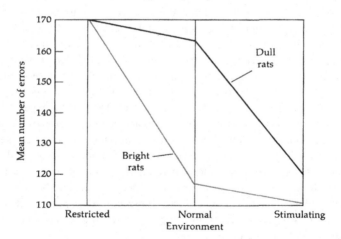

**FIGURE 17.3**  Results of an experiment with two strains of rats, one selected for "brightness," the other for "dullness." After many generations of selection, when raised in the same environment in which the selection was practiced ("normal"), bright rats made about 45 fewer errors than dull rats in the maze used for the tests. However, when the rats were raised in an impoverished ("restricted") environment, bright and dull rats made the same number of errors. When raised in an abundant ("stimulating") environment, the two strains performed nearly equally well (Ayala, 1982a). *After Cooper, R.M., Zubek, J.P., 1958. Effects of enriched and restricted early environments on the learning ability of bright and dull rats. Canadian Journal of Psychology 12 (3), 159–164.*

raised in an unfavorable environment of severe deprivation, where both strains averaged 170 errors. The differences also nearly disappeared when the rats were raised with abundant food and other favorable conditions. In this "optimal" environment, the dull rats reduced their average number of errors from 165 to 120. As with the cinquefoil plants, we see that a given *genotype* gives rise to different phenotypes in different environments, and that the *differences* in *phenotype* between two genotypes change from one environment to another—the *genotype* that is best in one environment may not be best in another.

## CLONING HUMANS?

In the second half of the 20th century, as dramatic advances were taking place in genetic knowledge as well as in the genetic technology often referred to as "genetic engineering," utopian proposals were advanced, at least as suggestions that should be explored and considered as possibilities once the technologies had sufficiently progressed. Some proposals suggested that persons of great intellectual or artistic achievement or of great virtue be cloned. If this were accomplished in large numbers, the genetic constitution of humankind would, it was argued, considerably improve (Plotz, 2005).

Such utopian proposals are grossly misguided. It should be apparent that, as stated earlier, it is not possible to clone a human individual. Seeking to multiply great benefactors of humankind, persons of great intelligence or character; we might obtain the likes of Stalin, Hitler, or Bin Laden. As the Nobel Laureate geneticist George W. Beadle asserted many years ago: "Few of us would have advocated preferential multiplication of Hitler's genes. Yet who can say that in a different cultural context Hitler might not have been one of the truly great leaders of men, or that Einstein might not have been a political villain" (Ayala, 1986). There is no reason whatsoever to expect that the genomes of individuals with excellent attributes would, when cloned, produce individuals similarly endowed with virtue or intelligence. Identical genomes yield, in different environments, individuals who may be quite different. Environments cannot be reproduced, particularly several decades apart, which would be the case when the *genotype* of the persons selected because of their eminent achievement might be cloned. These considerations by themselves would seem sufficient to reject human *cloning* on ethical and social grounds (Caplan and Arp, 2013; Levick, 2013).

Are there circumstances that would justify *cloning* a person because he or she wants it? One might think of a couple unable to have children, or a man or woman who does not want to marry, or two lesbian lovers who want to have a child with the *genotype* of one in an *ovum* of the other, or other special cases that might come to mind (Kitcher, 1996). It must be first pointed out that the *cloning* technology has not yet been developed to an extent that would make it possible to produce a healthy human individual by *cloning*. Second, and most important, the individual produced by *cloning* would be a very different person from the one whose *genotype* is cloned, as belabored previously.

Ethical, social, and religious values will come into play when seeking to decide whether a person might be allowed to be cloned. Most people are likely to disapprove. Indeed, many countries have prohibited human *cloning*. In 2004, the issue of *cloning* was raised in several countries where legislatures were considering whether research on embryonic stem cells should be supported or allowed. The Canadian parliament on March 12, 2004 passed legislation permitting research with stem cells from embryos under specific conditions, but human *cloning* was banned and the sale of sperm and payments to egg donors and surrogate mothers were prohibited. The French parliament on July 9, 2004 adopted a new *bioethics* law that allows embryonic stem-cell research but considers human *cloning* a "crime against the human *species*." Reproductive *cloning* experiments are punishable by up to 20 years in prison. Japan's Cabinet Council for Science and Technology Policy voted on July 23, 2004 to adopt policy recommendations that permit the limited *cloning* of human embryos for scientific research, but not the *cloning* of individuals. On January 14, 2001 the British government amended the Human Fertilisation and Embryology Act of 1990 by allowing embryo research on stem cells and *therapeutic cloning*. The Human Fertilisation and Embryology Act of 2008 explicitly prohibited reproductive *cloning* but allowed experimental stem-cell research for treating diabetes, Parkinson's disease, and Alzheimer's disease (Staff Writer, 2008; UK Statute Law Database, 2001). On February 3, 2014 the House of Commons voted to legalize the *gene therapy* technique known as MRT, or three-person in vitro fertilization, in which *mitochondria* from a donor's egg cell contribute to a couple's embryo (Callaway, 2015b; see also Cohen et al., 2015). In the United States there are currently no federal laws that ban *cloning* completely (UK Statute Law Database, 2001). Thirteen states (Arkansas, California, Connecticut, Iowa, Indiana, Massachusetts, Maryland, Michigan, North Dakota, New Jersey, Rhode Island, South Dakota, and Virginia) ban reproductive *cloning* and three states (Arizona, Maryland, and Missouri) prohibit use of public funds for research on reproductive *cloning* (NCSL, 2008). In early December 2015 the US National Academies of Science, Engineering, and Medicine, in concert with the Chinese Academy of Sciences and the Royal Society in London, hosted a large gathering of experts to address ethical and policy issues concerning the editing of human genomes in the early stages of *development* and in germ-line cells. In addition, a committee appointed by the US Institute of Medicine is exploring whether a consensus can be developed regarding the ethical and social implications of MRT.

## THERAPEUTIC CLONING

*Cloning* of embryonic cells, "stem cells," could have important health applications in organ transplantation, treating injured nerve cells, and other areas. In addition to SCNT, the method discussed earlier for *cloning* individuals, another technique is available, "induced pluripotent stem cells," although SCNT has proven to be much more effective and less costly. The objective is to obtain "pluripotent" stem cells that have the potential to differentiate in any of the three germ layers characteristic of humans and other animals: endoderm (lungs and interior lining of stomach and gastrointestinal tract), ectoderm (nervous systems and epidermal tissues), and mesoderm (muscle, blood, bone, and urogenital tissues). Stem cells, with more limited possibilities than pluripotent cells, can also be used for specific therapeutic purposes (Kfoury, 2007).

Stem-cell *therapy* consists of *cloning* embryonic or other stem cells to obtain pluripotent stem cells that can be used in regenerative medicine to treat or prevent all sorts of diseases and for the transplantation of organs. At present, bone marrow transplantation is a widely used form of stem-cell *therapy*; stem blood cells are used in the treatment of sickle-cell anemia, a lethal disease when untreated, which is very common in places where malaria is rife because heterozygous individuals are protected against infection by *Plasmodium falciparum*, the agent of malignant malaria, which has led to a high frequency of the sickle-cell *allele* (Elguero et al., 2015). One of the most promising applications of *therapeutic cloning* is the growth of organs for transplantation, using stem cells that have precisely the *genome* of the organ recipient. Two major transplantation hurdles would be overcome: the possibility of immune rejection, and the availability of organs from suitable donors. Another regenerative medical application that might be anticipated is the therapeutic growth of nerve cells. There are hundreds of thousands of individuals around the world paralyzed from the neck down and confined for life to a wheelchair as a consequence of damage to the spinal cord below the neck, often due to a car accident or a fall, that interrupted the transmission of nerve activity from the brain to the rest of the body and vice versa. A small growth of nerve cells sufficient to heal the wound in the spinal cord would have enormous health consequences for the injured persons and for society.

At present, the one *gene therapy* modification of the embryo that can be practiced is MRT, legalized in the United Kingdom by the House of Commons on February 3, 2014 (Callaway, 2015a,b), as mentioned earlier. Mutations in the mitochondrial *DNA* of about 1 in 6500 individuals account for a variety of severe and often fatal conditions, including blindness, muscular weakness, and heart failure (Gemmel and Wolff, 2015). These conditions are transmitted from the mother to her children through her *mitochondrial DNA (mtDNA)*. The technique consists in transplanting the *nucleus* of an *ovum* (or cell) from the mother with mitochondrial genetic impairments to an *ovum* from a donor in which the *nucleus* has been removed or otherwise neutralized. The *ovum* is then reimplanted (with or without previous fertilization by sperm from the father to be) in the mother. With MRT, the embryo possesses nuclear *DNA* from the mother and father, as well as *mtDNA* from a donor female who has healthy *mtDNA*. At present, MRT remains technically challenging, with a low rate of success and limited social acceptance. One complicating issue is that *mtDNA* replacement is not 100% successful; disease-causing *mutant mtDNA* persists in the developing embryo and may account for eventual diseases due to heteroplasmy, at least in some tissues. A second issue of concern is that *mtDNA* disorders often appear late in life. It remains unknown whether the benefits of MRT as currently practiced may persist in advanced age. An additional consideration concerns the gender of the child. If it is a man, he will not transmit his *DNA* to his progeny; if it is woman, she will transmit to her children the *mtDNA* from the donor woman. Some experts may therefore think that MRT is acceptable to produce male embryos, but not female embryos.

# Glossary

**Adaptation**   A structural or functional characteristic of an organism that allows it to cope better with its environment; the evolutionary process by which organisms become adapted to their environment.

**Adaptive value**   A measure of the reproductive efficiency of an organism (or *genotype*) compared with other organisms (or genotypes); also called *selective value*.

**Aesthetics**   The study of the nature of beauty, art, and taste; the creation and appreciation of beauty (also, *esthetics*).

**Allele**   Each of the two or more different forms of a *gene*, such as the alleles for *A*, *B*, and *O* at the *ABO* blood-group gene locus.

**Allopatric**   Referring to populations or species that live in different territories (see also *sympatric*).

**Altruism**   Acting for the benefit of another individual at one's own cost.

**Analogy (adj. analogous)**   Resemblance in function but not in structure due to independent evolutionary origin; eg, the wings of a bird and of an insect.

**Argument from design**   Traditional argument revived in the 1990s asserting that organisms are outcomes of intentional design and therefore evince the existence of a designer, namely God in the traditional argument.

**Atheism**   The denial of the existence of a deity.

**Autapomorphy**   A *derived trait* found only in a single *clade* of a cladogram.

**Autosome**   A *chromosome* other than a sex chromosome.

**Base pair**   Two nitrogenous bases that pair by hydrogen bonding in double-stranded *DNA* or *RNA*.

**Bioethics**   The study of ethical questions in biological research and medicine.

**Biogeography**   The geographic distribution of plants and animals.

**Biota**   All plants and animals of a given region or time.

**Bottleneck**   A period when a *population* becomes reduced to only a small number of individuals.

**Category**   Each of the levels in the Linnaean classification system (*kingdom, phylum, class, order, family, genus, species,* and their intermediates).

**Chromosome**   A thread-shaped structure visible in the cell's *nucleus* during cell division. Chromosomes contain most of the hereditary material or genes.

**Clade**   A complete group of organisms derived from a common ancestor, or the branches that separate in a cladistic event.

**Cladistics**   A classification system proposed by Hennig based on phylogenetic hypotheses and common ancestry that only admits speciation by means of *cladogenesis.*

**Cladogenesis**   An evolutionary process whereby a species gives rise to two different ones, after which it disappears. In a *cladogram,* the divergence point is known as a node.

**Cladogram**   The graphic representation of the branching relations between *species,* genera, families, and so on, which are represented as *clades.*

**Class**   A category formed by a set of *orders.*

**Cloning**   The precise replication of a molecule, cell, organ, or individual.

**Coalescence**   The convergence of the phylogenies of several species into their most recent common ancestor.

**Codon**   A group of three adjacent *nucleotides* in an *mRNA* molecule that code either for a specific *amino acid* or for *polypeptide* chain termination during *protein* synthesis.

**Condition, necessary**   An attribute that must be the case for an argument or description to be correct.

**Condition, sufficient**   An attribute that is enough to guarantee that an argument or description is correct.

**Convergence**   The parallel development of the same feature in unrelated organisms.

**Corroboration**   Evidence that supports a hypothesis or an argument that had been proposed earlier on the basis of some preexisting evidence.

**Court of law**   An official assembly or tribunal that decides cases on the basis of official legislation or the common law.

**Cranium**   The set of bones that constitute the skull except for the *mandible* (also known as the *brain case*).

**Creation**   The assertion that the world exists as an act of God.

**Creationism**   The belief that God is the author, ultimately, of everything that exists in the world and of all their attributes.

**Criterion of demarcation**   The distinctive feature of scientific knowledge is the testing of hypotheses by observation and experimentation.

**Cultural evolution**   The evolution of human distinctive characteristics, including language, art, morality, legal and political institutions, clothing, housing, agriculture, and industry.

**Deism**   The notion that God is the author of the universe and all its laws, but does not otherwise interfere in the world.

**Design** The attributes of organisms and everything in the world, which evince that they have been intentionally designed, or that give the appearance of having been designed.

**Development** The processes of growth and change in organisms that occur from their beginning until their maturity.

**DNA** Deoxyribonucleic acid; *nucleic acid* composed of units consisting of a deoxyribose sugar, a phosphate group, and the nitrogen bases adenine, guanine, cytosine, and thymine. The self-replicating genetic material of all living cells, it is made up of a double helix of two complementary strands of *nucleotides.*

**Dominant** A character that is manifest in the *phenotype* of heterozygous individuals.

**Drift** In populations, genetic and other changes that occur by chance (see *genetic drift*).

**Duplication** A chromosomal mutation characterized by two copies of a *chromosome* segment.

**Effective population size** The number of reproducing individuals in a population.

**Emergence** Higher level phenomena have attributes beyond those of the component lower level phenomena.

**Endemic** Applied to species restricted to a certain region or part of a region; in epidemiology, applied to diseases that are constantly present at relatively low levels in a particular *population.*

**Enzyme** A biochemical catalyst based on specialized *protein* molecules that speeds up biochemical processes.

**Epigenesis** Cellular processes that modify the expression of genes without changing their DNA sequence.

**Epistemology** Philosophical investigation of questions concerning the nature and justification of knowledge.

**Esthetics** The study of the nature of beauty, art, and taste; the creation and appreciation of beauty (also *aesthetics*).

**Ethics** The discipline dealing with what is morally good and bad in human behavior.

**Eugenics** The proposal that the human species can be improved over the generations by increasing the rate of reproduction of individuals or groups endowed with desirable characteristics, and reducing or eliminating the reproduction of those with undesirable features.

**Eukaryote** An organism whose cells contain a distinct *nucleus* as well as *mitochondria* and other *organelles.*

**Evil, problem of** How to account for the existence of physical and moral evil in the world, if this has been created by a benevolent God; also known as the *theodicy* problem.

**Evo-devo** The investigation of the processes of development and evolution of organisms and the interaction between the two.

**Evolution**   The hereditary change of organisms through time.

**Exaptation**   An adaptation that evolved for reasons other than its current biological role.

**Exon**   The *DNA* of a eukaryotic transcription unit whose transcript becomes a part of the *mRNA* produced by splicing out introns.

**Falsifiability**   The property of a proposition that could be shown to be false by empirical evidence or logical reasoning.

**Falsification**   Demonstration that a statement or theory is false on the ground of empirical evidence or logical reasoning.

**Family**   The category composed by a set of *genera*.

**Fitness**   The reproductive contribution of an organism or *genotype* to future generations.

**Flagellum**   A filiform appendage of bacteria, plants, and animals.

**Fossil**   Any preserved remains or traces of past life, more than about 10,000 years old, embedded in rock either as mineralized remains or impressions, casts, or tracks.

**Gamete**   A mature reproductive cell capable of fusing with a similar cell of opposite sex to give a *zygote*; also called a *sex cell*.

**Gene**   A genetic unit found on a specific *locus* of a *chromosome*. It consists of a sequence of *DNA* that codes for an *enzyme* or a *protein*, or that regulates activity of other genes.

**Genetic drift**   Chance fluctuations in gene frequency observed especially in small *populations*.

**Genome**   The genetic content of a cell; in *eukaryotes*, it sometimes refers to only one complete (*haploid*) *chromosome* set.

**Genotype**   The genetic information of an individual.

**Genotyping**   Procedure to obtain the genome of an individual and/or a species.

**Genus**   The category formed by a set of closely related *species*.

**Gradualism**   The view that evolutionary change is a gradual, more or less continuous process.

**Group selection**   Differential reproduction between groups of individuals.

**Habitat**   The natural home or environment of a plant or animal.

**Haploid**   Of cells, such as *gametes*, that in *eukaryotes* have half as many *chromosome* sets as the somatic cells.

**Hardy—Weinberg law**   Describes the genetic equilibrium in a population stating that genotypes, the genetic constitution of individual organisms, exist in certain frequencies that are a simple function of the allelic frequencies.

**Heterozygote**   An organism with two different *alleles* at a certain locus.

**High-throughput genetic sequencing**   See PEC.

**Hominin**   An individual belonging to the tribe *Hominini*.

**Hominini**   A tribe composed of current humans and their direct and lateral ancestors that are not also ancestral to chimpanzees.

**Hominoid**    An individual belonging to the superfamily *Hominoidea.*

**Hominoidea**    The superfamily composed of lesser apes, great apes, and humans, as well as their direct and lateral ancestors that are not also ancestral to *Old World monkeys.*

**Homo**    The genus to which the human species belongs. The species *Homo habilis, Homo naledi, Homo ergaster, Homo antecessor, Homo erectus, Homo heidelbergensis, Homo neanderthalensis,* and *Homo sapiens* are usually included within this genus. *Homo rudolfensis* is also included by some authors.

**Homologous (noun *homology*)**    A trait that is shared by two taxa, which inherited it from their closest common ancestor. Can also be used to refer to *chromosomes.*

**Homozygote**    A cell or organism having the same *allele* at a given locus on *homologous chromosomes* (adj. *homozygous*).

**Hybridization**    Genetic admixing between two *species.*

**Hypothetico-deductive method**    The claim that the scientific method consists of two stages: formulation of hypotheses and testing them by observation and experimentation.

**Immune system**    The bodily system that protects the body from foreign substances.

**Inbreeding**    Mating between relatives.

**Induction**    Making generalizations from particular instances.

**Inductionism**    The theory that scientific knowledge consists of generalizations derived from individual observations.

**Intelligent design**    The notion that the design of organisms and everything else in the world is due to the actions of an intelligent designer.

**Introgression**    Presence of some amount of genetic material coming from a different species.

**Intron**    A length of *DNA* within a functional gene in *eukaryotes,* separating two segments of coding DNA (*exons*).

**Inversion**    A chromosomal mutation characterized by the reversal of a *chromosome* segment.

**Irreducible complexity**    The claim that organs or other components of organisms can only come about if all parts are simultaneously present and thus the parts may not have evolved gradually or sequentially.

**Kin selection**    A form of altruism related to parental care.

**Levels of selection**    The different levels (such as genes, cells, individuals, populations) at which natural selection can operate.

**Locus**    A *gene's* specific place on a *chromosome;* sometimes used to refer to the gene itself (plural *loci*).

**Macroevolution**    Large-scale evolutionary change.

**Mendelism**    The study of biological heredity on the basis of discrete heritable components, as first formulated by Gregor Mendel.

**Messenger RNA (mRNA)** An RNA molecule whose *nucleotide* sequence is translated into an *amino acid* sequence on ribosomes during *polypeptide* synthesis.

**Microevolution** Small-scale evolutionary change.

**Missing links** Organisms not yet discovered that are intermediate between major groups of organisms.

**Mitochondria** Organelles in a eukaryotic cell that are involved in energy metabolism; each mitochondrion has its own small circular genome.

**Mitochondrial DNA (mtDNA)** Genetic information contained in *mitochondria* in the form of a single circular strand which is inherited only through the maternal line.

**Molecular clock** The estimated regularity of changes in *DNA* and *proteins* through time, which can be used to estimate the timing of evolutionary episodes.

**Molecular evolution** The study of evolutionary processes at the level of biological molecules.

**Morality** The study of good or bad human behavior.

**mtDNA** See *mitochondrial DNA*.

**Multiregional hypothesis** A hypothesis suggesting that the appearance of modern humans occurred by an independent evolution from earlier *hominins* in different geographical regions.

**Mutant** An *allele* different from the wild type, or an individual carrying such an allele.

**Mutation** An inheritable modification of genetic material.

**Natural selection** The differential reproduction of alternative *genotypes* due to variable *fitness*.

**Natural theology** Arguments seeking to demonstrate the existence and attributes of God based on the investigation of the natural world.

**Naturalism, methodological** The claim that the subject of science is to investigate natural phenomena, without claiming that they are all that exists or that can be investigated.

**Naturalism, ontological (or metaphysical)** The claim that natural phenomena studied by science are all that exist in the world.

**Neutral theory of molecular evolution** Asserts that most evolutionary changes at the level of molecules are neutral; that is, due to chance rather than to natural selection.

**Niche** The place of an organism in its environment, including the resources it exploits and its association with other organisms.

**Nitrogen base** An organic compound composed of a ring containing nitrogen. Used here to refer to each of the complementary molecules that keep the two *DNA* strands together transversally or form *RNA* strands (the bases are adenine, cytosine, guanine, thymine, and uracil).

**Node** The point where two *clades* diverge in a cladogram.

**Nomad**   One who continually moves from place to place to find food.

**Nominalism**   The claim that only individual things exist; universal or general classes are names used for convenience, not realities that exist as such in the world.

**Novelty**   A trait that has evolved in an organism or species but has not previously existed in any other organisms.

**Nucleic acid**   See *DNA* and *RNA*.

**Nucleotide**   A nucleic acid unit, composed of a sugar molecule, a nitrogen base, and a phosphate group. A set of three nucleotides constitutes a *triplet*, or *codon*, and each triplet codes for an *amino acid* or represents a stop signal during protein synthesis.

**Nucleus**   A membrane-enclosed *organelle* of *eukaryotes* that contains the *chromosomes*.

**Occipital torus**   A protuberance found on the posterior part of the *cranium*.

**Old World monkeys**   Relating to monkeys in all geographical areas except South and Central America, in the superfamily Cercopithecoidea.

**Ontogenetic**   Relative to *ontogeny*.

**Ontogeny**   Development of an organism after conception.

**Order**   The taxonomic category formed by a set of *families* and a division of a *class*.

**Organelle**   A functional membrane-enclosed body inside cells (eg, a *nucleus* or a *mitochondrion*).

**Orrorin**   The *genus* including late Miocene *hominin* remains found in the Tugen Hills region of Kenya.

**Orthologous genes or chromosomes**   Referring to *genes* or *chromosomes* of different *species* which are similar because they derive from a common ancestor.

**Out of Africa**   See *replacement hypothesis*.

**Ovum**   A female *gamete*.

**Paleolithic**   The Old Stone Age, the first and longest part of the Stone Age that began some 2.6 million years ago in Africa with the first recognizable stone tools belonging to the *Oldowan* industrial tradition and ended some 12,000–10,000 years ago.

*Pan*   The *genus* to which chimpanzees and bonobos belong.

**Paraphyletic**   A group of organisms including some, but not all, of the descendants of the group's common ancestor.

**Parsimony**   In evolution, the principle proposing that evolution has followed the most economical route, involving the assumption that closely related species (those that diverged more recently) will consistently have fewer differences than species that diverged longer ago.

**PCR (polymerase chain reaction)**   Technique for cloning genetic material, which allows millions of identical copies to be obtained from a single DNA molecule.

**PEC (primer extension capture)**   Also known as high-throughput sequencing. Technique that permits identification of short sequences of nitrogenous bases from very degraded DNA samples contaminated by the presence of microbial DNA.

**Phenetics**   A system of classification of organisms based principally on the similarity of morphological traits, also known as numerical *taxonomy*.

**Phenotype**   An organism's observable traits.

**Phlogiston**   The old hypothetical claim that fire is a natural substance.

**Phyletic**   Applied to a group of *species* with a common ancestor; a line of direct descent.

**Phylogenesis**   The process of evolution and differentiation of organisms.

**Phylogenetic tree**   The graphic representation of evolutionary relations among living and extinct organisms.

**Phylogeny**   The evolutionary history of a group of living or extinct organisms.

**Phylum**   The category formed by a set of *classes* (plural *phyla*).

**Platyrrhine**   A member of the *primate* infraorder Platyrrhini (*New World monkeys*).

**Pleiotropy**   When a gene has caused consequences in several phenotypic traits.

**Pleistocene**   The first epoch of the Quaternary period, which lasted from about 1.64 million to 10,000 years ago, and saw the radiation of the *genus Homo*.

**Plesiomorphy**   A trait that is already present in the ancestral group of the taxon being studied.

**Polymorphism**   The existence of alternative allelic forms at a *locus* within a *population*. Thus, in humans, there is a polymorphism for the ABO blood groups.

**Polypeptide**   A chain of *amino acids* covalently bound by peptide linkages.

**Polyphyletic**   The grouping of organisms derived from at least two different ancestral stocks.

*Pongo*   The *genus* to which orangutans belong.

**Population**   A set of individuals belonging to the same *species* that constitute an effective reproductive community.

**Primate**   The *order* to which the human *species* belongs, together with prosimians, tarsoids, and the rest of the anthropoids.

**Progress**   Advance or betterment; it can be biological or cultural.

**Prokaryote**   A microorganism that belongs to archaea or bacteria; it has single-strand DNA and lacks organelles.

**Prosimian**   Any *primate* in the suborder Prosimii (lemurs, lorises, and tarsiers).

**Protein**   A molecule composed of one or more *polypeptide* subunits and possessing a characteristic three-dimensional shape imposed by the sequence of its component *amino acid* residues.

**Punctuated evolution**   See rectangular evolution.

**Recessive**   An allele, or the corresponding trait, that is manifest only in *homozygotes*.

**Rectangular evolution**   Evolutionary change is discontinuous, with periods of stasis alternating with periods of rapid change; also known as punctuated evolution.

**Reductionism**   The claim that higher-level phenomena must be explained by lower-level laws or processes.

**Reductionism, epistemological**   The claim that higher-level phenomena can be fully explained by investigating lower-level processes.

**Reductionism, methodological**   The claim that the understanding of higher-level phenomena is benefitted by investigating their lower-level components or processes.

**Reductionism, ontological (or metaphysical)**   The claim that organisms are "nothing but" aggregations of their components and thus do not exhibit any distinctive attributes.

**Replacement hypothesis**   The proposal that the modern humans that had dispersed from Africa did not admix with earlier *populations* living in the territories they colonized.

**RNA**   Ribonucleic acid is a *nucleic acid* composed of units consisting of a ribose sugar, a phosphate group, and the nitrogen bases adenine, guanine, cytosine, and uracil.

**Selection**   See *natural selection*.

**Selective value**   See *adaptive value*.

**Sex cell**   See *gamete*.

**Sex chromosome**   A chromosome that differs between the two sexes and is involved in sex determination (see *autosome*).

**Simian**   Any member of the *primate* suborder Anthropoidea (monkeys, apes, and humans); a higher primate.

**Single nucleotide polymorphism**   See *SNP*.

**Single-species hypothesis**   The notion that all *hominin* specimens after *Homo habilis* can be adequately accommodated within a single *species*.

**Sister group**   Each one of the clades that separates at a node (see *clade*, *cladogenesis*, and *cladogram*).

**Skull**   Set formed by the cranial bones and the *mandible*.

**SNPs**   Single nucleotide polymorphisms. Alleles corresponding to a certain locus that differ in a single nitrogen base.

**Sociobiology**   The study of the genetic basis and evolutionary history of social organization in animals, including humans.

**Speciation**   The process of evolution of a new *species*.

**Species**   The basic unit of Linnaean classification, always expressed by two Latin names (such as *Homo sapiens*), the first of which specifies the *genus*; defined as groups of interbreeding natural *populations* that are reproductively isolated from other such groups.

**Stasis**   A condition when evolutionary change is not taking place.

**Symbiosis**   Interaction between different kinds of cells or different kinds of organisms.

**Symbol**   A word, behavior, or object that conveys meaning.

**Symbolism**   The capacity of creating and interpreting symbols.

**Sympatric**   Referring to species that share the same territory (see also *allopatric*).

**Synthetic theory**   The theory of evolution as it emerged in the second quarter of the 20th century by the integration of Mendelian genetics and Darwin's theory of natural selection.

**Systematics**   The discipline that studies the classification of organisms and their evolutionary relationships.

**Taxon**   A defined unit in the classification of organisms. For example, *Homo* is a taxon of the *genus* category; *Homo sapiens* is a taxon of the *species* category (plural *taxa*).

**Taxonomy**   The rules and procedures used in the classification of organisms.

**Teleological explanation**   Accounting for the presence of a feature or behavior of an organism in terms of the function or purpose it serves.

**Teleology**   Consideration of the function or purpose of organisms and their components.

**Teleonomy**   The same as teleology.

**Tertiary**   The first period of the Cenozoic era, from 65 to 1.64 million years ago.

**Tetrapod**   A four-footed animal: any amphibian, reptile, bird, or mammal.

**Theism**   The claim that there is a God who created the world and interacts with His creation.

**Theistic evolution**   The claim that God directs the evolutionary process.

**Theodicy**   Defense of God's goodness and omnipotence in spite of the existence of physical and moral evil in the world.

**Therapeutic cloning**   The treatment of a disorder or disease by cloning of an organ, cell, or gene.

**Therapy**   Medical treatment of a bodily, mental, or behavioral disorder.

**Triplet**   In genetics, set of three contiguous *DNA* or *RNA nucleotides* that specifies a particular *amino acid* in a *protein*, or indicates its end signal (also called *codon*).

**Utilitarianism**   Ethical and other philosophical themes that claim that the value of actions should be measured by their consequences on other human beings.

**Verifiability**   The confirmation that a hypothesis is true.

**Vertebrate**   An animal with a backbone or vertebral column.

**Wernicke's area**   The region of the human brain involved in the comprehension of speech, lying in the upper part of the temporal cortex and extending into the parietal cortex in the left cerebral hemisphere.

**Zygote**   The diploid cell formed by the union of egg and sperm nuclei in the cell.

# References

Abbott, A., 1992. Gene therapy. Italians first to use stem cells. Nature 356 (6369), 465.

Abi-Rached, L., Jobin, M.J., Kulkarni, S., McWhinnie, A., Dalva, K., Gragert, L., Babrzadeh, F., Gharizadeh, B., Luo, M., Plummer, F.A., Kimani, J., Carrington, M., Middleton, D., Rajalingam, R., Beksac, M., Marsh, S.G.E., Maiers, M., Guethlein, L.A., Tavoularis, S., Little, A.-M., Green, R.E., Norman, P.J., Parham, P., 2011. The shaping of modern human immune systems by multiregional admixture with archaic humans. Science 334, 89–94.

Abrahamov, B., 1986. Al-Kasim ibn Ibrahim's argument from design. Oriens 29, 259–284.

Adcock, G.J., Dennis, E.S., Easteal, S., Huttley, G.A., Jermin, L.S., Peacock, W.J., 2001. Mitochondrial DNA sequences in ancient Australians: implications for modern human origins. Proceedings of the National Academy of Sciences of the United States of America 98, 537–542.

Alexander, R., 1979. Darwinism and Human Affairs. University of Washington Press, Seattle.

Almond, G.A., Chodorow, M., Pearce, R.H. (Eds.), 1982. Progress and Its Discontents. University of California Press, Berkeley.

Altman, S., Wesolowski, D., Guerrier-Takada, C., Li, Y., 2005. RNase P cleaves transient structures in some riboswitches. Proceedings of the National Academy of Sciences of the United States of America 102, 11284–11289.

Alvarez, W., Alvarez, W., Asaro, F., Michel, H.V., 1980. Extraterrestrial cause for the cretaceous-tertiary extinction. Science 208, 1095–1108.

Aquinas, T., 1905. Of god and his creatures. In: Rickaby, J. (Ed.), Summa Contra Gentiles. Burns & Oates, London, pp. 241–368.

Aquinas, T., 1964. Existence and nature of god. In: Blackfriars (Ed.), Summa Theologiae. McGraw-Hill, New York, pp. 10–249.

Arias, E., 2010. United States life tables, 2006. National Vital Statistical Reports 58 (21), 1–40.

Armstrong, K., 1993. A History of God. The 4,000-Year Quest of Judaism, Christianity and Islam. Alfred Knopf, New York.

Armstrong, K., 2009. The Case for God. Alfred Knopf, New York.

Arp, R., 1998. Hume's mitigated skepticism and the design argument. American Catholic Philosophical Quarterly 72, 539–558.

Arp, R., 1999. The quinque via of Thomas Hobbes. History of Philosophy Quarterly 16, 367–394.

Attwater, J., Wochner, A., Holliger, P., 2013. In-ice evolution of RNA polymerase ribozyme activity. Nature Chemistry 5, 1011–1018.

Augustine., 1998. The city of god. In: Dyson, R. (Ed.), An Early Classic of Christian Theology. Cambridge University Press, Cambridge, UK.

Augustine., 2002. Work of St. Augustine. In: Rotelle, J.E. (Ed.), On Genesis, vol. 13. New City Press, Hyde Park, NY.

Avery, O.T., MacLeod, C.M., McCarthy, M., 1944. Studies on the chemical nature of the substance inducing transformation of pneumococcal types. Induction of transformation by a deoxyribonucleic fraction isolated from pneumococcus type III. Journal of Experimental Medicine 79, 137–158.

Avise, J., 2006. Evolutionary Pathways in Nature: A Phylogenetic Approach. Cambridge University Press, Cambridge, UK.

Avise, J.C., 2010. Inside the Human Genome. A Case for Non-Intelligent Design. Oxford University Press, Oxford, New York.

Ayala, F.J., 1968. Biology as an autonomous science. American Scientist 56, 207–221.

Ayala, F.J., 1969. An evolutionary dilemma: fitness of genotypes versus fitness of populations. Canadian Journal of Genetics and Cytology 11, 439–463.

Ayala, F.J., 1970. Teleological explanations in evolutionary biology. Philosophy of Science 37, 1–15.

Ayala, F.J., 1974. The concept of biological progress. In: Ayala, F.J., Dobzhansky, T. (Eds.), Studies in the Philosophy of Biology. University of California Press, Berkeley, pp. 339–355.

Ayala, F.J., 1977. Philosophical issues. In: Dobzhansky, T., Ayala, F.J., Stebbins, G.L., Valentine, J.W. (Eds.), Evolution. Freeman, San Francisco, pp. 474–516. Chapter 16.

Ayala, F.J., 1982a. Population and Evolutionary Genetics: A Primer. Benjamin Cummings, Menlo Park, CA.

Ayala, F.J., 1982b. The evolutionary concept of progress. In: Almond, G.A., Chodorow, M., Pearce, R.H. (Eds.), Progress and Its Discontents. University of California Press, Berkeley, pp. 106–124.

Ayala, F.J., 1983. Microevolution and macroevolution. In: Bendall, D.S. (Ed.), Evolution From Molecules to Men. Cambridge University Press, Cambridge, pp. 396–397.

Ayala, F.J., 1985. The theory of evolution: recent successes and challenges. In: McMullin, E. (Ed.), Evolution and Creation. Notre Dame Press, Notre Dame, Indiana, pp. 59–90.

Ayala, F.J., 1986. Whither mankind? The choice between a genetic twilight and a moral twilight. American Zoologist 26, 895–905.

Ayala, F.J., 1987a. Biological reductionism: the problems and some answers. In: Yates, F.E. (Ed.), Self-Organizing Systems: The Emergence of Order. Plenum Press, New York, London, pp. 315–324.

Ayala, F.J., 1987b. The biological roots of morality. Biology and Philosophy 2, 235–252.

Ayala, F.J., 1988. Can "progress" be defined as a biological concept? In: Nitecki, M.H. (Ed.), Evolutionary Progress. University of Chicago Press, Chicago, pp. 75–96.

Ayala, F.J., 1992a. DNA law. Journal of Molecular Evolution 35, 273–276.

Ayala, F.J., 1992b. Wistar's views. Journal of Molecular Evolution 35, 467–471.

Ayala, F.J., 1993. Junk science and DNA typing in the courtroom. Contention 2, 45–60.

Ayala, F.J., 1994. On the scientific method, its practice and pitfalls. History and Philosophy of the Life Sciences 16, 205–240.

Ayala, F.J., 1995. The distinctness of biology. In: Weinert, F. (Ed.), Laws of Nature. Essays on the Philosophical, Scientific and Historical Dimensions. Walter de Gruyter, Berlin, pp. 268–285.

Ayala, F.J., 1999. Adaptation and novelty: teleological explanations in evolutionary biology. History and Philosophy of the Life Sciences 21, 3–33.

Ayala, F.J., 2006. Darwin and Intelligent Design. Fortress Press, Minneapolis, MN.

Ayala, F.J., 2007. Darwin's Gift to Science and Religion. Joseph Henry Press, Washington, DC.

Ayala, F.J., 2008a. Darwin's gift to science and religion: commentaries and responses. Theology and Science 6, 179–196.

Ayala, F.J., 2008b. Reduction, emergence, naturalism, dualism, teleology: a précis. In: Cobb Jr., J.B. (Ed.), Back to Darwin. A Richer Account of Evolution. William B. Erdmans Publishing Company, Grand Rapids, MI and Cambridge, UK, pp. 76–87.

Ayala, F.J., 2010. What the biological sciences can and cannot contribute to ethics. In: Ayala, F.J., Arp, R. (Eds.), Contemporary Debates in Philosophy of Biology. Wiley-Blackwell, Chichester, West Sussex, UK, pp. 316–336.

Ayala, F.J., 2014. Introduction. In: Vargas, P., Zardoya, R. (Eds.), The Tree of Life. Evolution and Classification of Living Organisms. Sinauer Associates, Sunderland, MA.

Ayala, F.J., 2015. Cloning humans? Biological, ethical, and social considerations. Proceedings of the National Academy of Sciences of the United States of America 112, 8879–8886.

Ayala, F.J., 2016. Human evolution and progress. In: Tibayrenc, M., Ayala, F.J. (Eds.), On Human Nature. Elsevier (in press).

Ayala, F.J., Powell, J.R., 1972. Allozymes as diagnostic characters of sibling species in *Drosophila*. Proceedings of the National Academy of Sciences of the United States of America 69, 1094–1096.

Ayala, F.J., Dobzhansky, T. (Eds.), 1974. Studies in the Philosophy of Biology. University of California Press, Berkeley.

Ayala, F.J., Valentine, J., 1979. Evolving: The Theory and Processes of Organic Evolution. Benjamin Cummings, Menlo Park, CA.

Ayala, F.J., Kiger, J.A., 1984. Modern Genetics, second ed. Benjamin/Cummings, Menlo Park, California.

Ayala, F.J., Black, B., 1993a. The nature of science: a primer for the legal consumer of scientific information. Science and Courts 1, 1–21.

Ayala, F.J., Black, B., 1993b. Science and the courts. American Scientist 81, 230–239.

Ayala, F.J., Arp, R. (Eds.), 2010. Contemporary Debates in Philosophy of Biology. Wiley-Blackwell, Malden, MA.

Ayala, F.J., Mourão, C.A., Pérez-Salas, S., Richmond, R., Dobzhansky, T., 1970. Enzyme variability in the *Drosophila willistoni* group. I. Genetic differentiation among sibling species. Proceedings of the National Academy of Sciences of the United States of America 67, 225–232.

Ayala, F.J., Powell, J.R., Dobzhansky, T., 1971. Polymorphisms in continental and island populations of *Drosophila willistoni*. Proceedings of the National Academy of Sciences of the United States of America 68, 2480–2483.

Ayala, F.J., Powell, J.R., Tracey, M.L., Mourao, C.A., Perez-Salas, S., 1972. Enzyme variability in the *Drosophila willistoni* group. IV. Genic variation in natural populations of *Drosophila willistoni*. Genetics 70, 113–139.

Babbage, C., 1838. The Ninth Bridgewater Treatise. A Fragment. John Murray, London.

Bainbridge, J.W., Smith, A.J., Barker, S.S., Robbie, S., Henderson, R., Balaggan, K., Viswanathan, A., Holder, G.E., Stockman, A., Tyler, N., Petersen-Jones, S., Bhattacharya, S.S., Thrasher, A.J., Fitzke, F.W., Carter, B.J., Rubin, G.S., Moore, A.T., Ali, R.R., 2008. Effect of gene therapy on visual function in Leber's congenital amaurosis. New England Journal of Medicine 358, 2231–2239.

Baltimore, D., 1970. Viral RNA-dependent DNA polymerase in virions of RNA tumor viruses. Nature 226, 1209.

Baltimore, D., Berg, P., Botchan, M., Carroll, D., Charo, R.A., Church, G., Corn, J.E., Daley, G.Q., Doudna, J.A., Fenner, M., Greely, H.T., Jinek, M., Martin, G.S., Penhoet, E., Puck, J., Sternberg, S.H., Weissman, J.S., Yamamoto, K.R., 2015. A prudent path forward for genomic engineering and germline gene modification. Science 348, 36–38.

Bar-Yosef, O., Vandermeersch, B., 1993. El hombre moderno de Oriente medio. Investigación y Ciencia 201, 66–73.

Barash, D., 1977. Sociobiology and Behavior. Elsevier, New York.

Barkow, J., Cosmides, L., Tooby, J. (Eds.), 1992. The Adapted Mind: Evolutionary Psychology and the Generation of Culture. Oxford University Press, Oxford.

Beckner, M., 1959. The Biological Way of Thought. Columbia University Press, New York.

Behe, M., 1996. Darwin's Black Box: The Biochemical Challenge to Evolution. Free Press, New York.

Behe, M., 2007. The Edge of Evolution: The Search for the Limits of Darwinism. Free Press, New York.

Bell, C., 1833. The Hand, Its Mechanisms and Vital Endowments as Evincing Design. William Pickering, London.

Bell, E.A., Boehnke, P., Harrison, T.M., Mao, W.L., 2015. Potentially biogenic carbon preserved in a 4.1 billion-year-old zircon. Proceedings of the National Academy of Sciences of the United States of America 112, 14518–14521.

Benado, M., Aguilera, M., Reig, D.A., Ayala, F.J., 1979. Biochemical genetics of Venezuelan spiny rats of the *Proechimys guainae* and *Proechimys trinitatis* superspecies. Genetica 50, 89–97.

Berg, E.S., 1926 [1969]. Nomogenesis or Evolution Determined by Law (1926). MIT Press, London, Boston.

Bergson, H., 1907. L'Evolution créatrice (Creative Evolution). Alcan, Paris.

Bernard, C., 1865. Introduction à l'Étude de la Médecine Expérimentale. Éditions Garnier-Flammarion, Paris.

Bernardini, P.L., 2015. A note on the ethical value of aesthetics. Ndias Quarterly 3, 19.

Berra, T., 1990. Evolution and the Myth of Creationism. Stanford University Press, Stanford, CA.

Black, B., Ayala, F.J., Saffran-Brinks, C., 1994. Science and the law in the wake of *Daubert*: a new search for scientific knowledge. Texas Law Review 72, 715–802.

Blackmore, S., 1999. The Meme Machine. Oxford University Press, Oxford.

Blake, A., Truscianko, T. (Eds.), 1990. A.I. and the Eye. Wiley, New York.

Blaese, R.M., Culver, K.W., Miller, A.D., Carter, C.S., Fleisher, T., Clerici, M., Shearer, G., Chang, L., Chiang, Y., Tolstoshev, P., Greenblatt, J.J., Rosenberg, S.A., Klein, H., Berger, M., Mullen, C.A., Ramsey, W.J., Muul, L., Morgan, R.A., Anderson, W.F., 1995. T lymphocyte-directed gene therapy for ADA-SCID: initial trial results after 4 years. Science 270 (5235), 475–480.

Bordes, F., 1953. Nodules de typologie paléolithique I. Outils musteriens à fracture volontaire, vol. 1. Bulletin de la Societé Préhistorique Française 50, Paris.

Bordes, F., 1979. Typologie du Paléolithique ancien et moyen, vol. 1. CNRS, Paris.

Bowler, P.J., 2007. Monkey trials & gorilla sermons. In: Evolution and Christianity From Darwin to Intelligent Design. Harvard University Press, Cambridge, MA.

Bowler, P.J., 2009. Evolution. The History of an Idea. University of California Press, Berkeley.

Boyd, R., Richerson, P.J., 1985. Culture and the Evolutionary Process. University of Chicago Press, Chicago.

Boyd, R., Richerson, P.J., 2005. The Origin and Evolution of Cultures. Oxford University Press, Oxford, UK.

Bracher, P.J., 2015. Primordial soup that cooks itself. Nature Chemistry 7, 273–274.

Brasier, M.D., Antcliffe, J., Saunders, M., Wacey, D., 2015. Changing the picture of Earth's earliest fossils (3.5–1.9 Ga) with new approaches and new discoveries. Proceedings of the National Academy of Sciences of the United States of America 112, 4859–4864.

Brauer, M., Brumbaugh, D., 2001. Biology remystified: the scientific claims of the new creationist. In: Pennock, R. (Ed.), Intelligent Design Creationism and its Critics: Philosophical, Theological, and Scientific Perspectives. MIT Press, Cambridge, MA, pp. 289–334.

Broad, C.D., 1925. The Mind and Its Place in Nature. Kegan Paul, London.

Brock, A. (Ed.), 1999. The Molecular Origins of Life: Assembling Pieces of the Puzzle. Cambridge University Press, Cambridge.

Brosnan, S., de Waal, F., 2003. Monkeys reject unequal pay. Nature 425, 297–299.

Brown, W.P., 2010. The Seven Pillars of Creation. The Bible, Science and the Ecology of Wonder. Oxford University Press, Oxford, New York.

Brunet, M., 2010. Short note: the track of a new cradle of mankind in Sahelo-Saharan Africa (Chad, Libya, Egypt, Cameroon). Journal of African Earth Sciences 58, 680–683.

Brunet, M., Guy, F., Pilbeam, D., Mackaye, H.T., Likius, A., Ahounta, D., Beauvilain, A., Blondel, C., Bocherens, H., Boisserie, J.R., et al., 2002. A new hominid from the Upper Miocene of Chad, Central Africa. Nature 418, 145–151.

Brunet, M., Guy, F., Polbeam, D., Lieberman, D.E., Likius, A., Mackaye, H.T., Ponce de Leon, M.S., Zollikofer, C.N.E., Vignaud, P., 2005. New material of the earliest hominid from the Upper Miocene of Chad. Nature 434, 753–755.

Buckland, W., 1820. Vindiciae Geologicae. William Buckland Publisher, Oxford.

Buckland, W., 1823. Reliquiae Diluvianae. John Murray, London.

Buckland, W., 1836. Geology and Mineralogy. Considered with Reference to Natural Theology. William Pickering, London.

Bullen, K.E., 1976. Wegener, Alfred Lothar. In: Dictionary of Scientific Biography, vol. XIV. Scribner's Sons, New York, pp. 214–217.

Burnet, T., 1691 [1965]. Sacred Theory of the Earth. R. Norton, London.

Bury, J.B., 1955 [1932]. The Idea of Progress. An Inquiry into Its Growth and Origin. Dover Publications, New York.

Cain, A.J., Harrison, G.A., 1958. An analysis of the taxonomist's judgment of affinity. Journal of Zoology 131, 85–98.

Cairns, J., 1963. The chromosome of Escherichia coli. Cold Spring Harbor Symposia on Quantitative Biology 28, 43.

Callaway, E., 2015a. Oldest stone tools raise questions about their creator. Nature 520, 421.

Callaway, E., 2015b. World hails embryo vote. Nature 518, 145–146.

Caplan, A.L., Arp, R., 2013. Were it physically safe, would human reproductive cloning be acceptable? In: Caplan, A.L., Arp, R. (Eds.), Contemporary Debates in Bioethics. Wiley-Blackwell, New York, pp. 73–77.

Cartier, N., Aubourg, P., 2009. Hematopoietic stem cell transplantation and hematopoietic stem cell gene therapy in X-linked adrenoleukodystrophy. Brain Pathology 20, 857–862.

Cavalli-Sforza, L.L., Feldman, M.W., 1981. Cultural Transmission and Evolution. A Quantitative Approach, vol. 16. Princeton University Press, Princeton, NJ.

Cavalli-Sforza, L.L., Feldman, M.W., 2003. The application of molecular genetic approaches to the study of human evolution. Nature Genetics 33 (suppl.), 266–275.

Cavazzana-Calvo, M., Thrasher, A., Mavilio, F., 2004. The future of gene therapy. Nature 427, 779–781.

Cech, T.R., 1985. Self-splicing DNA: implications for evolution. International Review of Cytology 93, 3.

Cech, T.R., 1986. The generality of self-splicing RNA: relationship to nuclear mRNA splicing. Cell 44, 207–210.

Cech, T.R., 1987. The chemistry of self-splicing RNA and RNA enzymes. Science 236, 1532–1539.

Cech, T.R., 1993. Efficiency and versatility of catalytic RNA: implications for an RNA world. Gene 135, 33–36.

Cech, T.R., Bass, B.L., 1986. Biological catalysis by RNA. Annual Review of Biochemistry 55, 599–629.

Cela-Conde, C., Ayala, F.J., 2007. Human Evolution. Trails from the Past. Oxford University Press, Oxford.

Cela-Conde, C., Ayala, F.J., 2016. Human Evolution. Trails from the Past, second ed. Oxford University Press, Oxford.

Chakraborty, R., Kidd, K.K., 1991. The utility of DNA typing in forensic work. Science 254, 1735–1739.

Chambon, P., 1981. Split genes. Scientific American 244, 60–71.

Cheng, E.K., Yoon, A.H., 2005. Does Frye or Daubert Matter? A study of scientific admissibility standards. Virginia Law Review 91, 471–513.

Charlesworth, B., Lande, R., Slatkin, M., 1982. A new-Darwinian commentary on macroevolution. Evolution 36, 474–498.

Cideciyan, A.V., Hauswirth, W.W., Aleman, T.S., Kaushal, S., Schwartz, S.B., Boye, S.L., Windsor, E.A.M., Conlon, T.J., Sumaroka, A., Roman, A.J., Byrne, B.J., Jacobson, S.G., 2009. Vision 1 year after gene therapy for Leber's congenital amaurosis. New England Journal of Medicine 361, 725–727.

Claussen, J., Keck, D.D., Hiesey, W.M., 1940. Experimental Studies on the Nature of Species. I. The Effect of Varied Environment on Western North American Plants. Carnegie Institution of Washington Publication 520, Washington, DC.

Cohen, I.G., Savulescu, J., Adashi, E.Y., 2015. Transatlantic lessons in regulation of mitochondrial replacement therapy. Science 348, 178–180.

Condorcet, M.J.A.N.C., 1956 [1795]. Sketch for a Historical Picture of the Progress of the Human Mind. Noonday Press, New York.

Conway Morris, S., 1998. The Crucible of Creation. Oxford University Press, Oxford.

Conway Morris, S., 2003. Life's Solution: Inevitable Humans in a Lonely Universe. Cambridge University Press, Cambridge.

Cooper, R.M., Zubek, J.P., 1958. Effects of enriched and restricted early environments on the learning ability of bright and dull rats. Canadian Journal of Psychology 12, 159–164.

Copernicus, N., 1543. De revolutionibus orbium coelestium (On the Revolutions of the Heavenly Spheres). Johannes Petreius, Nuremberg.

Copp, D., 2006. The Oxford Handbook of Ethical Theory. Oxford University Press, Oxford.

Coppens, Y., 1991. L'évolution des hominidés, de leur locomotion et de leurs environnements. In: Coppens, Y., Senut, B. (Eds.), Origine(s) de la bipédie chez las hominidés. CNRS, Paris.

Coppens, Y., 1994. East side story: the origin of mankind. Scientific American 270, 62–69.

Cover Image, 2002. Science 295 (5557).

Coyne, J.A., 2015. Faith vs. Fact. Why Science and Religion Are Incompatible. Viking, New York.

Crick, F., 1979. Split genes and RNA splicing. Science 204, 264–271.

Cronin, J.E., Boaz, N.T., Stringer, C.B., Rak, Y., 1981. Tempo and mode in hominid evolution. Nature 292, 113–122.

Cronly-Dillon, J., 1991. Evolution of the Eye and Visual System. CRC Press, Boca Raton, FL.

Crow, J.F., 1958. Some possibilities for measuring selection intensities in man. Human Biology 30, 1–13.

Daeschler, E.B., Shubin, N.H., Jenkins Jr., F.A., 2006. A Devonian tetrapod-like fish and the evolution of the tetrapod body plan. Nature 440, 757–763.

Darwin, C., 1859. On the Origin of Species by Means of Natural Selection. John Murray, London.

Darwin, C., 1967 [1859]. On the Origin of Species. Atheneum, New York.

Darwin, C., 1871. The Descent of Man, and Selection in Relation to Sex (second ed., 1889). John Murray, London (Also: Appleton and Company, New York, 1971).

Darwin, C., 1958. In: Barlow, N. (Ed.), The Autobiography of Charles Darwin (1809–1882). Collins, London.

Darwin, E., 1794–1796. Zoonomia, or the Laws of Organic Life. J. Johnson in St. Paul's Church-Yard, London.

Darwin, F., 1887. The Life and Letters of Charles Darwin, Including an Autobiographical Chapter, vol. 3. Murray, London.

Darwin, F., 1903. More Letters of Charles Darwin, vol. 2. Murray, London.

Davenas, E., Beauvais, F., Amara, J., Oberbaum, M., Robinzon, B., Miadonnai, A., Tedeschi, A., Pomeranz, B., Fortner, P., Belon, P., Sainte-Laudy, J., Poitevin, B., Benveniste, J., 1988. Human basophil degranulation triggered by very dilute antiserum against IgE. Nature 333, 816–818.

Davidson, I., Noble, W., 1989. The archaeology of perception. Current Anthropology 30, 125–155.

Dawkins, R., 1976. The Selfish Gene. Oxford University Press, Oxford.

Dawkins, R., 1986. The Blind Watchmaker. Norton, New York.

Dawkins, R., 1992. Progress. In: Keller, E.F., Lloyd, E. (Eds.), Keywords in Evolutionary Biology. Harvard University Press, Cambridge, MA, pp. 263–272.

Dawkins, R., 1995. River Out of Eden. Harper Collins, New York.

Dawkins, R., 1996. Climbing Mount Improbable. W. W. Norton, New York.

Dawkins, R., 2004. The Ancestor's Tale: A Pilgrimage to the Darwin of Life. Weidenfeld & Nicholson, London.

Dawkins, R., 2006. The God Delusion. Bantam Press, London.

De Beer, G., 1964. Charles Darwin, A Scientific Biography. Doubleday, Garden City, New York.

DeDuve, C., 1996. The birth of complex cells. Scientific American 106, 36–40.

de Waal, F., 1996. Good Natured: The Origins of Right and Wrong in Humans and Other Animals. Harvard University Press, Cambridge, MA.

Dembski, W., 1995. The Design Inference: Eliminating Chance through Small Probabilities. Cambridge University Press, Cambridge, UK.

Dembski, W., 2002. No Free Lunch: Why Specified Complexity Cannot Be Purchased Without Intelligence. Rowman & Littlefield, Lanham, MD.

Dembski, W., Ruse, M. (Eds.), 2004. Debating Design: From Darwin to DNA. Cambridge University Press, Cambridge, UK.

Dennett, D., 1995. Darwin's Dangerous Idea. Simon and Schuster, New York.

DiMaggio, E.N., Campisano, C.J., Rowan, R., Dupont-Nivet, G., Deino, A.L., Bibi, F., Lewis, M.E., Souron, A., Garello, D., Werdelin, L., Reed, K.E., Arrowsmith, J.R., 2015. Late Pliocene fossiliferous sedimentary record and the environmental context of early *Homo* from Afar, Ethiopia. Science 347, 1355–1359.

Dobzhansky, T., 1937. Genetics and the Origin of Species (third ed., 1951). Columbia University Press, New York.

Dobzhansky, T., 1962. Mankind Evolving. The Evolution of the Human Species. Yale University Press, New Haven, CT and London.

Dobzhansky, T., 1967. The Biology of Ultimate Concern. New American Library, New York.

Dobzhansky, T., 1970. Genetics of the Evolutionary Process. Columbia University Press, New York.

Dobzhansky, T., 1972. Nothing in biology makes sense except in the light of evolution. The American Biology Teacher 35, 125–129.

Dobzhansky, T., Ayala, F.J., Stebbins, G.L., Valentine, J.W., 1977. Evolution. W.H. Freeman & Co., San Francisco.

Doebeli, M., Ispolatov, I., 2010. Continuously stable strategies as evolutionary branching points. Journal of Theoretical Biology 266, 529–535.

Doench, J.G., Zhang, F., 2014. Genome-scale CRISPR-Cas9 knockout screening in human cells. Science 343, 84–87.

Doolittle, R., 1993. The evolution of vertebrate blood coagulation: a case of Yin and Yang. Thrombosis Haemostasis 70, 24–28.

Doolittle, W.F., 2000. Phylogenetic classification and the universal tree. Science 284, 2124–2129.

Doronichev, V., Golovanova, L., 2003. Bifacial tools in the lower and middle Paleolithic of the Caucasus and their contexts. In: Soresi, M., Dibble, H.L. (Eds.), Multiple Approaches to the Study of Bifacial Technologies. Museum of Archaeology and Anthropology, Philadelphia, PA, pp. 77–107.

Doudna, J., 2015. My whirlwind year with CRISPR. Nature 528, 469–471.

Doudna, J.A., Charpentier, E., 2014. The new frontier of genome engineering with CRISPR-Cas9. Science 346, 1077–1086.

Downes, J.A., 1978. Feeding and mating in the insectivorous Ceratopogoninae (Diptera). Memoirs of the Entomological Society of Canada 104, 1–62.

Draper, J.W., 1875. History of the Conflict between Religion and Science. Henry S. King & Co., London.

Dupré, J., 2001. Human Nature and the Limits of Science. Oxford University Press, Oxford.

Dupré, J., 2012. Process of Life. Essays in the Philosophy of Biology. Oxford University Press, Oxford.

Edelman, G.M., 1974. The problem of molecular recognition by a selective system. In: Ayala, F.J., Dobzhansky, T. (Eds.), Studies in the Philosophy of Biology. Macmillan, London, pp. 45–56.

Edelman, G.M., 1987. Neural Darwinism: The Theory of Neuronal Group Selection. Basic Books, New York.

Edwards v. Aguilar, 1987. 482 U.S. 578.

Ehrlich, P.R., 2000. Human Natures: Genes, Cultures, and the Human Prospect. Island Press, Washington, DC.

Ehrlich, P.R., Ehrlich, A.H., 2008. The Dominant Animal. Human Evolution and the Environment. Island Press, Washington, DC.

Eldredge, N., 1971. The allopatric model and phylogeny in Paleozoic invertebrates. Evolution 25, 156–167.

Eldredge, N., 2005. Darwin. Norton, New York.

Eldredge, N., Gould, S.J., 1972. Punctuated equilibria: an alternative to phyletic gradualism. In: Schopf, T.J.M. (Ed.), Models in Paleobiology. Freeman, Cooper, Co., San Francisco.

Elgar, M.A., 1992. Sexual cannibalism in spiders and other invertebrates. In: Elgar, M.A., Crespi, B.J. (Eds.), Cannibalism: Ecology and Evolution Among Diverse Taxa. Oxford University Press, Oxford.

Elguero, E., Délicat-Loembet, L., Rougeron, V., Arnathau, C., Roche, B., Becquart, P., Gonzalez, J.-P., Nkoghe, D., Sica, L., Leroy, E., Durand, P., Ayala, F.J., Ollomo, B., Renaud, F., Prugnolle, F., 2015. Malaria continues to select for sickle cell trait in Central

Africa. Proceedings of the National Academy of Sciences of the United States of America 112, 7051–7054.

Embley, T.M., Williams, T.A., 2015. Evolution: steps on the road to eukaryotes. Nature 521, 169–170.

Erwin, D.H., 2006. Extinction: How Life Nearly Died 250 Million Years Ago. Princeton University Press, Princeton, NJ.

Erwin, D.H., Valentine, J.W., 2013. The Cambrian Explosion. The Construction of Animal Biodiversity. Roberts and Co., Greenwood Village, CO.

Felsenstein, J., 2007. Has natural selection been refuted? The arguments of William Dembski. National Center for Science Education Reports 27, 21–26.

Fischer, A., Hacein-Bey-Abina, S., Cavazzana-Calvo, M., 2010. 20 years of gene therapy for SCID. Nature Immunology 11, 457–460.

Fitch, W.M., Margoliash, E., 1967. Science 155, 279–284.

Forrest, B., Gross, P., 2004. Creationism's Trojan Horse: The Wedge of Intelligent Design. Oxford University Press, New York.

Friedmann, T., Roblin, R., 1972. Gene therapy for human genetic disease? Science 175, 949–955.

Gagnon, K.T., Corey, D.R., 2015. Stepping toward therapeutic DRISPR. Proceedings of the National Academy of Sciences of the United States of America 112, 15536–15537.

Gazzaniga, M.S., 2005. The Ethical Brain. Dana Press, New York.

Gazzaniga, M.S., 2008. Human. The Science Behind What Makes Us Unique. HarperCollins, New York.

Gemmel, N., Wolff, J.N., 2015. Mitochondrial replacement therapy: cautiously replace the master manipulator. BioEssays 37, 584–585.

Gene Therapy Clinical Trials Worldwide Database, 2014. The Journal of Gene Medicine. wiley. com. http://www.abedia.com/wiley/index.html.

Genome Editing, 2015. Nature outlook, supplement. Nature 528, S1–S17.

Ghiselin, M.T., 1969. The Triumph of the Darwinian Method. University of California Press, Berkeley.

Ghiselin, M.T., 1974. The Economy of Nature and the Evolution of Sex. University of California Press, Berkeley.

Giere, R., Bickle, J., Mauldin, R., 2005. Understanding Scientific Reasoning. Wadsworth Publishing, Belmont, CA.

Gilbert, S.F., 1986. Developmental Biology, fifth ed. 1997. Sinauer, Sunderland, MA.

Gilbert, S.F., The Swarthmore College Evolution and Development Seminar, 2007. The aerodynamics of flying carpets. In: Comfort, N. (Ed.), The Panda's Black Box. Johns Hopkins University Press, Baltimore, MD, pp. 40–62.

Gill, P., Jeffreys, A.J., Werrett, D.J., 1985. Forensic application of DNA "fingerprints." Nature 318, 577–579.

Gingerich, P.D., 1976. Paleontology and phylogeny: patterns of evolution at the species level in early Tertiary mammals. America Journal of Science 276, 1–28.

Ginsberg, M., 1944. Moral Progress. Frazer Lecture at the University of Glasgow. Glasgow University Press, Glasgow.

Godfrey-Smith, P., 2014. Philosophy of Biology. Princeton University Press, Princeton, NJ.

Goldstein, M., Goldstein, L.F., 1978. How We Know. Plenum Press, New York.

Gott, R., Duggan, S., 2003. Understanding and Using Scientific Evidence: How to Critically Evaluate Data. Sage Publications, Thousand Oaks, CA.

Goudge, T.A., 1961. The Ascent of Life. A Philosophical Study of the Theory of Evolution. University of Toronto Press, Toronto.

Gould, S.J., 1980. Is a new general theory of evolution emerging? Paleobiology 6, 137—161.

Gould, S.J., 1982a. The meaning of punctuated equilibrium and its role in validating a hierarchical approach to macroevolution. In: Milkman, R. (Ed.), Perspectives in Evolution. Sinauer, Sunderland, MA.

Gould, S.J., 1982b. Darwinism and the expansion of evolutionary theory. Science 216, 380—387.

Gould, S.J., 1988. On replacing the idea of progress with an operational notion of directionality. In: Nitecki, M.H. (Ed.), Evolutionary Progress. University of Chicago Press, Chicago, pp. 319—338.

Gould, S.J., 1997. Full House. The Spread of Excellence From Plato to Darwin. Harmony, New York.

Gould, S.J., 1999. Rock of Ages. Science and Religion in the Fullness of Life. Ballantine Publishing Group, New York.

Gould, S.J., 2002. The Structure of Evolutionary Theory. Harvard University Press, Cambridge, MA.

Gould, S.J., Lewontin, R.C., 1979. The spandrels of San Marco and the Pan glossian paradigm. Proceedings of the Royal Society of London B 205, 581—598.

Graur, D., Li, W.-S., 2000. Fundamentals of Molecular Evolution, second ed. Sinauer, Sunderland, MA.

Green, R.E., Krause, J., Briggs, A.W., Maricic, T., Stenzel, U., Kircher, M., Patterson, N., Li, H., Zhai, W., Fritz, M.H.-Y., Hansen, N.F., Durand, E.Y., Malaspinas, A.-S., Jensen, J.D., Marques-Bonet, T., Alkan, C., Prüfer, K., Meyer, M., Burbano, H.A., Good, J.M., Schultz, R., Aximu-Petri, A., Butthof, A., Höber, B., Höffner, B., Siegemund, M., Weihmann, A., Nusbaum, C., Lander, E.S., Russ, C., Novod, N., Affourtit, J., Egholm, M., Verna, C., Rudan, P., Brajkovic, D., Kucan, D.Z., Gušic, I., Doronichev, V.B., Golovanova, L.V., Lalueza-Fox, C., de la Rasilla, M., Fortea, J., Rosas, A., Schmitz, R.W., Johnson, P.L.F., Eichler, E.E., Falush, D., Birney, E., Mullikin, J.C., Slatkin, M., Nielsen, R., Kelso, J., Lachmann, M., Reich, D., Pääbo, S., 2010. A draft sequence of the Neandertal genome. Science 328, 710—722.

Greene, J.D., Sommerville, R.B., Nystrom, L.E., Darley, J.M., Cohen, J.D., 2001. An fRMI investigation of emotional engagement in moral judgment. Science 293, 2105—2108.

Grene, M., 1974. The Understanding of Nature. Essays in the Philosophy of Biology. Reidel, Boston.

Haeckel, E., 1866. Generelle Morphologie der Organismen, vol. 2. Reimer, Berlin.

Haeckel, E., 1896. The Evolution of Man, vol. 2. Appleton, New York.

Haidt, J., 2007. The new synthesis in moral psychology. Science 316, 998—1002.

Haile-Selassie, Y., 2001. Late Miocene hominids from the Middle Awash, Ethiopia. Nature 412, 178—181.

Haile-Selassie, Y., Asfaw, B., White, T.D., 2004. Hominid cranial remains from Upper Pleistocene deposits at Aduma, Middle Awash, Ethiopia. American Journal of Physical Anthropology 123, 1—10.

Haile-Salassie, Y., Saylor, B.Z., Deino, A., Levin, N.E., Alene, M., Latimer, B.M., 2012. A new hominin foot from Ethiopia shows multiple Pliocene bipedal adaptations. Nature 483, 565—569.

Hallam, A., 1978. How rare is phyletic gradualism and what is its evolutionary significance? Evidence from Jurassic bivalves. Paleobiology 4, 16—25.

Hamilton, W.D., 1964. The genetical evolution of social behavior. Journal of Theoretical Biology 7, 1—52.

Harrison, G.A.S. (Ed.), 1993. Human Adaptation. Oxford University Press, Oxford UK.

Haught, J.F., 1998. Darwin's gift to theology. In: Russell, J.R., Stoeger, W.R., Ayala, F.J. (Eds.), Evolutionary and Molecular Biology: Scientific Perspectives on Divine Action. Vatican Observatory Press, Vatican City State and Center for Theology and the Natural Sciences, Berkeley, CA, pp. 393–418.

Hauser, M., 2006. Moral Minds: How Nature Designed Our Universal Sense of Right and Wrong. Harper, New York.

Hegel, G.W.F., 1970 [1817]. Philosophy of Nature. Oxford University Press, Oxford.

Hempel, C.G., 1965. Aspects of Scientific Explanation. Free Press, New York.

Hennig, W., 1950. Grundzüge einer Theorie der phylogenetischen Systematik. Aufbau, Berlin.

Hennig, W., 1966. Phylogenetic systematics. Annual Review of Entomology 10, 97–116.

Herper, M., 2014. Gene Therapy's Big Comeback. Forbes magazine March 26, 2014. http://www.forbes.com/sites/matthewherper/2014/03/26/once-seen-as-too-scary-editing-peoples-genes-with-viruses-makes-a-618-million-comeback/.

Herrick, G.J., 1956. The Evolution of Human Nature. The University of Texas Press, Austin.

Higham, T., Douka, K., Wood, R., Ramsey, C.B., Brock, F., Basell, L., Camps, M., Arrizabalaga, A., Baena, J., Barroso-Ruíz, C., Bergman, C., Boitard, C., Boscato, P., Caparrós, M., Conard, N.J., Draily, C., Froment, A., Galván, B., Gambassini, P., Garcia-Moreno, A., Grimaldi, S., Haesaerts, P., Holt, B., Iriarte-Chiapusso, M.-J., Jelinek, A., Jordá Pardo, J.F., Maíllo-Fernández, J.-M., Marom, A., Maroto, J., Menéndez, M., Metz, L., Morin, E., Moroni, A., Negrino, F., Panagopoulou, E., Peresani, M., Pirson, S., de la Rasilla, M., Riel-Salvatore, J., Ronchitelli, A., Santamaria, D., Semal, P., Slimak, L., Soler, J., Soler, N., Villaluenga, A., Pinhasi, R., Jacobi, R., 2014. The timing and spatiotemporal patterning of Neanderthal disappearance. Nature 512, 306–309.

Highfield, R., November 16, 2007. Dolly Creator Prof. Ian Wilmut Shuns Cloning. Daily Telegraph. http://www.telegraph.co.uk/news/science/science-news/3314696/Dolly-creator-Prof-Ian-Wilmut-shuns-cloning.html.

Hitchens, C., 2007. God Is Not Great. How Religion Poisons Everything. Hachette Book Group, New York.

Hoagland, H., Burhoe, R.W. (Eds.), 1962. Evolution and Man's Progress. Columbia University Press, New York.

Hodge, C., 1874. What Is Darwinism. Scribner, Armstrong & Co., New York.

Hooke, R., 1665. Micrographia, or, Some Physiological Descriptions of Minute Bodies Made My Magnifying Glasses with Observations and Inquiries Thereupon. Martyn & Allestry, London.

Horridge, G., 1987. Evolution of visual processing and the construction of seeing systems. Proceedings of the Royal Society of London B 230, 279–292.

Hoving, T., 1993. Making the Mummies Dance: Inside the Metropolitan Museum of Art. Simon & Schuster, New York.

Huber, P.W., 1991. Galileo's Revenge: Junk Science in the Courtroom. Basic Books, New York.

Hull, D., 1973. Darwin and His Critics. Harvard University Press, Cambridge, Massachusetts.

Hull, D., 1974. Philosophy of Biological Science. Prentice-Hall, Englewood Cliffs, New Jersey.

Hull, D., 1988. Science as a Process. University of Chicago Press, Chicago.

Hull, D.L., 1992. God of the Galapagos. Nature 352, 485–486.

Hume, D., 1902 [1739]. An Enquiry Concerning Human Understanding. Clarendon Press, Oxford.

Hume, D., 1978 [1740]. Treatise of Human Nature. Oxford University Press, Oxford.

Hume, D., 1779/2006. Dialogues concerning natural religion. In: Coleman, D. (Ed.), Dialogues Concerning Natural Religion and Other Writings. Cambridge University Press, Cambridge, UK, pp. 3–104.

Hume, D., Smith, N.K., 1935. Dialogues Concerning Natural Religion. Clarendon Press, Oxford.

Hung, E., 1996. The Nature of Science: Problems and Perspectives. Wadsworth Publishing, Belmont, CA.

Huxley, J.S., 1942. Evolution, the Modern Synthesis. Harper, New York.

Huxley, J.S., 1953. Evolution in Action. Harper, New York.

Huxley, T.H., Huxley, J.S., 1947. Touchstone for Ethics. Harper, New York.

International Human Genome Sequencing Consortium, 2001. Initial sequencing and analysis of the human genome. Nature 409, 860–921.

Isaac, G.L., 1969. Studies of early cultures in East Africa. World Archaeology 1, 1–28.

Jablonski, D., 1999. The Future of the Fossil Record. Science 284, 2114–2116.

Jabr, F., March 11, 2013. Will Cloning Ever Save Endangered Animals? Scientific American. http://www.scientificamerican.com/article/cloning-endangered-animals/.

Jacob, F., 1988. The Statue Within: An Autobiography. Basic Books, New York.

Jeffreys, A.J., Wilson, V., Thein, S.L., 1985. Individual-specific "fingerprints" of human DNA. Nature 316, 76–79.

Jobling, M.A., Hurles, M.E., Tyler-Smith, C., 2004. Human Evolutionary Genetics. Origins, Peoples and Disease. Garland Science, New York.

John Paul II, P., 1996. The address of Pope John Paul II to the members of the Pontifical Academy of Sciences appeared in L'Osservatore Romano on October 23, 1996, in its French original, and on October 30, 1996, in English. Both texts are reproduced. In: Russell, J.R., Stoeger, W.R., Ayala, F.J. (Eds.), Evolutionary and Molecular Biology: Scientific Perspectives on Divine Action. Vatican Observatory Press, Vatican City State and Center for Theology and the Natural Sciences, Berkeley, CA, 1998, pp. 2–9.

Johnson, P., 1993. Darwin on Trial. InterVarsity Press, Downers Grove, IL.

Johnson, P., 2002. The Wedge of Truth: Splitting the Foundations of Naturalism. InterVarsity Press, Downers Grove, IL.

Jones, H., Cohen, I. (Eds.), 1963. Science Before Darwin: A Nineteenth-Century Anthology. A. Deutsch, London.

Joyce et al., 1984. [note 3, Nature 515, p. 348, November 20, 2014].

Judson, O., 2002. Dr. Tatiana's Sex Advice to All Creation. Holt, New York.

Jurs, A., DeVito, S., 2013. The stricter standard: an empirical assessment of Daubert's effect on civil defendants. Catholic University Law Review 62, 679–726.

Kandel, E.R., 2003. The molecular biology of memory storage: a dialogue between genes and synapses. In: Jörnwal, H. (Ed.), Nobel Lecture 2000. World Scientific Publishing Co., Singapore.

Kellogg, D.E., 1975. The role of phyletic change in the evolution of Pseudocubus vema (Radiolaria). Paleobiology 1, 359–370.

Kfoury, C., 2007. Therapeutic cloning: promises and issues. McGill Journal of Medicine 10, 112–120.

Kimura, M., 1961. Natural selection as the process of accumulating genetic information in adaptive evolution. Genetical Research 2, 127–140.

Kimura, M., 1983. The Neutral Theory of Molecular Evolution. Cambridge University Press, London.

King, T.J., 1996. Nuclear transplantation in amphibia. In: Prescott, D.M. (Ed.), Methods in Cell Physiology, vol. 2. Academic Press, New York, pp. 1–36.

Kirk, G., Raven, J., Schofield, P., 1983. The Presocratic Philosophers: A Critical History With a Selection of Texts. Cambridge University Press, Cambridge, UK.

Kirsch, J., 2004. God Against the Gods. The History of the War between Monotheism and Polytheism. Viking, New York.

Kitcher, P., 1985. Vaulting Ambition: Sociobiology and the Quest for Human Nature. MIT Press, Cambridge, MA.

Kitcher, P., 1996. The Lives to Come. The Genetic Revolution and Human Possibilities. Simon & Schuster, New York.

Kitzmiller, v. Dover Area School District, Case 4:04-cv-02688-JEJ Document 342 Filed 12/20/ 2005. Available at: https://upload.wikimedia.org/wikipedia/commons/8/8d/Kitzmiller_v._ Dover_Area_School_District.pdf.

Klocker, H., 1968. God and the Empiricists. The Bruce Publishing Company, New York.

Knoll, A.H., 2003. Life on a Young Planet. The First Three Billion Years of Evolution on Earth. Princeton University Press, Princeton, NJ.

Knoll, A.H., Bargoorn, E.S., 1977. Archean microfossils showing cell division from the Swaziland system of South Africa. Science 198, 396–398.

Koch, R., 1890. An address on bacteriological research. British Medical Journal 2, 380.

Krause, J., Fu, Q., Good, J.M., Viola, B., Shunkov, M.V., Derevianko, A.P., Pääbo, S., 2010. The complete mitochondrial DNA genome of an unknown hominin from southern Siberia. Nature 464, 894–897.

Lamarck, J.-B., 1809. Philosophie Zoologique. Duminil-Lesueur, Paris (English Translation: Elliot, H., 1914. The Zoological Philosophy. Macmillan, London.).

Lamphier, E., Urnov, F., Haecker, S.E., Werner, M., Smolenski, J., 2015. Don't edit the human germ line. Nature 519, 410–411.

Lander, E., 1992. DNA fingerprinting: science, law, and the ultimate identifier. In: Kevles, D.J., Hood, L. (Eds.), The Code of Codes. Harvard University Press, Cambridge, MA, pp. 191–210.

Lange, M. (Ed.), 2007. Philosophy of Science. An Anthology. Blackwell, Oxford.

Larson, M.H., Gilbert, L.A., Wang, X., Lim, W.A., Weissman, J.S., Qi, L.S., 2013. CRISPR interference (CRISPRi) for sequence-specific control of gene expression. Nature Protocols 8, 2180–2196.

Lawrence, S.E., 1992. Sexual cannibalism in the praying mantis, *Mantis religiosa*. A field study. Animal Behaviour 43, 569–583.

Leakey, M.D., 1971. Olduvai Gorge. In: Excavations in Beds I and II 1960–1963, vol. 3. Cambridge University Press, Cambridge.

Leakey, M.D., 1975. Cultural patterns in the Olduvai sequence. In: Butzer, K.W., Isaac, G.L. (Eds.), After the Australopithecines. Mouton, The Hague, pp. 477–493.

Ledford, H., 2015. CRISPR, the disruptor. Nature 522, 20–24.

Leroi, A.M., 2014. The Lagoon. How Aristotle Invented Science. Viking, New York.

Levick, S.E., 2013. Were it physically safe, human reproductive cloning would not be acceptable. In: Caplan, A.L., Arp, R. (Eds.), Contemporary Debates in Bioethics. Wiley-Blackwell, New York, pp. 90–97.

Levinton, J.S., Simon, C.M., 1980. A critique of the punctuated equilibria model and implications for the detection of speciation in the fossil record. Systematic Zoology 29, 130–142.

LeWitt, P.A., Rezai, A.R., Leehey, M.A., Ojemann, S.G., Flaherty, A.W., Eskandar, E.N., Kostyk, S.K., Thomas, K., Sarkar, A., Siddiqui, M.S., Tatter, S.B., Schwalb, J.M., Poston, K.L., Henderson, J.M., Kurlan, R.M., Richard, I.H., Van Meter, L., Sapan, C.V., During, M.J., Kaplitt, M.G., Feigin, A., 2011. AAV2-GAD gene therapy for advanced Parkinson's disease: a double-blind, sham-surgery controlled, randomised trial. The Lancet Neurology 10, 309–319.

Lewontin, R.C., 1968. The concept of evolution. In: Sills, D.L. (Ed.), International Encyclopedia of the Social Sciences, vol. 5. Macmillan Co. and Free Press, London and New York.

Lewontin, R.C., 1991. Biology as Ideology. The Doctrine of DNA. Anansi, Toronto.

Lewontin, R.C., Hartl, D.L., 1991. Population genetics in forensic DNA typing. Science 254, 1745–1750.

Li, W.-S., 1997. Molecular Evolution. Sinauer, Sunderland, MA.

Li, W.-H., Saunders, M.A., 2005. The chimpanzee genome. Nature 437, 50–51.

Lieb, B., Dimitrova, K., Kang, H.-S., Braun, S., Gebauer, W., Martin, A., Hanelt, B., Saenz, S.A., Adema, C.M., Markl, J., 2006. Red blood with blue-blood ancestry: intriguing structure of a snail hemoglobin. Proceedings of the National Academy of Sciences of the United States of America 103, 12011–12016.

Lieberman, P., 1998. Eve Spoke. Human Language and Human Evolution. Norton, New York.

Liu, R., Ochman, H., 2007. Stepwise formation of the bacterial flagellar system. Proceedings of the National Academy of Sciences of the United States of America 104, 7116–7121.

Liu, W., Martinon-Torres, M., Cai, Y.-J., Xing, S., Tong, H.-W., Pei, S.-W., Sier, M.J., Wu, X.-H., Edwards, R.L., Cheng, H., Li, Y.-Y., Yang, X.-X., Bermúdez de Castro, J.M., Wu, X.-J., 2015. The earliest unequivocally modern humans in southern China. Nature 526, 696–699.

Lordkipanidze, D., Ponce de León, M.S., Margvelashvili, A., Rak, Y., Rightmire, G.P., Vekua, A., Zollikofer, C.P.E., 2013. A complete skull from Dmanisi, Georgia, and the evolutionary biology of early *Homo*. Science 342, 326–331.

Lotka, A.J., 1945. The law of evolution as a maximal principle. Human Biology 17, 167–194.

Lovejoy, A.O., 1960 [1936]. The Great Chain of Being. A Study in the History of an Idea. Harvard University Press, Cambridge, MA.

Lucrecius Carus, T., 1743. Of the Nature of Things. Daniel Brown, London.

Lyell, C., 1830–1833. Principles of Geology. John Murray, London.

Lynch, M., 2007. The Origins of Genome Architecture. Sinauer, Sunderland, MA.

Maddox, J., Randi, J., Stewart, W.W., 1988. High-dilution' experiments a delusion. Nature 334, 287–291.

Maienschein, J., Ruse, M. (Eds.), 1999. Biology and the Foundations of Ethics. Cambridge University Press, Cambridge.

Maimonides, M. [Felshin, M.] 1956. Moses Maimonides (Rambam). Book Guild, New York.

Martineau, H., 1832–1834. In: Illustrations of Political Economy, third ed., vol. 9. Charles Fox, London.

Mascie-Taylor, C.G.N. (Ed.), 1993. The Anthropology of Disease. Oxford University Press, Oxford, UK.

Maynard Smith, J., Szathmány, E., 1995. The Major Transitions in Evolution. Freeman, New York.

Mayr, E., 1942. Systematics and the Origin of Species. Columbia University Press, New York.

Mayr, E., 1963. Animal Species and Evolution. Harvard University Press, Cambridge, MA.

Mayr, E., 1964. Introduction. In: Darwin, C. (Ed.), On the Origin of Species. Harvard University Press, Cambridge, Massachusetts.

Mayr, E., 1965. Cause and effect in biology. In: Lerner, D. (Ed.), Cause and Effect. Free Press, New York, pp. 33–50.

Mayr, E., 1969. Principles of Systematic Zoology, sixth ed. McGraw-Hill, New York.

Mayr, E., 1974. Teleological and teleonomic, a new analysis. In: Cohen, R.S., Wartofsky, M.W. (Eds.), Boston Studies in the Philosophy of Science, vol. XIV. Reidel, Boston, pp. 91–117.

Mayr, E., 1982. The Growth of Biological Thought. Harvard University Press, Cambridge, MA.

Mayr, E., 1988. Toward a New Philosophy of Biology. Harvard University Press, Cambridge, MA.

Mayr, E., 1998. The multiple meanings of teleological. History and Philosophy of Life Sciences 20, 35—40.

Mayr, E., 2001. What Evolution Is. Basic Books, New York.

Mayr, E., Pohl, B., Peters, D.S., 2005. A well-preserved Archaeopteryx specimen with theropod features. Science 310, 1483—1486.

McCormick, C.T., 1954. Handbook of the Law of Evidence. West Publishing Company, St. Paul, MN.

McLaren, A., 2000. Cloning: pathways to a pluripotent future. Science 288, 1775—1780.

McLean v. Arkansas Board of Education, 1982. 529 F. Supp. 1255.

Medawar, P.B., 1967. The Art of the Soluble. Methuen, London.

Medawar, P.B., Medawar, J.S., 1983. Aristotle to Zoos. Harvard University Press, Cambridge, MA.

Medvedev, Z.A., 1969. In: Lerner, I.M. (Ed.), The Rise and Fall of T.D. Lysenko. Columbia University Press, New York.

Mellars, P., 2006a. Going east: new genetic and archaeological perspectives on the modern human colonization of Eurasia. Science 313, 796—800.

Mellars, P., 2006b. Why did modern human populations disperse from Africa ca. 60,000 years ago? A new model. Proceedings of the National Academy of Sciences of the United States of America 103, 9381—9386.

Mendel, G., 1866. Experiments in Plant Hybridization. Reprinted in Sinnott, E.W., Dunn, L.C., Dobzhansky, T., 1958. Principles of Genetics, fifth ed. McGraw-Hill, New York, pp. 419—443.

Meselson, M., Stahl, F., 1958. The replication of DNA in *Escherichia coli*. Proceedings of the National Academy of Sciences of the United States of America 44, 671—682.

Michener, C.D., Sokal, R.R., 1957. A quantitative approach to a problem of classification. Evolution 11, 130—162.

Mill, J.S., 1974 [1843]. In: Robson, J.M. (Ed.), A System of Logic Rationative and Inductive, vol. 2. University of Toronto Press, Toronto.

Miller, H., 1858. The Testimony of the Rocks; or, Geology in its Bearings on the Two Theologies, Natural and Revealed. Gould and Lincoln, Boston.

Miller, K., 1999. Finding Darwin's God: A Scientist's Search for Common Ground. HarperCollins, New York.

Miller, K., 2004. The flagellum unspun: the collapse of "irreducible complexity". In: Dembski, W., Ruse, M. (Eds.), Debating Design: From Darwin to DNA. Cambridge University Press, Cambridge, UK, pp. 81—97.

Miller, S.L., 1953. Production of amino acids under possible primitive earth conditions. Science 117, 528—529.

Monod, J., 1970. Le hasard et la necessite. Editions du Seuil, Paris.

Monod, J., 1972. Chance and Necessity. Vintage Books, New York.

Moore, G., 1903. Principia Ethica. Cambridge University Press, Cambridge.

Moore, J.A., 1993. Science as a Way of Knowing. The Foundations of Modern Biology. Harvard University Press, Cambridge, MA.

Moore, J.A., 2002. From Genesis to Genetics. The Case of Evolution and Creationism. University of California Press, Berkeley.

Moore, J.R., 1979. The Post-Darwinian Controversies: A Study of the Protestant Struggle to Come to Terms with Darwin in Great Britain and America 1870—1900. Cambridge University Press, Cambridge.

Moorehead, A., 1969. Darwin and the Beagle. Harper and Row, New York.

Moorehead, P.S., Kaplan, M.M. (Eds.), 1969. Mathematical Challenges to the Neo-Darwinian Interpretation of Evolution. The Wistar Institute Press, Philadelphia.

Moorhead, P.S., Kaplan, M.M. (Eds.), 1967. Mathematical Challenges to the Neo-Darwinian Interpretation of Evolution. The Wistar Institute Press, Philadelphia.

Morange, M., 2015. What history tells us XXXIX. CRISPR-Cas: from a prokaryotic immune system to a universal genome editing tool. Journal of Biosciences 40, 829–832.

More, H., 1662. An antidote against atheism. In: A Collection of Several Philosophical Writings of Dr. Henry More. Flesher and Morden, London.

Muller, H.J., 1950. Our load of mutations. American Journal of Human Genetics 2, 111–176.

Musgrave, I., 2004. Evolution of the bacterial flagellum. In: Young, M., Edis, T. (Eds.), Why Intelligent Design Fails. Rutgers University Press, New Brunswick, NJ, pp. 58–84.

Nagel, E., 1961. The Structure of Science. Problems in the Logic of Scientific Explanation. Harcourt, Brace & World, New York.

Nagel, E., 1965. Types of causal explanation in science. In: Lerner, D. (Ed.), Cause and Effect. Free Press, New York, pp. 24–25.

National Academy of Sciences, 1989. On Being a Scientist. National Academy Press, Washington, DC.

National Academy of Sciences and Institute of Medicine, 2008. Science, Evolution, and Creationism. National Academies Press, Washington, DC.

NCSL, 2008. Human Cloning Laws (2008). National Conference of State Legislatures (NCSL). http://www.ncsl.org/research/health/human-cloning-laws.aspx.

Nei, M., 1987. Molecular Evolutionary Genetics. Columbia University Press, New York.

Nesse, R.M., Williams, G.C., 1994. Why We Get Sick. Random House, New York.

Nevo, E., Shaw, C.R., 1972. Genetic variation in a subterranean mammal, *Spalax ehrenbergi*. Biochemical Genetics 7, 235–241.

Newton, I., 1687. Philosophiæ Naturalis Principia Mathematica (Mathematical Principles of Natural Philosophy). London.

Nieuwenfijdt, B., 1718/2007. In: Chamberlayne, J. (Trans.), The Religious Philosopher, or the Right Use of Contemplating the Works of the Creator in the Most Amazing Structure of the Heavens with All Their Furniture. Kessinger Publishing, LLC, Whitefish, MT.

Nitecki, M.H. (Ed.), 1988. Evolutionary Progress. University of Chicago Press, Chicago.

Noll, M.P., Petraglia, M.D., 2003. Acheulean bifaces and early human behavioral patterns in East Africa and South India. In: Soressi, M., Dibble, H.L. (Eds.), Multiple Approaches to the Study of Bifacial Technologies. University of Pennsylvania Museum of Archaeology and Anthropology, Philadelphia, pp. 31–53.

Orgel, L.E., 1994. The origin of life on earth. Scientific American 271, 77–83.

Osborn, H.F., 1934. Aristogenesis, the creative principle in the origin of species. American Naturalist 68, 193–235.

Owen, R., 1843. Lectures on the Comparative Anatomy and Physiology of the Invertebrate Animals. Longman, London.

The Oxford English Dictionary, 1933. Oxford University Press, Oxford.

Pääbo, S., 2014. Neanderthal man. In: Search of Lost Genomes. Basic Books, New York.

Paley, W., 1785. The Principles of Moral and Political Philosophy. Exshaw, Dublin.

Paley, W., 1794. A View of the Evidences of Christianity. Collected Works, 1819. Rivington, London.

Paley, W., 1802. Natural Theology, or Evidences of the Existence and Attributes of the Deity Collected From the Appearances of Nature. American Tract Society, New York.

Pallen, M., Matzke, N., 2006. From the origin of species to the origin of the bacterial flagella. Nature Reviews Microbiology 4, 784–790.

Pascal, B., 1669. Pensées. Chez Guillaume Desprez, Paris.

Patel, B.H., Percivalle, C., Ritson, D.J., Duffy, C.D., Sutherland, J.D., 2015. Common origins of RNA, protein and lipid precursors in a cyanosulfidic protometabolism. Nature Chemistry 7, 301−307.

Peacocke, A.R., 1998. Biological evolution: a positive appraisal. In: Russell, J.R., Stoeger, W.R., Ayala, F.J. (Eds.), Evolutionary and Molecular Biology: Scientific Perspectives on Divine Action. Vatican Observatory Press, Vatican City State and Center for Theology and the Natural Sciences, Berkeley, CA, pp. 357−376.

Pennisi, E., 2013. The CRISPR craze. Science 341, 833−836.

Pennock, R. (Ed.), 2001. Intelligent Design Creationism and Its Critics: Philosophical, Theological, and Scientific Perspectives. MIT Press, Cambridge, MA.

Pennock, R., 2002. Tower of Babel: The Evidence Against the New Creationism. MIT Press, Cambridge, MA.

Perakh, M., 2004a. Unintelligent Design. Prometheus Books, New York.

Perakh, M., 2004b. There is a free lunch after all: Dembski's wrong answers to irrelevant questions. In: Young, M., Edis, T. (Eds.), Why Intelligent Design Fails: A Scientific Critique of the New Creationism. Rutgers University Press, New Brunswick, NJ, pp. 153−171.

Pigliucci, M., Müller, G.B., 2010. Evolution − The Extended Synthesis. MIT Press, Cambridge, MA.

Pinker, S., 2002. The Blank Slate: The Modern Denial of Human Nature. Viking, New York.

Pinker, S., 2015. The untenability of faitheism. Current Biology 25, R635−R653.

Pittendrigh, C.S., 1958. Adaptation, natural selection and behavior. In: Roe, A., Simpson, G.G. (Eds.), Behavior and Evolution. Yale University Press, New Haven, CT, pp. 390−416.

Pius XII, 1950. Humani Generis (Of the Human Race). Libreria Editrice Vaticana, Rome.

Plato, 1997a. Timaeus. In: Cooper, J. (Ed.), Plato: Complete Works. Hackett Publishing Company, Indianapolis, IN.

Plato, 1997b. Phaedo. In: Cooper, J. (Ed.), Plato: Complete Works. Hackett Publishing Company, Indianapolis, IN.

Pleins, J.D., 2013. The Evolving God. Charles Darwin on the Naturalness of Religion. Bloomsbury, New York and London.

Plummer, T., 2004. Flaked stones and old bones: biological and cultural evolution at the dawn of technology. Yearbook of Physical Anthropology 47, 118−164.

Plotz, D., 2005. The Genius Factory. Unravelling the Mysteries of the Nobel Prize Sperm Bank. Simon & Schuster, London.

Pomeranz Krummel, D., Altman, S., 1999a. Multiple binding modes of substrate to the catalytic RNA subunit of RNase P from *Escherichia coli*. RNA 5, 1021−1033.

Pomeranz Krummel, D., Altman, S., 1999b. Verification of phylogenetic predictions in vivo and the importance of the tetraloop motif in a catalytic RNA. Proceedings of the National Academy of Sciences of the United States of America 96, 11200−11205.

Pool, R., 1988. Unbelievable results spark a controversy. Science 241, 407.

Popper, K.R., 1959. The Logic of Scientific Discovery. Routeledge, London.

Popper, K.R., 1962. Conjectures and Refutations. The Growth of Scientific Knowledge. Basic Books, New York.

Popper, K.R., 1963. Conjectures and Refutations: The Growth of Scientific Knowledge. Routledge and Kegan Paul, London.

Popper, K.R., 1974. Scientific reduction and the essential incompleteness of all science. In: Ayala, F.J., Dobzhansky, T. (Eds.), Studies in the Philosophy of Biology. Reduction and Related Matters. University of California Press, Berkeley and Los Angeles, pp. 259−284.

Provine, W., 1988. Evolution and the foundation of ethics. MBL Science 3, 25−29.

Quammen, D., 2007. The Reluctant Mr. Darwin: An Intimate Portrait of Charles Darwin and the Making of His Theory of Evolution. W.W. Norton, New York.

Randall, J.H., 1960. Aristotle. Columbia University Press, New York.

Rasmussen, M., Guo, X., Wang, Y., Lohmueller, K.E., Rasmussen, S., Albrechtsen, A., Skotte, L., Lindgreen, S., Metspalu, M., Jombart, T., Kivisild, T., Zhai, W., Eriksson, A., Manica, A., Orlando, L., De La Vega, F.M., Tridico, S., Metspalu, E., Nielsen, K., Ávila-Arcos, M.C., Moreno-Mayar, J.V., Muller, C., Dortch, J., Gilbert, M.T.P., Lund, O., Wesolowska, A., Karmin, M., Weinert, L.A., Wang, B., Li, J., Tai, S., Xiao, F., Hanihara, T., van Driem, G., Jha, A.R., Ricaut, F.-X., de Knijff, P., Migliano, A.B., Romero, I.G., Kristiansen, K., Lambert, D.M., Brunak, S., Forster, P., Brinkmann, B., Nehlich, O., Bunce, M., Richards, M., Gupta, R., Bustamante, C.D., Krogh, A., Foley, R.A., Lahr, M.M., Balloux, F., Sicheritz-Pontén, T., Villems, R., Nielsen, R., Wang, J., Willerslev, E., 2011. An Aboriginal Australian genome reveals separate human dispersals into Asia. Science 334, 94–98.

Raup, D.M., 1978. Cohort analysis of generic survivorship. Paleobiology 4, 1–15.

Raup, D.M., 1991. Extinction: Bad Genes or Bad Luck. Norton, New York.

Ray, J., 1691. The Wisdom of God Manifested in the Works of Creation. Wernerian Club, London.

Redi, F., 1685. Bacco in Toscana (Baccus in Tuscany). Per Piero Matini, Firenze.

Reich, D., Patterson, N., Kircher, M., Delfin, F., Nandineni, M.R., Pugach, I., Ko, A.M., Ko, Y.C., Jinam, T.A., Phipps, M.E., Saitou, N., Wollstein, A., Kayser, M., Pääbo, S., Stoneking, M., 2011. Denisova admixture and the first modern human dispersals into Southeast Asia and Oceania. American Journal of Human Genetics 89, 516–528.

Rensch, B., 1947. Evolution Above the Species Level. Columbia University Press, New York.

Reznick, D.N., Mateos, M., Springer, M.S., 2002. Independent origins and rapid evolution of the placenta in the fish genus Poeciliopsis. Science 298, 1018–1020.

Reznick, D.N., 2010. The Origin Then and Now. An Interpretive Guide to the Origin of Species. Princeton UniversityPress, Princeton, NJ.

Richards, R.J., 1988. The moral foundations of the idea of evolutionary progress: Darwin, Spencer, and the neo-Darwinians. In: Nitecki, M.H. (Ed.), Evolutionary Progress. University of Chicago Press, Chicago, pp. 129–148.

Richerson, P.J., Boyd, R., 2005. Not by Genes Alone: How Culture Transformed Human Evolution. University of Chicago Press, Chicago.

Richerson, P.J., Boyd, R., Henrich, J., 2010. Gene-culture coevolution in the age of genomics. Proceedings of the National Academy of Sciences of the United States of America 107 (Suppl. 2), 8985–8992.

Rivera, M.C., Jain, R., Moore, J.E., Lake, J.A., 1998. Genomic evidence for two functionally distinct gene classes. Proceedings of the National Academy of Sciences of the United States of America 95, 6239–6244.

Rivera, M.C., Lake, J.A., 2004. The ring of life provides evidence for a genome fusion origin of eukaryotes. Nature 431, 152–155.

Roberts, M., 2004. Intelligent design: some geological, historical, and theological questions. In: Dembski, W., Ruse, M. (Eds.), Debating Design: From Darwin to DNA. Cambridge University Press, Cambridge, UK, pp. 275–293.

Romano, M., Cifelli, R.L., 2015. 100 years of continental drift. Science 350, 915–916.

Rosenberg, A., 1985. The Structure of Biological Science. Cambridge University Press, Cambridge.

Ross, D., 1949. Aristotle, fifth ed. Barnes and Noble, New York.

Ruse, M., 1973. The Philosophy of Biology. Hutchinson, London.

Ruse, M., 1988. Philosophy of Biology Today. State University of New York Press, Albany, NY.

Ruse, M., 1995. Evolutionary Naturalism. Routledge, London.

Ruse, M., 1996. Monad to Man. The Idea of Progress in Evolutionary Biology. Harvard University Press, Cambridge, MA.

Ruse, M., 2000. The Evolution Wars: A Guide to the Controversies. ABC-CLIO, Santa Barbara, CA.

Ruse, M., 2001. Can a Darwinian Be a Christian? The Relationship Between Science and Religion. Cambridge University Press, Cambridge, UK.

Ruse, M., 2006. Darwinism and Its Discontents. Cambridge University Press, Cambridge.

Ruse, M., 2008a. Evolution and Religion. A Dialogue. Rowman & Littlefield, Lanham, MD.

Ruse, M. (Ed.), 2008b. The Oxford Handbook of Philosophy of Biology. Oxford University Press, Oxford.

Ruse, M., 2010. The biological sciences can act as a ground for ethics. In: Ayala, F.J., Arp, R. (Eds.), Contemporary Debates in Philosophy of Biology. Wiley-Blackwell, Malden, MA, pp. 297—315.

Ruse, M., 2012. The Philosophy of Human Evolution. Cambridge University Press, Cambridge.

Ruse, M., 2015. Atheism. What Every One Needs to Know. Oxford University Press, Oxford.

Ruse, M., Richards, R.J. (Eds.), 2009. The Cambridge Companion to the "Origin of Species". Cambridge University Press, Cambridge.

Ruse, M., Wilson, E.O., 1985. The evolution of ethics. New Scientist 108, 50—52.

Russell, R.J., 2006. Cosmology, Evolution, and Resurrection Hope. Pandora Press, Kitchener, Ontario, Canada.

Russell, R.J., 2007. Physics, cosmology, and the challenge to consequentialist natural theology. In: Murphy, N., Russell, R.J., Stoeger, W.R. (Eds.), Physics and Cosmology: Scientificc Perspectives on the Problem of Natural Evil. Vatican Observatory Press, Vatican City State and Center for Theology and the Natural Sciences, Berkeley, pp. 109—130.

Sacks, O., February 19, 2015. My Own Life. Oliver Sacks on Learning He Has Terminal Cancer. The Opinion Pages, A25, The New York Times.

Salvini-Plawen, L.V., Mayr, E., 1977. On the evolution of photoreceptors and eyes. Evolutionary Biology 10, 207—263.

Sankararaman, S., Patterson, N., Li, H., Pääbo, S., Reich, D., 2012. The date of interbreeding between Neandertals and modern humans. PLoS Genetics 8 (10), e1002947. http://dx.doi.org/10.1371/journal.pgen.1002947.

Schick, K.D., Toth, N., 1993. Making Silent Stones Speak. Simon & Schuster, New York.

Schopf, W.J., 1993. Microfossils of the early archean apex chert: new evidence of the antiquity of life. Science 260, 640—646.

Schopf, W.J., 1999. Cradle of Life. Princeton University Press, Princeton, NJ.

Schrödinger, E., 1944 [1992]. What Is Life? MIT Press, Cambridge, MA.

Schwartz, J., 1992. The Creative Moment. Harper Collins, New York.

Scott, E.C., 2005. Evolution vs. Creationism. University of California Press, Berkeley, CA.

Scott, E.C., 2009. Evolution vs. Creationism, second ed. University of California Press, Berkeley.

Sczepanski, J.T., Joyce, G.F., 2014. A cross-chiral RNA polymerase ribozyme. Nature 515, 440—442.

Searle, J.R., 1998. How to study consciousness scientifically. Brain Research 26, 379—387.

Segawa, Y., Suga, H., Iwabe, N., Oneyama, C., Akagi, T., Miyata, T., Okada, M., 2006. Functional development of Src tyrosine kinases during evolution from a unicellular ancestor to multicellular animals. Proceedings of the National Academy of Sciences of the United States of America 103, 12021—12026.

Senut, B., Pickford, M., Gommery, D., Mein, P., Cheboi, K., Coppens, Y., 2001. First hominid from the Miocene (Lukeino Formation, Kenya). Comptes Rendus de l'Académie des Sciences — Series IIA — Earth and Planetary Science 322, 137—144.

Sepkoski Jr., J.J., 1987. Environmental trends in extinction during the Paleozoic. Science 235, 64—66.

Sepkoski, J.J., 1997. Biodiversity: past, present, and future. Journal of Paleontology 71, 533—539.

Service, R.F., 2015. Origin-of-life puzzle cracked. Science 347, 1298.

Shalem, O., Sanjana, N.E., Hartenian, E., Shi, X., Scott, D.A., Mikkelsen, T., Heckl, D., Ebert, B.L., Root, D.E., Doench, J.G., Zhang, F., 2014. Genome-scale CRISPR-Cas9 knockout screening in human cells. Science 343, 83—87.

Shelke, S.A., Piccirilli, J.A., 2014. Nature 515, 347—348.

Sheridan, C., 2011. Gene therapy finds its niche. Nature Biotechnology 29, 121—128.

Shields, P.G., Kind, A.J., Campbell, K.H.S., Waddington, D., Wilmut, I., Colman, A., Schnieke, A.E., 1999. Analysis of telomere lengths in cloned sheep. Nature 399, 316—317.

Shubin, N.H., 2008. Your Inner Fish: A Journey into the 3.5-Billion-Year History of the Human Body. Pantheon, New York.

Shubin, N.H., Daeschler, E.B., Jenkins Jr., F.A., 2006. The pectoral fin of *Tiktaalik roseae* and the origin of the tetrapod limb. Nature 440, 764—771.

Simpson, G.G., 1944. Tempo and Mode in Evolution. Columbia University Press, New York.

Simpson, G.G., 1949. The Meaning of Evolution. Yale University Press, New Haven.

Simpson, G.G., 1953. The Major Features of Evolution. Columbia University Press, New York.

Simpson, G.G., 1961. Principles of Animal Taxonomy. Columbia University Press, New York.

Simpson, G.G., 1964. This View of Life. Harcourt, Brace and World, New York.

Skoglund, P., Jakobsson, M., 2011. Archaic human ancestry in East Asia. Proceedings of the National Academy of Sciences of the United States of America 108, 18301—18306.

Slack, G., 2007. The Battle over the Meaning of Everything. Wiley, New York.

Slaymaker, I.M., Gao, L., Zetsche, B., Scott, D.A., Yan, W.X., Zhang, F., 2016. Rationally engineered Cas9 nucleases with improved specificity. Science 351, 84—88.

Smith, A., 1937 [1776]. The Wealth of Nations. Modern Library, New York.

Sneath, P.H., 1957. The application of computers to taxonomy. Journal of General Microbiology 17, 201—226.

Sober, E., 1993. Philosophy of Biology. Westview Press, Boulder and San Francisco.

Sober, E., 2007. What is wrong with intelligent design? The Quarterly Review of Biology 82, 3—6.

Sober, E., Wilson, D., 1998. Unto Others: The Evolution and Psychology of Unselfish Behavior. Harvard University Press, Cambridge.

Sokal, R.R., Sneath, P.H.A., 1963. Principles of Numerical Taxonomy. Freeman, San Francisco.

Sonleitner, F., 2006. Intelligent design is not testable. Meeting of the Geological Society of America 38, 10.

Speiser, S.M., March 15, 1993. Peer panels can stomp out truth. National Law Journal 15.

Spencer, H., 1852. Progress: its law and cause. Westminster Review 67, 244—267. Reprinted in 1868. Essays: Scientific, Political and Speculative, Williams and Norgate, London, vol. 1, pp. 1—67.

Spencer, H., 1857. Progress: its law and cause. Popular Science Literature 17, 233—285. J. Fitzgerald Publisher, New York.

Spencer, H., 1862. First Principles. Williams & Norgate, London.

Spencer, H., 1893. The Principles of Ethics. Williams and Norgate, London.

Staff Writer, August 10, 2001. In the News: Antinori and Zavos. Times Higher Education. http://www.timeshighereducation.co.uk/164313.article.

Staff Writer, October 23, 2008. MPs Support Embryology Proposals. BBC News Online, London. http://news.bbc.co.uk/2/hi/uk_news/politics/7682722.stm.

Stanley, S.M., 1979. Punctuated equilibrium and evolutionary stasis. Paleobiology 7, 156−166.

Stanley, S.M., 1982. Macroevolution and the fossil record. Evolution 36, 460−473.

Stebbins, G.L., 1950. Variation and Evolution in Plants. Columbia University Press, New York.

Stebbins, G.L., 1969. The Basis of Progressive Evolution. University of North Carolina Press, Chapel Hill.

Stebbins, G.L., Ayala, F.J., 1981. Is a new evolutionary synthesis necessary? Science 213, 967−971.

Stebbins, G.L., Ayala, F.J., 1985. The evolution of Darwinism. Scientific American 253, 72−82.

Strimling, P., Enquist, M., Eriksson, K., 2009. Repeated learning makes cultural evolution unique. Proceedings of the National Academy of Sciences of the United States of America 106, 13870−13874.

Strong, A.H., 1885. Systematic Theology, vol. 3. Fleming Revell, Westwood, NJ, 1907.

Swinburne, R., 1994. The Christian God. Oxford University Press, Oxford, UK.

Taubes, G., 1993. Bad Science: The Short Life and Weird Times of Cold Fusion. Random House, New York.

Taylor, J.H., Woods, P.S., Hughes, W.L., 1957. The organization and duplication of chromosomes as revealed by autoradiographic studies using tritium-labelled thymidine. Proceedings of the National Academy of Sciences of the United States of America 43, 122−128.

Teilhard de Chardin, P., 1959. The Phenomenon of Man. Harper, New York.

Teilhard de Chardin, P., 1960. The Divine Milieu. Harper & Row, New York.

Temin, H.M., Mizutani, S., 1970. RNA-dependent DNA polymerase in virions of *Rous sarcoma* virus. Nature 226, 1211.

Templeton, A.R., 2002. Out of Africa again and again. Nature 416, 45−51.

Templeton, A.R., 2005. Haplotype trees and modern human origins. Yearbook of Physical Anthropology 48, 33−59.

Thoday, J.M., 1953. Components of Fitness. Symposia of the Society for the Study of Experimental Biology 7 (Evolution), 96−113.

Thoday, J.M., 1958. Natural selection and biological progress. In: Barnet, S.A. (Ed.), A Century of Darwin. Allen and Unwin, London.

Thoday, J.M., 1970. Genotype versus population fitness. Canadian Journal of Genetics and Cytology 12, 674−675.

Thompson, P., 1988. The Structure of Biological Themes. State University of New York Press, Albany, NY.

Tillich, P., 1959. Theology of Culture. Oxford University Press, New York.

Tomasello, M., 1999. The Cultural Origins of Human Cognition. Harvard University Press, Cambridge, MA.

Tooby, J., Cosmides, L., 1992. The psychological foundations of culture. In: Barkow, J.H., Cosmides, L., Tooby, J. (Eds.), The Adapted Mind. Oxford University Press, New York, pp. 19−136.

Travis, J., 2015. Making the cut. CRISPR genome-editing technology shows its power. Science 350, 1456−1457.

UK Statute Law Database, 2001. Text of the Human Fertilisation and Embryology (Research Purposes) Regulations 2001 (No. 188). legislation.gov.uk. http://www.legislation.gov.uk/uksi/2001/188/contents/made.

Ussery, D., 2004. Darwin's transparent box: the biochemical evidence for evolution. In: Young, M., Edis, T. (Eds.), Why Intelligent Design Fails. Rutgers University Press, New Brunswick, NJ, pp. 48–57.

Ussher, J., 1650–1654. Annals of the Old and New Testament. London.

Valentine, J.W., 1973. Evolutionary Paleoecology of the Marine Biosphere. Prentice Hall, Upper Saddle River, NJ.

Valentine, J.W., 1985. Phanerozoic Diversity Patterns. Profiles in Macroevolution. Princeton University Press, Princeton, NJ.

Valentine, J.W., 2004. On the Origin of Phyla. University of Chicago Press, Chicago.

van der Oost, J., Westra, E.R., Jackson, R.N., Wiedenheft, B., 2014. Unravelling the structural and mechanistic basis of CRISPR-Cas systems. Nature Reviews Microbiology 12, 479–492.

van Erp, P.B., Bloomer, G., Wilkinson, R., Wiedenheft, B., 2015a. The history and market impact of CRISPR RNA-guided nucleases. Current Opinion Virology 121, 85–90.

van Erp, P.B.G., Jackson, R.N., Carter, J., Golden, S.M., Bailey, S., Wiedenheft, B., 2015b. Mechanism of CRISPR-RNA guided recognition of DNA targets in *Escherichia coli*. Nucleic Acids Research 43, 8381–8391.

Vaquero Turcios, J., 1995. Maestros subterráneos. Celeste Ediciones, SA, Madrid.

Vermeij, G.J., 1987. Evolution and Escalation: An Ecological History of Life. Princeton University Press, Princeton, NJ.

Villmoare, B., Kimbel, W.H., Seyoum, C., Campisano, C.J., DiMaggio, E.N., Rowan, J., Braun, D.R., Arrowsmith, J.R., Reed, K.E., 2015. Early *Homo* at 2.8 Ma from Ledi-Geraru, Afar, Ethiopia. Science 347, 1352–1355.

Voltaire, F.-M.A., 1967. Voltaire, François-Marie Arouet de. In: Encyclopedia of Philosophy, vol. 8. Macmillan, London, pp. 262–270.

Vrba, E.S., 1980. Evolution, species, and fossils: how does life evolve? South African Journal Science 76, 61–84.

Waddington, C.H., 1960. The Ethical Animal. Allen and Unwin, London.

Wallace, A.R., 1858. On the tendency of varieties to depart indefinitely from original type; instability of varieties supposed to prove the permanent distinctness of species. Journal of the Proceedings of the Linnean Society of London (Zoology) 3, 53–62.

Ward, K., 2008. The Big Questions in Science and Religion. Templeton Foundation, West Conshohocken, PA, pp. 79–80.

Watson, J.D., 1968. The Double Helix. Atheneum, New York.

Watson, J.D., Crick, F.H.C., 1953. Molecular structure of nucleic acid. A structure for deoxyribose nucleic acid. Nature 171, 737–738.

Watson, J.D., Tooze, J., 1981. The DNA Story. A Documentary History of Gene Cloning. Freeman, San Francisco.

Webster's New Collegiate Dictionary, tenth ed. 1998. Merriam-Webster, Springfield, MA.

Weinberg, S., 1999. A Designer Universe? This article is based on a talk given in April 1999 at the Conference on Cosmic Design of the American Association for the Advancement of Science in Washington, DC Retrieved January 28, 2016. http://www.physlink.com/Education/essay_weinberg.cfm.

Weir, B.S., 1992. Independence of VNTR alleles defined as fixed bins. Genetics 130, 873–887.

Westra, E.R., Buckling, A., Fineran, P.D., 2014. CRISPR-Cas systems: beyond adaptive immunity. Nature Reviews Microbiology 12, 317–326.

White, M.J.D., 1978. Modes of Speciation. W.H. Freeman, San Francisco.

Whiten, A., 2005. The second inheritance system of chimpanzees and humans. Nature 437, 52–55.

Whiten, A., Goodall, J., McGrew, W.C., Nishida, T., Reynolds, V., Sugiyama, Y., Tutin, G.E.G., Wrangham, R.W., Boesch, C., 1999. Cultures in chimpanzees. Nature 399, 682—685.

Whiten, A., Horner, V., de Waal, F., 2005. Conformity to cultural norms of tool use in chimpanzees. Nature 437, 737—740.

Wicken, J.S., 1987. Evolution, Thermodynamics, and Information. Extending the Darwinian Program. Oxford University Press, New York.

Williams, G.C., 1966. Adaptation and Natural Selection. Princeton University Press, Princeton, NJ.

Wilson, E.O., 1975. Sociobiology, the New Synthesis. Belknap Press, Cambridge.

Wilson, E.O., 1978. On Human Nature. Harvard University Press, Cambridge, MA.

Wilson, E.O., 1998. Consilience: The Unity of Knowledge. Knopf, New York.

Wilson, E.O., 2012. The Social Conquest of Earth. Norton/Liveright, New York.

Wilson, E.O., 2014. The Meaning of Human Existence. Liveright, New York.

Wimsatt, W.C., 1972. Teleology and the logical structure of function statements. Studies in the History and Philosophy of Science 3, 1—80.

Woese, C., 1987. Bacterial evolution. Microbiology Reviews 51, 221—271.

Woese, C.R., 2000. Interpreting the universal phylogenetic tree. Proceedings of the National Academy of Sciences of the United States of America 97, 8392—8396.

Wolpert, D., 2003. William Dembski's treatment of no free lunch theorems is written in Jell-O. Mathematical Reviews 12, 2003b, 00012.

Wolpert, D.H., Macready, W.G., 1997. No free lunch theorems for optimization. IEEE Transactions on Evolutionary Computation 1, 67—82.

Wolpoff, M.H., Spuhler, J.N., Smith, F.H., Radovic, J., Pope, J., Frayer, D.W., Eckhardt, R., Clark, G., 1988. Modern human origins. Science 241, 772—774.

Wolpoff, M.H., Hawks, J., Frayer, D.W., Hunley, K., 2001. Modern human ancestry at the peripheries: a test of the replacement theory. Science 291, 293—297.

Wolpoff, M.H., Senut, B., Pickford, M., Hawks, J., 2002. Shahelanthrpus or Sahelpithecus? Nature 419, 581—582.

Wynn, T., 2007. Archaeology and cognitive evolution. Behavioral and Brain Sciences 25 (3), 389—402.

Young, Y., Edis, T. (Eds.), 2004. Why Intelligent Design Fails. Rutgers University Press, New Brunswick, NJ.

Zimmerman, C., October 15, 2015. Editing of Pig DNA May Lead to More Organs for People. New York Times. http://www.nytimes.com/2015/10/20/science/editing-of-pig-dna-may-lead-to-more-organs-for-people.html.

# Index

Printed in the United States
By Bookmasters